# Advanced Computational Methods for Oncological Image Analysis

# Advanced Computational Methods for Oncological Image Analysis

Editors

**Leonardo Rundo**
**Carmelo Militello**
**Vincenzo Conti**
**Fulvio Zaccagna**
**Changhee Han**

MDPI • Basel • Beijing • Wuhan • Barcelona • Belgrade • Manchester • Tokyo • Cluj • Tianjin

*Editors*
Leonardo Rundo
University of Cambridge
UK

Carmelo Militello
Italian National Research Council (IBFM-CNR)
Italy

Vincenzo Conti
University of Enna KORE
Italy

Fulvio Zaccagna
University of Toronto
Canada

Changhee Han
LPIXEL Inc.
Japan

*Editorial Office*
MDPI
St. Alban-Anlage 66
4052 Basel, Switzerland

This is a reprint of articles from the Special Issue published online in the open access journal *Journal of Imaging* (ISSN 2313-433X) (available at: https://www.mdpi.com/journal/jimaging/special_issues/oncological_image_analysis).

For citation purposes, cite each article independently as indicated on the article page online and as indicated below:

LastName, A.A.; LastName, B.B.; LastName, C.C. Article Title. *Journal Name* **Year**, *Volume Number*, Page Range.

**ISBN 978-3-0365-2554-9 (Hbk)**
**ISBN 978-3-0365-2555-6 (PDF)**

# Contents

Journal of *Imaging*

MDPI

*Editorial*

# Advanced Computational Methods for Oncological Image Analysis

Leonardo Rundo [1,2,*], Carmelo Militello [3,*], Vincenzo Conti [4], Fulvio Zaccagna [5,6] and Changhee Han [7]

1 Department of Radiology, University of Cambridge, Cambridge CB2 0QQ, UK
2 Department of Information and Electrical Engineering and Applied Mathematics (DIEM), University of Salerno, 84084 Fisciano, Italy
3 Institute of Molecular Bioimaging and Physiology, Italian National Research Council (IBFM-CNR), 90015 Cefalù, Italy
4 Faculty of Engineering and Architecture, University of Enna KORE, 94100 Enna, Italy; vincenzo.conti@unikore.it
5 Department of Biomedical and Neuromotor Sciences, University of Bologna, 40138 Bologna, Italy; fulvio.zaccagna@unibo.it
6 IRCCS Istituto delle Scienze Neurologiche di Bologna, Functional and Molecular Neuroimaging Unit, 40139 Bologna, Italy
7 Saitama Prefectural University, Saitama 343-8540, Japan; mykallis13@gmail.com
* Correspondence: lrundo@unisa.it (L.R.); carmelo.militello@ibfm.cnr.it (C.M.)

The Special Issue "Advanced Computational Methods for Oncological Image Analysis", published for the *Journal of Imaging*, covered original research papers about state-of-the-art and novel algorithms and methodologies, as well as applications of computational methods for oncological image analysis, ranging from radiogenomics to deep learning. Interesting review articles were also considered.

Nowadays, the amount of heterogeneous biomedical data is constantly increasing, owing to the advances in image acquisition modalities and high-throughput technologies [1–3]. In particular, this trend applies to oncological image analysis [4]. Cancer is the second most common cause of death worldwide and encompasses highly variable clinical and biological scenarios. Some of the current clinical challenges are (i) early disease diagnosis and (ii) precision medicine, which allows for treatments targeted at specific clinical cases. The ultimate goal is to optimize the clinical workflow by combining accurate diagnosis with the most suitable therapies [5].

The automated analysis of these large-scale datasets creates new compelling challenges that require advanced computational methods, ranging from classic machine learning techniques [6,7] to deep learning [8,9].

The developed reliable computer-assisted methods (i.e., artificial intelligence), together with clinicians' unique knowledge, can be used to properly handle typical issues in evaluation/quantification procedures (i.e., operator dependence and time-consuming tasks) [10]. These technological advances can significantly improve result repeatability in disease diagnosis and act as a guide towards appropriate cancer care. Indeed, the need for applying machine learning and computational intelligence techniques to effectively perform image processing operations —such as segmentation, co-registration, classification, and dimensionality reduction, and multi-omics data integration—has steadily increased.

This Special Issue collects 13 papers related to oncological image analysis, including 10 original contributions and 3 review articles.

In the last few years, the role of medical image computing and quantification has been remarkably growing. Several areas have benefited from these advances, including oncology, since the advancement of computational techniques provides a technological bridge between radiology and oncology. This aspect could significantly accelerate the adoption of precision medicine. Regarding medical imaging focusing on traditional image analysis

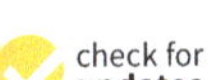
check for updates

**Citation:** Rundo, L.; Militello, C.; Conti, V.; Zaccagna, F.; Han, C. Advanced Computational Methods for Oncological Image Analysis. *J. Imaging* **2021**, *7*, 237. https://doi.org/10.3390/jimaging7110237

Received: 9 November 2021
Accepted: 9 November 2021
Published: 12 November 2021

tasks—such as registration, fusion, and segmentation—in recent years we have witnessed the advances of model-based medical image processing for biomarker development [11].

Among sex-related cancers, breast cancer for women and prostate cancer for men are major causes of disease and death.

Concerning breast cancer, methods to predict its risk or to stratify women in different risk levels could help achieve early diagnosis and consequently, mortality reduction Literature reviews are useful in providing a comprehensive vision of computer-assisted approaches to support the clinical process, especially for young scientists [12,13]. In particular, [14] reviews extraction methods of textural features from mammograms, where machine learning and deep learning algorithms are used to infer knowledge from the features and assess breast cancer risk. The accurate diagnosis of breast cancer is very challenging due to the increasing disease complexity, such as changes in treatment procedures and patient population samples. Improving the performance with suitable diagnosis techniques could lead to personalized care and treatment, thus reducing and controlling cancer recurrence [15].

Even though magnetic resonance (MR) has a better capability to differentiate soft tissues, mammography is the primary imaging modality used for the screening and early detection of breast cancer. The analysis of mammography images starts with detecting regions of interest around tumors. Those regions are then delimited through segmentation and classified as probably benign or malignant tumors. Meanwhile, the manual detection and delimitation of masses in images is time consuming and error prone. Therefore, integrated computer-aided detection systems have been proposed to assist radiologists in the process [16].

Along with the well-known imaging modalities, such as MR, CT, PET, US, which are now consolidated and used in clinical routine, recently new modalities have emerged that exploit techniques initially born in non-clinical contexts, such as microwaves [17,18]. When the aim is to reconstruct the dielectric/conductivity profile of the tissue under examination, "quantitative" algorithms must be adopted. In these cases, the reconstructions are basically optimized iteratively to consider the non-linearity. Among linear imaging methods, commonly addressed as radar approaches, beam forming (BF) is probably the most popular in microwave breast imaging. Basically, it consists of time-shifting the signals received over the measurement aperture to isolate signals scattered from (and hence to focus at) a particular synthetic focal point belonging to the imaged spatial area [17]. Microwave-based tomography is a model-based imaging modality that approximately reconstructs the actual internal spatial distribution of a breast's dielectric properties over a reconstruction model consisting of discrete elements. Breast tissue types are characterized by their dielectric properties, so the complex permittivity profile could help distinguish different tissue types [18].

Prostate cancer is one of the most diagnosed cancers in men and can often cause bone metastases. In this case, the most common imaging technique for screening, diagnosis, and the follow-up of disease evolution is bone scintigraphy, due to its high sensitivity and widespread availability in nuclear medicine facilities. To date, the assessment of bone scans relies solely on the interpretation of an expert physician who visually assesses the scan. This time-consuming task is also subjective, due to the lack of well-established criteria to identify bone metastases and quantify them using a straightforward and universally accepted procedure. The aim of the work in [19] was to provide the physician with a fast, precise, and reliable tool to quantify bone scans and evaluate disease progression/response to treatment.

Immunotherapy is one of the most significant breakthroughs in cancer treatment. Unfortunately, only a few patients respond positively to the treatment. Moreover, to date, no efficient biomarkers exist for discriminating patients eligible for this treatment in an early stage. To help overcome these limitations, the development of tools for discriminating between patients with high chances of response and those with disease progression is needed [20].

Among tumors, brain lesions are one of the foremost reasons for the rise in mortality among children and adults. A brain tumor is a mass of tissue that propagates out of control of the normal forces that regulate growth inside the brain [21]. The quantitative analysis of brain tumors provides valuable information for understanding tumor characteristics and planning better treatment. The manual segmentation of brain tumors is a challenging and time-consuming task. The accurate segmentation of lesions requires multiple image modalities with varying contrasts. As a result, manual segmentation, which is arguably the most accurate segmentation method, would be impractical for more extensive studies. Moreover, automated brain tumor classification on MRI is non-invasive, so that it avoids biopsy and makes the diagnosis process safer. The effort of the research community to propose automatic brain tumor segmentation and classification methods has been tremendous. As a result, ample literature exists on segmentation using region growing, traditional machine learning and deep learning methods [22,23]. Similarly, a number of tasks have been successfully conducted in the area of brain tumor classification into their respective histological type.

Structural and metabolic imaging are fundamental for diagnosis, treatment and follow-up in oncology. Beyond the well-established diagnostic imaging applications, ultrasounds are currently emerging in clinical practice as a non-invasive technology for therapy. Indeed, the sound waves can increase the temperature inside the target solid tumors, leading to the apoptosis or necrosis of neoplastic tissues. The MR-guided focused ultrasound surgery (MRgFUS) technology represents a valid application of this ultrasound property, mainly used in oncology and neurology [24]. Patient safety during MRgFUS treatments was investigated because temperature increases during the treatment are not always accurately detected by MRI-based referenceless thermometry methods. For these reasons, in-depth studies about these aspects are needed to monitor temperature and improve safety during MRgFUS treatments.

Deep learning approaches represent state-of-the-art techniques in many clinical scenarios, allowing for excellent performance. In the clinical setting, the main problem derives from their black-box approach (i.e., the nature of neural networks)—understanding and interpreting their internal mechanisms are difficult. Moreover, they require a training phase on large-scale datasets. These drawbacks undermine their immediate clinical feasibility. Apart from that, deep learning architectures, specifically convolutional neural networks (CNNs), are well-established in image analysis, processing, and representation. They can optimize feature design tasks that are essential to automatically analyze different types of medical images [25–27]. Various approaches have been developed using CNN architectures, aiming to support the clinical routine, such as tumor segmentation [16], skin melanoma prediction [28], and the estimation of the immunotherapy treatment response [20].

**Acknowledgments:** The guest editors, and all the editorial staff, would like to express their gratitude to the authors, who have chosen to publish their articles in this Special Issue in the *Journal of Imaging*, as well as to the reviewers whose precious support in their evaluation of the manuscripts allowed us to select only high-quality contributions. We appreciate their demonstrated professionalism and effort despite the COVID-19 pandemic.

## References

1. Castiglioni, I.; Rundo, L.; Codari, M.; Di Leo, G.; Salvatore, C.; Interlenghi, M.; Gallivanone, F.; Cozzi, A.; D'Amico, N.C.; Sardanelli, F. AI Applications to Medical Images: From Machine Learning to Deep Learning. *Phys. Med.* **2021**, *83*, 9–24. [CrossRef]
2. Rundo, L.; Militello, C.; Vitabile, S.; Russo, G.; Sala, E.; Gilardi, M.C. A Survey on Nature-Inspired Medical Image Analysis: A Step Further in Biomedical Data Integration. *Fund. Inform.* **2019**, *171*, 345–365. [CrossRef]
3. Badr, E. Images in Space and Time. *ACM Comput. Surv.* **2021**, *54*, 345–365. [CrossRef]
4. Bi, W.L.; Hosny, A.; Schabath, M.B.; Giger, M.L.; Birkbak, N.J.; Mehrtash, A.; Allison, T.; Arnaout, O.; Abbosh, C.; Dunn, I.F.; et al. Artificial Intelligence in Cancer Imaging: Clinical Challenges and Applications. *CA Cancer J. Clin.* **2019**, *69*, 127–157. [CrossRef] [PubMed]
5. Topol, E.J. High-Performance Medicine: The Convergence of Human and Artificial Intelligence. *Nat. Med.* **2019**, *25*, 44–56. [CrossRef]

6. Conti, V.; Militello, C.; Rundo, L.; Vitabile, S. A Novel Bio-Inspired Approach for High-Performance Management in Service-Oriented Networks. *IEEE Trans. Emerg. Top. Comput.* **2020**. [CrossRef]
7. Zaccagna, F.; Ganeshan, B.; Arca, M.; Rengo, M.; Napoli, A.; Rundo, L.; Groves, A.M.; Laghi, A.; Carbone, I.; Menezes, L.J. CT Texture-Based Radiomics Analysis of Carotid Arteries Identifies Vulnerable Patients: A Preliminary Outcome Study. *Neuroradiology* **2021**, *63*, 1043–1052. [CrossRef]
8. Han, C.; Rundo, L.; Murao, K.; Nemoto, T.; Nakayama, H. Bridging the Gap Between AI and Healthcare Sides: Towards Developing Clinically Relevant AI-Powered Diagnosis Systems. In Proceedings of the 16th IFIP WG 12.5 International Conference, AIAI 2020, Neos Marmaras, Greece, 5–7 June 2020; pp. 320–333.
9. Litjens, G.; Kooi, T.; Bejnordi, B.E.; Setio, A.A.A.; Ciompi, F.; Ghafoorian, M.; van der Laak, J.A.W.M.; van Ginneken, B.; Sánchez, C.I. A Survey on Deep Learning in Medical Image Analysis. *Med. Image Anal.* **2017**, *42*, 60–88. [CrossRef]
10. Rundo, L.; Pirrone, R.; Vitabile, S.; Sala, E.; Gambino, O. Recent Advances of HCI in Decision-Making Tasks for Optimized Clinical Workflows and Precision Medicine. *J. Biomed. Inform.* **2020**, *108*, 103479. [CrossRef]
11. Marias, K. The Constantly Evolving Role of Medical Image Processing in Oncology: From Traditional Medical Image Processing to Imaging Biomarkers and Radiomics. *J. Imaging* **2021**, *7*, 124. [CrossRef] [PubMed]
12. Michael, E.; Ma, H.; Li, H.; Kulwa, F.; Li, J. Breast Cancer Segmentation Methods: Current Status and Future Potentials. *Biomed Res. Int.* **2021**, *2021*, 9962109. [CrossRef] [PubMed]
13. Rezaei, Z. A Review on Image-Based Approaches for Breast Cancer Detection, Segmentation, and Classification. *Expert Syst. Appl.* **2021**, *182*, 115204. [CrossRef]
14. Mendes, J.; Matela, N. Breast Cancer Risk Assessment: A Review on Mammography-Based Approaches. *J. Imaging* **2021**, *7*, 98. [CrossRef]
15. Ibrahim, S.; Nazir, S.; Velastin, S.A. Feature Selection Using Correlation Analysis and Principal Component Analysis for Accurate Breast Cancer Diagnosis. *J. Imaging* **2021**, *7*, 225. [CrossRef]
16. Viegas, L.; Domingues, I.; Mendes, M. Study on Data Partition for Delimitation of Masses in Mammography. *J. Imaging* **2021**, *7*, 174. [CrossRef]
17. Cuccaro, A.; Dell'Aversano, A.; Ruvio, G.; Browne, J.; Solimene, R. Incoherent Radar Imaging for Breast Cancer Detection and Experimental Validation against 3D Multimodal Breast Phantoms. *J. Imaging* **2021**, *7*, 23. [CrossRef]
18. Kurrant, D.; Omer, M.; Abdollahi, N.; Mojabi, P.; Fear, E.; LoVetri, J. Evaluating Performance of Microwave Image Reconstruction Algorithms: Extracting Tissue Types with Segmentation Using Machine Learning. *J. Imaging* **2021**, *7*, 5. [CrossRef]
19. Providência, L.; Domingues, I.; Santos, J. An Iterative Algorithm for Semisupervised Classification of Hotspots on Bone Scintigraphies of Patients with Prostate Cancer. *J. Imaging* **2021**, *7*, 148. [CrossRef]
20. Rundo, F.; Banna, G.L.; Prezzavento, L.; Trenta, F.; Conoci, S.; Battiato, S. 3D Non-Local Neural Network: A Non-Invasive Biomarker for Immunotherapy Treatment Outcome Prediction. Case-Study: Metastatic Urothelial Carcinoma. *J. Imaging* **2020**, *6*, 133. [CrossRef]
21. Biratu, E.S.; Schwenker, F.; Debelee, T.G.; Kebede, S.R.; Negera, W.G.; Molla, H.T. Enhanced Region Growing for Brain Tumor MR Image Segmentation. *J. Imaging* **2021**, *7*, 22. [CrossRef]
22. Magadza, T.; Viriri, S. Deep Learning for Brain Tumor Segmentation: A Survey of State-of-the-Art. *J. Imaging* **2021**, *7*, 19. [CrossRef]
23. Biratu, E.S.; Schwenker, F.; Ayano, Y.M.; Debelee, T.G. A Survey of Brain Tumor Segmentation and Classification Algorithms. *J. Imaging* **2021**, *7*, 179. [CrossRef] [PubMed]
24. Militello, C.; Rundo, L.; Vicari, F.; Agnello, L.; Borasi, G.; Vitabile, S.; Russo, G. A Computational Study on Temperature Variations in MRgFUS Treatments Using PRF Thermometry Techniques and Optical Probes. *J. Imaging* **2021**, *7*, 63. [CrossRef] [PubMed]
25. Sandeep Kumar, E.; Satya Jayadev, P. Deep learning for clinical decision support systems: A review from the panorama of smart healthcare. In *Studies in Big Data*; Springer: Cham, Switzerland, 2020; pp. 79–99. ISBN 9783030339654.
26. Choi, G.H.; Yun, J.; Choi, J.; Lee, D.; Shim, J.H.; Lee, H.C.; Chung, Y.-H.; Lee, Y.S.; Park, B.; Kim, N.; et al. Development of Machine Learning-Based Clinical Decision Support System for Hepatocellular Carcinoma. *Sci. Rep.* **2020**, *10*, 14855. [CrossRef] [PubMed]
27. Rundo, L.; Han, C.; Nagano, Y.; Zhang, J.; Hataya, R.; Militello, C.; Tangherloni, A.; Nobile, M.S.; Ferretti, C.; Besozzi, D.; et al. USE-Net: Incorporating Squeeze-and-Excitation Blocks into U-Net for Prostate Zonal Segmentation of Multi-Institutional MRI Datasets. *Neurocomputing* **2019**, *365*, 31–43. [CrossRef]
28. Manzo, M.; Pellino, S. Bucket of Deep Transfer Learning Features and Classification Models for Melanoma Detection. *J. Imaging* **2020**, *6*, 129. [CrossRef]

Journal of
*Imaging*

*Article*

# The Constantly Evolving Role of Medical Image Processing in Oncology: From Traditional Medical Image Processing to Imaging Biomarkers and Radiomics

**Kostas Marias** [1,2]

[1] Department of Electrical and Computer Engineering, Hellenic Mediterranean University, 71410 Heraklion, Greece; kmarias@hmu.gr
[2] Computational Biomedicine Laboratory (CBML), Foundation for Research and Technology—Hellas (FORTH), 70013 Heraklion, Greece

**Abstract:** The role of medical image computing in oncology is growing stronger, not least due to the unprecedented advancement of computational AI techniques, providing a technological bridge between radiology and oncology, which could significantly accelerate the advancement of precision medicine throughout the cancer care continuum. Medical image processing has been an active field of research for more than three decades, focusing initially on traditional image analysis tasks such as registration segmentation, fusion, and contrast optimization. However, with the advancement of model-based medical image processing, the field of imaging biomarker discovery has focused on transforming functional imaging data into meaningful biomarkers that are able to provide insight into a tumor's pathophysiology. More recently, the advancement of high-performance computing, in conjunction with the availability of large medical imaging datasets, has enabled the deployment of sophisticated machine learning techniques in the context of radiomics and deep learning modeling. This paper reviews and discusses the evolving role of image analysis and processing through the lens of the abovementioned developments, which hold promise for accelerating precision oncology, in the sense of improved diagnosis, prognosis, and treatment planning of cancer.

**Keywords:** medical imaging; imaging biomarkers; radiomics; deep learning

check for
**updates**

**Citation:** Marias, K. The Constantly Evolving Role of Medical Image Processing in Oncology: From Traditional Medical Image Processing to Imaging Biomarkers and Radiomics. *J. Imaging* **2021**, *7*, 124. https://doi.org/10.3390/imaging7080124

Academic Editors: Leonardo Rundo, Carmelo Militello, Vincenzo Conti, Fulvio Zaccagna and Changhee Han

Received: 7 July 2021
Accepted: 22 July 2021
Published: 23 July 2021



## 1. Introduction

To better understand the evolution of medical image processing in oncology, it is necessary to explain the importance of measuring tumor appearance from medical images. Medical image processing approaches contain useful diagnostic and prognostic information that can add precision in cancer care. In addition, because biology is a system of systems, it is reasonable to assume that image-based information may convey multi-level pathophysiology information. This has led to the establishment of many sophisticated predictive and diagnostic image-based biomarker extraction approaches in cancer. In more detail, medical image processing efforts are focused on extracting imaging biomarkers able to decipher the variation within individuals in terms of imaging phenotype, enabling the identification of patient subgroups for precision medicine strategies [1]. From the very beginning, the main prerequisite for clinical use was that quantitative biomarkers must be precise and reproducible. If these conditions are met, imaging biomarkers have the potential to aid clinicians in assessing the pathophysiologic changes in patients and better planning personalized therapy. This is important, as in clinical practice subjective characterizations might be used (e.g., average heterogeneity, speculated mass, necrotic core) which can decrease the precision of diagnostic processes.

Based on the above considerations, the extraction of quantitative parameters characterizing size, shape, texture, and activity can enhance the role of medical imaging in assisting in diagnosis or therapy response assessment. However, in clinical practice, only simpler image metrics (e.g., linear) are often used in oncology, especially in the evaluation of solid tumor response to therapy (e.g., a longer lesion diameter in RECIST). Both RECIST and WHO evaluation criteria rely on anatomical image measurements, mainly in CT or MRI data, and were originally developed mainly for cytotoxic therapies. Such linear measures suffer from high intra/inter-observed variability, which in some cases can compromise the accurate assessment of tumor response, since some studies report inter-observer RECIST variability of up to 30% [2]. Several studies have shown that 3D quantitative response assessments are better correlated with disease progression than those based on 1D linear measurements [3]. Nevertheless, traditional tumor quantification approaches based on linear or 3D tumor measures have experienced substantial difficulties in assessing response to newer oncology therapies, such as targeted, anti-angiogenic treatments and immunotherapies [2]. Size-based tumor assessments do not always represent tumor response to therapy, since, for example, tumors may display internal necrosis formation, with or without reduction in lesion size (as in traditional cytotoxic treatments). Even if RECIST criteria are constantly updated to address these issues, as in the case of Immune RECIST [4], such approaches still do not take into consideration a tumor's image structure and texture over time. In addition the size and the location of metastases have been reported to play a significant role in assessing early tumor shrinkage and depth of response [5]. To address these limitations, medical image processing has provided over the last few decades the means to extract tumor texture and size descriptors for obtaining more detailed (e.g., pixel-based) descriptors of tissue structure and for discovering feature patterns connected to disease or response. In this paper, it is argued that the evolution of medical image processing has been a gradual process, and the diverse factors that contributed to unprecedented progress in the field with the use of AI are explained. Initially, simplistic approaches to classify benign and malignant masses, e.g., in mammograms, were based on traditional feature extraction and pattern recognition methods. Functional tomographic imaging such as PET gave rise to more sophisticated, model-based approaches from which quantitative markers from tissue properties could be extracted in an effort to optimize diagnosis, treatment stratification, and personalize response criteria. Lastly, the advancement of artificial intelligence enabled the more exhaustive search of imaging phenotype descriptors and led to the increased performance of modern diagnostic and predictive models.

## 2. Traditional Image Analysis: The First Efforts towards CAD Systems

In the 1990s, one of the first challenges in medical image analysis was to facilitate the interpretation of mammograms in the context of national screening programs for breast cancer. In the United Kingdom, the design of the first screening program was undertaken by a working group under Sir Patrick Forrest, whose report was accepted by the government in 1986. As a consequence, the UK screening program was established for women between 50 and 64 in 1990 [6]. The implementation of such screening programs throughout Europe led to the establishment of specialist breast screening centers and the formal training of both radiographers and radiologists. X-ray mammography proved to be a cost-effective imaging modality for national screening, and population screening led to smaller and usually non-palpable masses being increasingly detected.

As a result, the radiologist's task became more complex, since the interpretation of a mammogram is challenging, due to the projective nature of mammography, while at the same time the need for early and accurate detection of cancer became pressing. To address these needs, medical image analysis became an active field of research in the early nineties, giving rise to numerous research efforts into cancer and microcalcification detection, as well as mammogram registration for improving the comparison of temporal mammograms. Figure 1 depicts the temporal mammogram registration concept towards CAD systems that would facilitate comparison and aid clinicians in early diagnose of cancer in screening mammography [7]. When the ImageChecker system was certified by the FDA for screening mammography in 1998, R2 Technology became the first company to employ computer-assisted diagnosis (CAD) for mammography, and later for digital mammography as well.

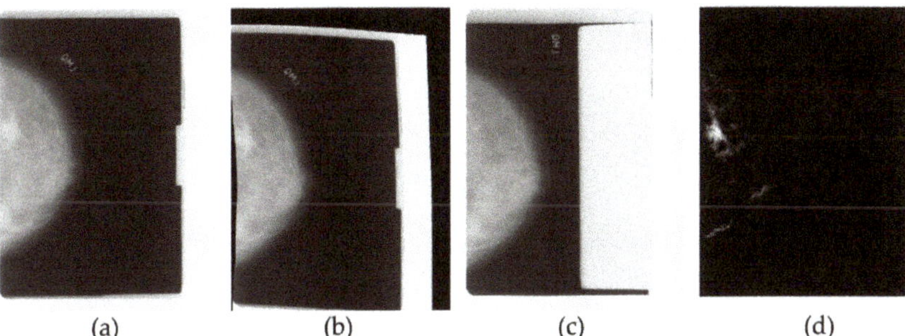

**Figure 1.** Traditional medical image processing on temporal mammograms. From left to right: the most recent mammogram (**a**) is registered to the previous mammogram (**b**), which is shown in (**c**). After registration there is one predominant region of significant difference in the subtraction image (**d**), which corresponds to a mass developed in the breast.

However, early diagnostic decision support systems suffered from low precision, which in turn could potentially lead to a negative impact in the number of unnecessary biopsies. In a relevant study [8], the positive predictive values of the interpretations worsened from 100%, 92.7%, and 95.5%, to 86.4%, 97.3%, and 91.1%, when mammograms were analyzed by three independent observers, with and without the CAD. This limitation was representative of the low generalizability of such cancer detection tools in these early days. At the same time the lack of more sophisticated imaging modalities hampered the research efforts towards predicting therapy response and optimizing therapy based on imaging data.

## 3. Quantitative Imaging Based on Models

With the advent of more sophisticated imaging modalities enabling functional imaging, medical image analysis efforts shifted towards the quantification of tissue properties. This opened new horizons in CAD systems towards translating image signals to cancer tissue properties such as perfusion and cellularity and developing more intuitive imaging biomarkers for several cancer imaging applications. For example, in the case of MRI, complex phenomena that occur after excitation are amenable to mathematical modeling, taking into consideration tissue interactions within the tumor microenvironment. In the context of evaluating a model-based approach, the model can be regarded reliable when the predicted data converges on the observed signal intensities and at the same time provides useful insights to radiologists and oncologists. MRI perfusion and diffusion imaging has been the main focus of such modeling efforts, not least due to fact that MRI is ionizing radiation-free.

Diffusion weighted MRI (DWI-MRI) is based on sequences sensitized to microscopic water mobility by means of strong gradient pulses and can provide quantitative information on tumor environment and architecture. Diffusivity can be assessed in the intracellular, extracellular, and intravascular spaces. Apparent diffusion coefficient (ADC) per pixel values derived from DWI-MRI theoretically have an inverse relationship to tumor cell density. In addition, with the introduction of the intravoxel incoherent motion (IVIM) model, both cellularity and microvascular perfusion information could be assessed after parametric modeling [9]. Figure 2 presents a parametric map of the stretching parameter $\alpha$ from the DWI-MRI stretched-exponential model (SEM), revealing highly heterogeneous parts of a dedifferentiated liposarcoma (DDLS) of Grade 3 [9].

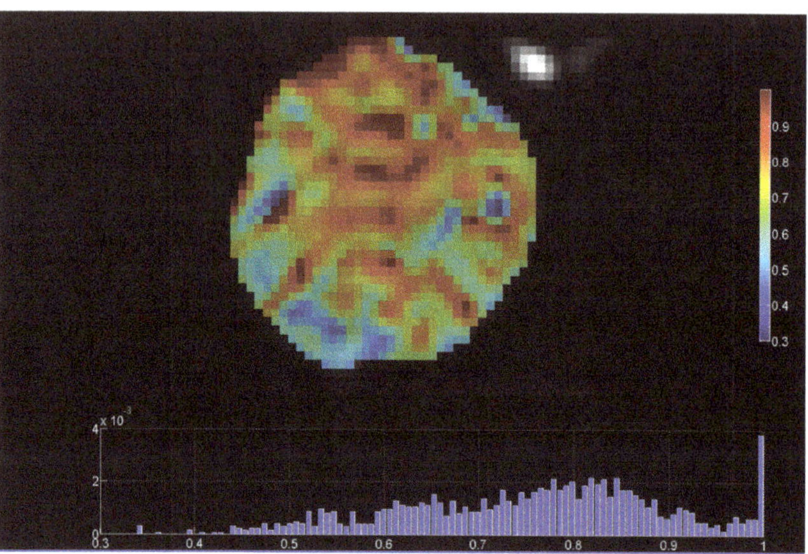


**Figure 2.** DWI-MRI stretched-exponential (SEM) DWI-MRI parametric map, revealing highly heterogeneous parts of a dedifferentiated liposarcoma (with permission from the department of Medical Imaging, Heraklion University Hospital). Heterogeneity index $\alpha$ ranges from 0 to 1, with lower values of $\alpha$ indicating microstructural heterogeneity.

DWI-MRI has been tested in most solid tumors for discriminating malignant from benign lesions, to automatize tumor grading, and to predict treatment response and post-treatment monitoring [10].

However, there is still a lack of standardization and generalization of these results, as well as validation against histopathology. While in clinical routine, in-depth DWI-MRI biomarker validation is difficult, recent pre-clinical studies have found that derived parametric maps can serve as a non-invasive marker of cell death and apoptosis in response to treatment [11]. To this end, they also confirmed significant correlations of ADC with immunohistochemistry measurements of cell density, cell death, and apoptosis.

In a similar fashion, in dynamic contrast-enhanced MRI (DCE-MRI), T1-weighted sequences are acquired before, during, and after the administration of a paramagnetic contrast agent (CA). Tissue-specific information about pathophysiology can be inferred from the dynamics of signal intensity in every pixel of the studied area. Usually this is performed by visual or semi-quantitative analysis from the signal time curves in selected regions of interest. However, with the use of pharmacokinetic modeling, e.g., between the intravascular and the extravascular extracellular space, it became possible to map signal intensities per pixel to CA concentration and then fit model parameters describing, e.g., interstitial space and transfer constant (ktrans). This enabled the generation of parametric maps, e.g., for ktrans providing more quantitative representation of tumor perfusion and heterogeneity within the tumor image region of interest. Although promising, e.g., for assessing treatment efficacy, such approaches have found limited use in clinical practice, not least due to the low reported reproducibility of model parameter estimation. One aspect of this problems is presented in the example shown in Figure 3, where the use of image-driven methods based on multiple-flip angles produces a parametric map of a tumor with different contrast compared to the one produced with the Fritz–Hansen population based AIF [12]. This issue has several implications, including for the accuracy of assessing breast cancer response to neoadjuvant chemotherapy [13].

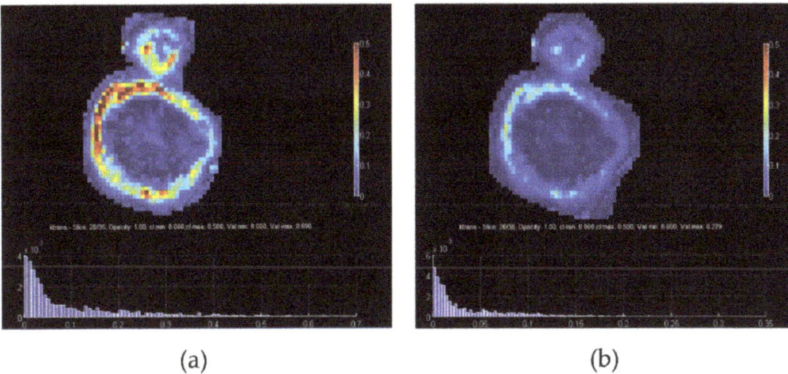

(a) (b)

**Figure 3.** (**a**) ktrans map of a tumor from PK analysis using AIF measured directly from the MR image, while for the conversion from signal to CA concentration the multiple flip angles method (mFAs) was used, (**b**) ktrans map of the same tumor using a population based AIF from Fritz and Hansen.

In conclusion, the clinical translation of DWI and DCE MRI is hampered by low repeatability and reproducibility across several studies in oncology. To address this problem initiatives such as the Quantitative Imaging Biomarkers Alliance (QIBA) propose clinical and technological requirements for quantitative DWI and DCE-derived imaging biomarkers, as well as image acquisition, processing, and quality control recommendations aimed at improving reproducibility error, precision, and accuracy [14]. It is argued that this active area of medical image processing has not yet reached its full potential and still represents a complementary approach to AI driven methods, towards CAD systems for promoting precision oncology. In addition, the exploitation of multimodality imaging strategies (e.g., PET/MRI) can provide added value through the combination of anatomical and functional information.

## 4. Radiomics and Deep Learning Approaches in Oncology through the Cancer Continuum

Traditional cancer medical image analysis was for decades based on human-defined features, usually inspired by low-level image properties, such as intensity, contrast, and a limited number of texture measures. Such methods were successfully used. e.g., in cancer subclassification, but it was hard to capture the high-level, complex patterns that an expert radiologist uses to define the presence or absence of cancer [1].

However, with the advancement of machine learning and the availability of more powerful, high-performance computational infrastructures, it became possible to exhaustively analyze the texture and shape content of medical images in an effort to decipher high-level pathophysiology patterns. At the same time the evolution of texture representation and feature extraction, through a growing number of techniques during the last decades, played a catalytic role in better capturing tumor appearance through medical image analysis [15]. Last but not least, the need to decipher the imaging phenotype in cancer became even more pressing, due to the fact that the vast majority of visible phenotypic variation is now considered attributable to non-genetic determinants in chronic and age-associated disorders [1].

All these factors played a central role in the advancement of radiomics, where in analogy to genomics high-throughput feature extraction followed by ML enabled the development of significant discriminatory and predictive signatures, based on imaging phenotype. Radiomics have been enhanced with deep learning techniques, offering an alternative approach to medical image feature extraction by the learning of complex, high-level features in an automated fashion from a large number of medical images that contain variable instances of a particular tumor. Figure 4 illustrates the main AI/radiomics applications that can assist clinicians in adding precision in the management of cancer patients.

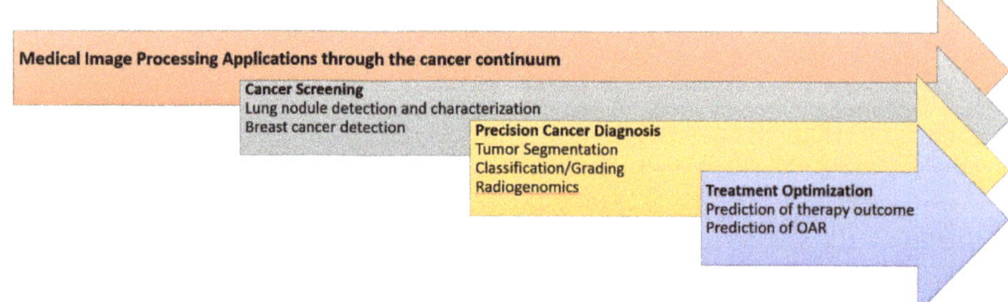


**Figure 4.** The main medical image processing applications enhanced with AI/radiomics towards precision oncology.

### 4.1. Cancer Screening

Recent advancements in AI driven medical image processing can have a positive impact in national cancer screening programs, alleviating the heavy workload of radiologists and aiding clinicians to reduce the number of missed cancers and to detect them at an earlier stage. Compared to the initial efforts mentioned in previous sections, recent AI-driven image processing can exceed the limits of human vision and potentially reduce the number of cancers missed in screening, as well as cope with inter-observer variability.

Regarding lung cancer screening, early nodule detection and classification is of paramount importance for improving patient outcomes and quality of life. Despite the existence of such screening programs the majority of lung cancers are detected in the later stages, leading to increased mortality and low 5-year survival rate [16]. To this end, radiomics and deep-learning-based methods have shown encouraging results towards precision pulmonary nodule evaluation [17]. A very interesting recent example is reported by Ardill et al., who developed a deep learning algorithm that uses a patient's current and prior computed tomography volumes to predict the risk of lung cancer. Their model achieved a state-of-the-art performance (94.4% area under the curve) on 6716 cases and performed similarly on an independent clinical validation set of 1139 cases. When prior computed tomography imaging was not available, their model outperformed all six radiologists, with absolute reductions of 11% in false positives and 5% in false negatives [18].

Regarding breast cancer screening technologies, it is argued that AI may provide the means to limit the inherent drawbacks of mammography and enhance diagnostic performance and robustness. In a prospective clinical study, a commercially available AI algorithm was evaluated as an independent reader of screening mammograms, and adequate diagnostic performance was reported [19].

### 4.2. Precision Cancer Diagnosis

During the last decades CAD-driven precision diagnosis has been the holy grail of medical image processing research efforts. However, the clinical interest in such applications has significantly grown only recently with the advancement of AI-driven efforts to generalize performance across diverse datasets. AI systems have reported unprecedented performance regarding the segmentation and classification of cancer. A recent study reported increased performance in segmenting and classifying brain tumors into meningioma, glioma, and pituitary tumors [20].

In addition, a growing number of studies are concerned with automated tumor grading, which is a prerequisite for optimal therapy planning. Yang et al. presented a retrospective glioma grading study (grade II and grade III concerning low grade glioma and high grade glioma) on one hundred and thirteen glioma patients and used transfer learning with AlexNet and GoogLeNet architectures, achieving up to 0.939 AUC [21].

At the same time, the quest to decode imaging phenotype has given rise to efforts to correlate imaging features with molecular and genetic markers in the context of radiogenomics [22]. This promising field of research can provide surrogate molecular information directly from medical images and is not prone to biopsy sampling errors, as the whole tumor can be analyzed. In a recent study, MRI radiomics were able to predict IDH1 mutation with an AUC of up to 90% [23].

### 4.3. Treatment Optimization

There are many challenging problems in optimizing treatment for cancer patients, such as accurate segmentation of organs at risk (OAR) in radiotherapy and prediction of neoadjuvant chemotherapy response. Intelligent processing of medical images has opened new horizons to address these clinical needs. In the case of nasopharyngeal carcinoma radiotherapy planning, a deep learning organs-at-risk (OAR) detection and segmentation network provides useful insights for clinicians for the accurate delineation of OARs [24]. Regarding prediction of neoadjuvant chemotherapy, the use of image-based algorithms to predict outcome has the potential to add precision, not least due to the fact that depending on tumor subtype the outcome can differ significantly. To this end, recent studies report promising preliminary results in applying AI to predict breast cancer neoadjuvant therapy response. Vulchi et al. reported improved prediction of response to HER2-targeted neoadjuvant therapy based on deep learning of DCE-MRI data [25]. Notably, the AUC dropped from 0.93 to 0.85 in the external validation cohort.

## 5. Radiomics Limitations Regarding Clinical Translation

While promising, radiomics methodologies are still in a translational phase and thorough clinical validation is needed towards clinical translation. To this end, when these technologies are tested and reviewed, a number of important limitations becomes apparent. In a recent review on MRI based radiomics in nasopharyngeal cancer [26], the authors reviewed the state of the art and used a radiomic quality score assessment (RQS). Several limitations were highlighted in the reviewed studies, including the absence of a validation cohort in 21% of them, as well as the lack of external validation in 92% of them. In another RQS based evaluation study on radiomics and radio-genomics papers, the RQS was low regarding clinical utility, test-retest analysis, prospective study, and open science [27]. It was also very interesting that no single study used phantoms to assess the robustness of radiomics features or performed a cost-effectiveness analysis. In a similar fashion, lack of feature robustness assessment and external validation was reported in studies regarding prostate cancer [28], while the main reported shortcomings in the quality of the MRI lymphoma radiomics studies regarded inconsistencies in the segmentation process and the lack of temporal data to increase model robustness [29]. All these recent studies clearly indicate that, although medical image processing in oncology has evolved significantly, the clinical translation of radiomics is still hampered by the lack of extensive, high quality validation studies. In addition, the lack of standardization in radiomics extraction remains a problem, which is currently being investigated by several studies, with respect to different software packages [30] and the reproducibility of standardized radiomics features using multi-modality patient data [31].

## 6. Discussion

Contrary to common belief, medical image processing has been evolving for the last few decades and its main application is cancer image analysis. Traditional medical image processing was founded on classical image processing and computer vision principles, focusing on low-level feature extraction and simple classification tasks, e.g., benign vs. malignant, or in the geometrical alignment of temporal images and the segmentation of tumors for volumetric analyses. This early stage in the 1990s was an important milestone for further development, since several radiologists and oncologists understood the future potential and helped in the creation of a multidisciplinary community on medical image analysis and processing. More importantly, it laid the foundations of radiomics by proposing the shape and textural analysis of tumors as useful patterns for detection, segmentation, and classification. However, the main limitation was the high degree of fragmentation in such efforts, the limited computational resources, and the very low availability of cancer image data; usually being mammograms or MRIs.

Functional imaging was another important milestone for medical image computing, since the idea of transforming dynamic image signals to tissue properties paved the way for the discovery of reliable and reproducible image biomarkers for oncology. To achieve this goal, non-conventional medical image processing was deployed based on compartmental models to link the imaging phenotype with microscopic tumor environment properties, based on diffusion and perfusion. Such model-based approaches include compartment pharmacokinetic models for DCE-MRI and the IVIM model for DWI-MRI, often requiring laborious pre-processing to transform the original signal to quantitative parametric maps able to convey perfusion and cellularity information to the clinician. It is argued that this is still an evolving research field and that the potential for clinical translation is significant, especially since techniques such as DWI-MRI do not involve ionizing radiation or the administration of contrast agent. That said, significant standardization efforts are still required in order to converge on stable imaging protocols and model implementations that will guarantee reproducible parametric maps and robust cancer biomarkers. Another limitation when comparing to modern radiomics/deep learning efforts is that the processing of such functional data with compartmental models is a very demanding task, requiring a deeper understanding of imaging protocols, as well as of numerical analysis methods for model fitting.

The gradual advancements of high-performance computing and machine learning and neural networks have revolutionized research in the field, especially during the last decade. The field of radiomics has extended the cancer medical image processing concepts regarding texture and shape descriptors to massive feature extraction and modeling. Such radiomics approaches have also been enhanced by convolutional neural networks, which outperformed the traditional image analysis methods in tasks such as lesion segmentation, while introducing more sophisticated predictive, diagnostic, and correlative pipelines towards precision diagnostics, therapy optimization, and synergistic radio-genomic biomarker discovery. The availability of open access computational tools for machine and deep learning, in combination with public cancer image resources such as the Cancer Imaging Archive (TCIA), has led to an unprecedented number of publications, AI start-ups, and accelerated discussions for the establishment of AI regulatory processes and clinical translation of such technologies. At the same time, the main limitation of these impressive technologies has been their low explainability, which came as a tradeoff for the impressive performances in oncological applications throughout the cancer continuum. Low explainability also contributed to reduced trust in these models, while the vast number of features explored made generalization difficult, especially due to the large variability of image quality and imaging protocols across vendors and clinical sites.

Medical image processing is still evolving and will continue to provide useful tools and methodological concepts for improving cancer image analysis and interpretation. Data science approaches focusing on radiomics have paved the way for accelerating precision oncology [32]. However, most of the efforts to date only use imaging data, which limits the performance of diagnostic and prognostic tools. To this end, novel data integration paradigms, exploiting both imaging and multi-omics data, is a very promising field for future research [33]. Recent studies have started exploring the synergy of deep learning with quantitative parametric maps. In [34], the authors present a deep learning method to predict good responders of locally advanced rectal cancer trained on apparent diffusion coefficient (ADC) parametric scans from different vendors. The fusion of standard imaging representations with parametric maps, as well as integrative diagnostic approaches [35] involving medical image and other cancer related data, hold promise for increasing accuracy and trustworthiness.

**Funding:** This research received no external funding.

**Institutional Review Board Statement:** Not applicable.

**Informed Consent Statement:** Not applicable.

**Conflicts of Interest:** The author declares no conflict of interest.

## References

1. Oakden-Rayner, L.; Carneiro, G.; Bessen, T.; Nascimento, J.C.; Bradley, A.P.; Palmer, L.J. Precision Radiology: Predicting longevity using feature engineering and deep learning methods in a radiomics framework. *Sci. Rep.* **2017**, *7*, 1648. [CrossRef] [PubMed]
2. Cai, W.-L.; Hong, G.-B. Quantitative image analysis for evaluation of tumor response in clinical oncology. *Chronic Dis. Transl. Med.* **2018**, *4*, 18–28. [CrossRef] [PubMed]
3. Duran, R.; Chapiro, J.; Frangakis, C.; De Lin, M.; Schlachter, T.R.; Schernthaner, R.E.; Wang, Z.; Savic, L.J.; Tacher, V.; Kamel, I.R.; et al. Uveal melanoma metastatic to the liver: The role of quantitative volumetric contrast-enhanced MR imaging in the assessment of early tumor response after transarterialchemo. *Transl. Oncol.* **2014**, *7*, 447–455. [CrossRef] [PubMed]
4. Aykan, N.F.; Özatlı, T. Objective response rate assessment in oncology: Current situation and future expectations. *World J. Clin. Oncol.* **2020**, *11*, 53–73. [CrossRef] [PubMed]
5. Froelich, M.F.; Petersen, E.L.; Heinemann, V.; Nörenberg, D.; Hesse, N.; Gesenhues, A.B.; Modest, D.P.; Sommer, W.H.; Hofmann, F.O.; Stintzing, S.; et al. Impact of Size and Location of Metastases on Early Tumor Shrinkage and Depth of Response in Patients With Metastatic Colorectal Cancer: Subgroup Findings of the Randomized, Open-Label Phase 3 Trial FIRE-3/AIO KRK-0306. *Clin. Colorectal Cancer* **2020**, *19*, 291–300.e5. [CrossRef] [PubMed]
6. Sasieni, P. Evaluation of the UK breast screening programmes. *Ann. Oncol.* **2003**, *14*, 1206–1208. [CrossRef]
7. Marias, K.; Behrenbruch, C.; Parbhoo, S.; Seifalian, A.; Brady, M. A registration framework for the comparison of mammogram sequences. *IEEE Trans. Med. Imaging* **2005**, *24*, 782–790. [CrossRef]
8. Funovics, M.; Schamp, S.; Lackner, B.; Wunderbaldinger, P.; Lechner, G.; Wolf, G. Computerassistierte diagnose in der mammographie: Das R2 imagechecker- system in der detektion spikulierter lasionen. *Wien. Med. Wochenschr.* **1998**, *148*, 321–324.
9. Manikis, G.C.; Nikiforaki, K.; Lagoudaki, E.; de Bree, E.; Maris, T.G.; Marias, K.; Karantanas, A.H. Differentiating low from high-grade soft tissue sarcomas using post-processed imaging parameters derived from multiple DWI models. *Eur. J. Radiol.* **2021**, *138*, 109660. [CrossRef]
10. Messina, C.; Bignone, R.; Bruno, A.; Bruno, A.; Bruno, F.; Calandri, M.; Caruso, D.; Coppolino, P.; De Robertis, R.; Gentili, F.; et al. Diffusion-Weighted Imaging in Oncology: An Update. *Cancers* **2020**, *12*, 1493. [CrossRef]
11. Fliedner, F.P.; Engel, T.B.; El-Ali, H.H.; Hansen, A.E.; Kjaer, A. Diffusion weighted magnetic resonance imaging (DW-MRI) as a non-invasive, tissue cellularity marker to monitor cancer treatment response. *BMC Cancer* **2020**, *20*, 134. [CrossRef] [PubMed]
12. Fritz-Hansen, T.; Rostrup, E.; Larsson, H.B.W.; Søndergaard, L.; Ring, P.; Henriksen, O. Measurement of the arterial concentration of Gd-DTPA using MRI: A step toward quantitative perfusion imaging. *Magn. Reson. Med.* **1996**, *36*, 225–231. [CrossRef]
13. Woolf, D.K.; Taylor, N.J.; Makris, A.; Tunariu, N.; Collins, D.J.; Li, S.P.; Ah-See, M.-L.; Beresford, M.; Padhani, A.R. Arterial input functions in dynamic contrast-enhanced magnetic resonance imaging: Which model performs best when assessing breast cancer response? *Br. J. Radiol.* **2016**, *89*, 20150961. [CrossRef] [PubMed]
14. Shukla-Dave, A.; Obuchowski, N.A.; Chenevert, T.L.; Jambawalikar, S.; Schwartz, L.H.; Malyarenko, D.; Huang, W.; Noworolski, S.M.; Young, R.J.; Shiroishi, M.S.; et al. Quantitative imaging biomarkers alliance (QIBA) recommendations for improved precision of DWI and DCE-MRI derived biomarkers in multicenter oncology trials. *J. Magn. Reson. Imaging* **2019**, *49*, e101–e121. [CrossRef]
15. Liu, L.; Chen, J.; Fieguth, P.; Zhao, G.; Chellappa, R.; Pietikäinen, M. From BoW to CNN: Two Decades of Texture Representation for Texture Classification. *Int. J. Comput. Vis.* **2019**, *127*, 74–109. [CrossRef]
16. Svoboda, E. Artificial intelligence is improving the detection of lung cancer. *Nature* **2020**, *587*, S20–S22. [CrossRef] [PubMed]
17. Binczyk, F.; Prazuch, W.; Bozek, P.; Polanska, J. Radiomics and artificial intelligence in lung cancer screening. *Transl. Lung Cancer Res.* **2021**, *10*, 1186–1199. [CrossRef]
18. Ardila, D.; Kiraly, A.P.; Bharadwaj, S.; Choi, B.; Reicher, J.J.; Peng, L.; Tse, D.; Etemadi, M.; Ye, W.; Corrado, G.; et al. End-to-end lung cancer screening with three-dimensional deep learning on low-dose chest computed tomography. *Nat. Med.* **2019**, *25*, 954–961. [CrossRef]
19. Salim, M.; Wåhlin, E.; Dembrower, K.; Azavedo, E.; Foukakis, T.; Liu, Y.; Smith, K.; Eklund, M.; Strand, F. External Evaluation of 3 Commercial Artificial Intelligence Algorithms for Independent Assessment of Screening Mammograms. *JAMA Oncol.* **2020**, *6*, 1581. [CrossRef] [PubMed]
20. Díaz-Pernas, F.J.; Martínez-Zarzuela, M.; Antón-Rodríguez, M.; González-Ortega, D. A Deep Learning Approach for Brain Tumor Classification and Segmentation Using a Multiscale Convolutional Neural Network. *Healthcare* **2021**, *9*, 153. [CrossRef]
21. Yang, Y.; Yan, L.-F.; Zhang, X.; Han, Y.; Nan, H.-Y.; Hu, Y.-C.; Hu, B.; Yan, S.-L.; Zhang, J.; Cheng, D.-L.; et al. Glioma Grading on Conventional MR Images: A Deep Learning Study With Transfer Learning. *Front. Neurosci.* **2018**, *12*. [CrossRef]
22. Trivizakis, E.; Papadakis, G.Z.; Souglakos, I.; Papanikolaou, N.; Koumakis, L.; Spandidos, D.A.; Tsatsakis, A.; Karantanas, A.H.; Marias, K. Artificial intelligence radiogenomics for advancing precision and effectiveness in oncologic care (Review). *Int. J. Oncol.* **2020**, *57*, 43–53. [CrossRef] [PubMed]

23. Choi, Y.; Nam, Y.; Lee, Y.S.; Kim, J.; Ahn, K.-J.; Jang, J.; Shin, N.-Y.; Kim, B.-S.; Jeon, S.-S. IDH1 mutation prediction using MR-based radiomics in glioblastoma: Comparison between manual and fully automated deep learning-based approach of tumor segmentation. *Eur. J. Radiol.* **2020**, *128*, 109031. [CrossRef] [PubMed]
24. Liang, S.; Tang, F.; Huang, X.; Yang, K.; Zhong, T.; Hu, R.; Liu, S.; Yuan, X.; Zhang, Y. Deep-learning-based detection and segmentation of organs at risk in nasopharyngeal carcinoma computed tomographic images for radiotherapy planning. *Eur Radiol.* **2019**, *29*, 1961–1967. [CrossRef]
25. Vulchi, M.; El Adoui, M.; Braman, N.; Turk, P.; Etesami, M.; Drisis, S.; Plecha, D.; Benjelloun, M.; Madabhushi, A.; Abraham, J. Development and external validation of a deep learning model for predicting response to HER2-targeted neoadjuvant therapy from pretreatment breast MRI. *J. Clin. Oncol.* **2019**, *37*, 593. [CrossRef]
26. Spadarella, G.; Calareso, G.; Garanzini, E.; Ugga, L.; Cuocolo, A.; Cuocolo, R. MRI based radiomics in nasopharyngeal cancer: Systematic review and perspectives using radiomic quality score (RQS) assessment. *Eur. J. Radiol.* **2021**, *140*, 109744. [CrossRef]
27. Park, J.E.; Kim, D.; Kim, H.S.; Park, S.Y.; Kim, J.Y.; Cho, S.J.; Shin, J.H.; Kim, J.H. Quality of science and reporting of radiomics in oncologic studies: Room for improvement according to radiomics quality score and TRIPOD statement. *Eur. Radiol.* **2020**, *30*, 523–536. [CrossRef]
28. Stanzione, A.; Gambardella, M.; Cuocolo, R.; Ponsiglione, A.; Romeo, V.; Imbriaco, M. Prostate MRI radiomics: A systematic review and radiomic quality score assessment. *Eur. J. Radiol.* **2020**, *129*, 109095. [CrossRef]
29. Wang, H.; Zhou, Y.; Li, L.; Hou, W.; Ma, X.; Tian, R. Current status and quality of radiomics studies in lymphoma: A systematic review. *Eur. Radiol.* **2020**, *30*, 6228–6240. [CrossRef]
30. McNitt-Gray, M.; Napel, S.; Jaggi, A.; Mattonen, S.A.; Hadjiiski, L.; Muzi, M.; Goldgof, D.; Balagurunathan, Y.; Pierce, L.A.; Kinahan, P.E.; et al. Standardization in Quantitative Imaging: A Multicenter Comparison of Radiomic Features from Different Software Packages on Digital Reference Objects and Patient Data Sets. *Tomography* **2020**, *6*, 118–128. [CrossRef] [PubMed]
31. Zwanenburg, A.; Vallières, M.; Abdalah, M.A.; Aerts, H.J.W.L.; Andrearczyk, V.; Apte, A.; Ashrafinia, S.; Bakas, S.; Beukinga, R.J.; Boellaard, R.; et al. The Image Biomarker Standardization Initiative: Standardized Quantitative Radiomics for High-Throughput Image-based Phenotyping. *Radiology* **2020**, *295*, 328–338. [CrossRef]
32. Capobianco, E.; Dominietto, M. From Medical Imaging to Radiomics: Role of Data Science for Advancing Precision Health. *J. Pers. Med.* **2020**, *10*, 15. [CrossRef]
33. Rundo, L.; Militello, C.; Vitabile, S.; Russo, G.; Sala, E.; Gilardi, M.C. A Survey on Nature-Inspired Medical Image Analysis: A Step Further in Biomedical Data Integration. *Fundam. Inform.* **2019**, *171*, 345–365. [CrossRef]
34. Zhu, H.-T.; Zhang, X.-Y.; Shi, Y.-J.; Li, X.-T.; Sun, Y.-S. A Deep Learning Model to Predict the Response to Neoadjuvant Chemoradiotherapy by the Pretreatment Apparent Diffusion Coefficient Images of Locally Advanced Rectal Cancer. *Front. Oncol.* **2020**, *10*. [CrossRef]
35. Chaddad, A.; Daniel, P.; Sabri, S.; Desrosiers, C.; Abdulkarim, B. Integration of Radiomic and Multi-omic Analyses Predicts Survival of Newly Diagnosed IDH1 Wild-Type Glioblastoma. *Cancers* **2019**, *11*, 1148. [CrossRef]

Journal of
*Imaging*

MDPI

*Review*

# Breast Cancer Risk Assessment: A Review on Mammography-Based Approaches

**João Mendes and Nuno Matela** *

Faculdade de Ciências, Instituto de Biofísica e Engenharia Biomédica, Universidade de Lisboa, 1749-016 Lisboa, Portugal; jpmendes@fc.ul.pt
* Correspondence: nmatela@fc.ul.pt

**Abstract:** Breast cancer affects thousands of women across the world, every year. Methods to predict risk of breast cancer, or to stratify women in different risk levels, could help to achieve an early diagnosis, and consequently a reduction of mortality. This paper aims to review articles that extracted texture features from mammograms and used those features along with machine learning algorithms to assess breast cancer risk. Besides that, deep learning methodologies that aimed for the same goal were also reviewed. In this work, first, a brief introduction to breast cancer statistics and screening programs is presented; after that, research done in the field of breast cancer risk assessment are analyzed, in terms of both methodologies used and results obtained. Finally, considerations about the analyzed papers are conducted. The results of this review allow to conclude that both machine and deep learning methodologies provide promising results in the field of risk analysis, either in a stratification in risk groups, or in a prediction of a risk score. Although promising, future endeavors in this field should consider the possibility of the implementation of the methodology in clinical practice.

**Keywords:** breast cancer; risk assessment; machine learning; deep learning; texture; mammography



**Citation:** Mendes, J.; Matela, N. Breast Cancer Risk Assessment: A Review on Mammography-Based Approaches. *J. Imaging* **2021**, *7*, 98. https://doi.org/10.3390/jimaging7060098

Academic Editors: Antoine Vacavant, Leonardo Rundo, Carmelo Militello, Vincenzo Conti, Fulvio Zaccagna and Changhee Han

Received: 14 April 2021
Accepted: 9 June 2021
Published: 12 June 2021



## 1. Introduction

One in eight women will be diagnosed with breast cancer (BC) in their lifetime, with one in thirty-nine women dying from this disease just in the USA. In the same country, in 2020, approximately 42,170 women were expected to die from BC and it was anticipated that approximately 30% of the cancers detected in women will be BC [1]. Around 95% of cancers are due to genetic mutations that result from environmental or lifestyle factors, where the remaining percentage is related to inherited genes—with BRCA1/BRCA2 genes being responsible for most of cases of BC [2,3].

BC diagnosis occurs either during a common screening program, before symptoms appear, or after women noticing some breast changes. Screening programs are important for an early detection of BC—that is, in a more treatable stage—resulting in a decrease in mortality [1,4].

The criterion that defines if a woman is eligible for screening is, normally, only her age. Different countries have different recommendations on which age is the best to start screening; the USA states that women from age 45 to 54 should have a mammography once a year, while 55+ plus women should have a mammography once every two years. On the other hand, the UK NHS says that only women between 50 to 71 should be screened, and only once every three years [5,6].

Although there are multiple screening programs, they might not serve all women. Some younger women may be at higher risk of developing breast cancer than women in their fifties and, despite that, these women are not eligible for screening. With that in mind, the perfect screening program should not consider age as the only risk factor that determines when to screen women.

The question resides in what risk factors are not being considered when choosing the best screening option. Age is one of the best documented risk factors, with the incidence of BC being extremely low before the age of 30 and having a linearly increase until the age of 80 [7]. Body Mass Index has also been shown to be a potential risk factor for the development of BC but only in post-menopausal ages [7,8]. Prior history of neoplastic or hyperplastic breast disease also presents itself as a risk factor for the development of BC. When it comes to family history, a woman who had a first-degree relative with BC when was 50 years or older, is almost twice at risk of developing breast cancer than a woman with no family history of BC [7]. Early menarche, late first full-term pregnancy and late menopause are three major risk factors for breast cancer [9]. Normally, the earlier the age of the first menarche, the higher the cancer risk. The fact that both women with early menarche and later menopause are at higher risk of BC, can lead to the conclusion that prolonged exposure to estrogen is also a risk factor for this disease [9]. Longer duration of the breastfeeding period is associated with a diminished risk of breast cancer, in comparison with women that had shorter breastfeeding periods. Use of oral contraceptives also puts women at higher risk of developing BC [10]. As it was previously discussed, the existence of the BRCA1/BRCA2 mutated gene in women karyotype puts them at higher risk of BC, compared to women who do not possess that gene [11]. Besides these risk factors, in 1976 Wolfe, started studying the association between breast parenchyma patterns and breast cancer. Wolfe showed that a prominent duct pattern helps to classify a woman as having higher risk than average for developing breast cancer. Wolfe also stated that it is possible to predict which women will develop breast cancer and which are less likely to develop it based only on the parenchymal pattern [12–15].

Many descriptors of these texture patterns have been documented. Mammographic density is one of those descriptors, normally represented numerically by percent mammographic density (% PD), that is also highly associated with an increased risk of breast cancer [16–18]. In fact, women with 60–70% PD are at four to five times higher risk than women with fatty breasts. Dense breasts are not only at higher risk of developing breast cancer as are also more prone to more aggressive tumors.

Screening programs all around the world use mammography, that can be acquired in a cranio-caudal (CC) and/or in a mediolateral-oblique (MLO) view, as a standard method for diagnosis, but although widely used, mammography has both benefits and harms. The aim for an early detection of this disease started in the beginning of the 20th century with awareness campaigns, but a decrease in BC mortality was only observed when the first mammographic screening started. On the bright side of mammography screening, life-threatening cancers will be detected early, improving prognosis, and consequently, decreasing risk of mortality. Studies point out that BC mortality rates, decreased at least 20% [19] thanks to an increase in mammographic screening—some studies even point out a reduction ranging from 30–50% [20]. Besides that, since cancer can be detected in an early stage, the available treatment can be less invasive and, consequently, have lower costs. The treatment will also be less intense, resulting in fewer time off of work, and, consequently, smaller money losses.

One of the problems associated with mammography is the rate of false positives. In Europe, the risk of having a false-positive result, for women in the range of 50–69 years having biennial screening, is about 20%. More alarming are the results in the United States, where all screened women will experience one false-positive in their life. These false-positive results have an impact in women lives, especially in day-to-day well-being and in costs concerning healthcare. But the presence of false positive is not the only downside of mammography. A summary of the benefits and harms of mammography in 1000 women with a screening every two years showed that 200 of them will experience a false positive, 30 will have a biopsy due to the false positive result, 15 will be over-diagnosed and three will develop interval cancers – the name given to a cancer that appears between two consecutive mammograms. These interval cancers may have been developed between the two mammograms, however, around 35% of them were already present in the previous mammogram but were overlooked. This means that the patient received a false negative result that can occur because, in mammography, there is an overlap of tissue that can obscure the presence of cancers [21]. Since the population being screened is mainly composed of asymptomatic women, it is expected that with increased screening, it will also be seen an increase in cancer incidence. Life-threatening cancers will be detected early, improving prognosis, which is clearly a point in favor of mammography screening, however, cancers that would never be detected and that, in theory, were not harmful for the woman who presents it, will also be diagnosed. This is called overdiagnosis. Overdiagnosis leads to an ethical dilemma since there is a probability for the patient to live longer with cancer than with the treatment, and this decision-making process could lead to an increased anxiety state of the patient [21]. Another important aspect to consider, related to this type of screening, is the relation between mammography and dense breasts. The sensitivity of mammography decreases in women who have dense breasts (30–64% vs. 76–98% in women with fatty breasts) [19], which occurs because cancers have attenuation coefficients closer to dense tissue. Actually, a study from 1999 [22] showed that there was a significant trend between breast density and the appearance of false positives. Since it is known that breast density is an important risk factor for the development of BC, the fact that mammography does not perform so well in dense breasts should be of great concern. As seen, there are multiple downsides to mammography and, yet it continues to be the standardized screening method. However, in 2014, the Swiss Medical Board stated that the harms produced by these screening programs outweighed the benefits and, therefore, they recommended Switzerland to stop all the mammography screening programs [23].

New technologies that allow a risk stratification, in line with the current medicine paradigm of preventive and personalized care, could help overcoming some discussed problems associated with the screening programs.

A review of Artificial Intelligence (AI) in the field of Breast imaging was already performed by Le and his colleagues [24]. In this work, a brief introduction to Artificial Intelligence is performed, concerning commonly used terminology and widely used algorithms. Applications of computer aided detection (CAD) systems in mammography screening are explored, like the automatic detection of breast cancer, or the distinction between malignant and benign lesions. Software based on AI for breast density classification assessment is also addressed. The authors, beside describing deep learning approaches in mammography, proceed to address relationships and applications of AI to digital breast tomosynthesis, ultrasound and MRI. Finally, the implementation of AI-CAD systems in clinical practice, the limitations of these systems, obstacles to its implementation and future applications are discussed.

In the current work, Section 2 explains the methods by which this review was performed in terms of inclusion criteria. Section 3 presents the results of this review, with each included paper being analyzed in terms of proposed goals, methodology used, and results obtained. Finally, in Section 4 a conclusion and a discussion about future endeavors is made.

## 2. Methods

The review done here aims to present a global picture in what is already done in the field of breast cancer risk assessment through computerized methods, using mammograms. A search in Google Scholar using different Boolean operators and the keywords—breast cancer risk, mammography, machine learning (ML), features, parenchyma/texture patterns—was performed. This search produced eight-hundred and forty-two matches that were screened through title and/or abstract. In order to be considered for this review, papers should meet the following inclusion criteria:

(1) Aim for a risk assessment, either by differentiating risk groups, predicting a risk value, or proposing new methods for the said assessment, using Machine/Deep Learning (DL) tools.

(2) The methodology used should consider textural features with/without epidemiological factors.
(3) Mammography images should be used for feature extraction, when that procedure is done.
(4) All papers' publication date should be within the 2000–2020 range.

Papers cited by the accepted manuscripts were also screened through the previously referred criteria. Only the articles that better served the scope of this work were considered, which resulted in 11 included manuscripts. Figure 1 represents a flowchart of the methodology used.

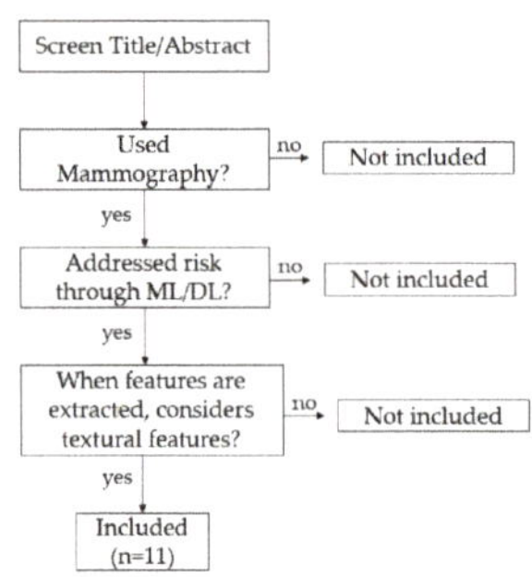


**Figure 1.** Paper Inclusion flowchart, for a search within the range 2000–2020 for publication date.

## 3. Results

### 3.1. Risk Assessment Using a Single Region of Interest

In the start of the millennium, Huo et al. [18], investigated feature selection in breast parenchymal patterns, for BC risk assessment. The specific aim of the study was to classify women, based on their mammograms, into high or low-risk groups. To label the training data, the authors used Gail's model, which asks as input some epidemiologic information: age; age at the first menarche; age at full first-time birth; number of first-degree relatives with BC; and number of previous breast biopsies; then, with that information, calculates a probability for developing BC [25]. This method, although very used, has its limitations, for example, it cannot be used in younger women, and it is unable to predict risk for women with the BRCA1/BRCA2 gene.

To be included in the low-risk group, besides having a risk lower than 10%, women could not have any family history of BC. Mammograms from women with the BRCA1/2 mutation were considered high-risk. An important step in dividing the dataset needs to be referred, as it happens in other similar researches. The age of mutation carriers and the "low-risk" group tends to be different, and, in order to avoid bias due to that difference, an aged-match dataset was constructed, and risk analysis was also conducted in this "sub-dataset".

Once the datasets were divided, mammograms proceeded to be pre-processed feature extraction being conducted in a pre-defined region of interest (ROI). Intensity-based features, statistical measures based in absolute pixel value; co-occurrence (GLCM) features [26], metrics that describe pixel pairs co-occurrences throughout the image; and two Fourier analysis features were extracted—a description of these features can be found in the referred manuscript. Once the extraction step was done, each feature was analyzed, through receiver operating characteristic (ROC) analysis, in order to access their discriminative capacity between high-risk and low-risk groups. The area under the curve (AUC) for each individual feature ranged from $0.53 \pm 0.09$ (minimum gray-level) to $0.87 \pm 0.05$ (skewness). After this evaluation, a feature selection method was applied so as to reduce the dimensionality of the problem and increase computational efficiency. This was achieved using stepwise selection followed by linear discriminant analysis. Discriminant Analysis chose intensity-based and co-occurrence features for the best set of features in the task of differentiating the two considered risk groups. Curiously, the chosen features were the ones that presented better discriminative capacity in the individual ROC analysis for the age-matched group. The linear discriminant analysis approach presented an AUC higher than any of the features alone—0.91.

Besides proving the usefulness of the texture features to characterize the difference between low and high-risk women, some interesting conclusions can made from this study considering features' average values for each group: the textural patterns from high-risk women tend to be coarser and lower in contrast; skewness measure should have negative values for high-risk women; and all the remaining intensity-based features should have higher values for high-risk women.

To ensure that a good parenchyma characterization could be done through one projection, the authors made a correlation study between CC-L and MLO-L, and between CC-L and CC-R views for each feature, which provided positive results. A limitation of this study might also be the fact that some of the low-risk women may have the mutation without knowing, which clearly affects features' discriminative capacity.

Li et al. [27] studied, in 2004, the effect of ROI size and location for feature extraction in BC risk analysis. The performance of the size and location of each region was evaluated in the task of differentiating high-risk (mutation carriers) and low-risk women. An aged-matched group between mutation carriers and low-risk women was created and used for risk analysis in this study. Researchers designated five different ROI locations, as depicted in Figure 2, identified by the letters A, B, C, D and E.

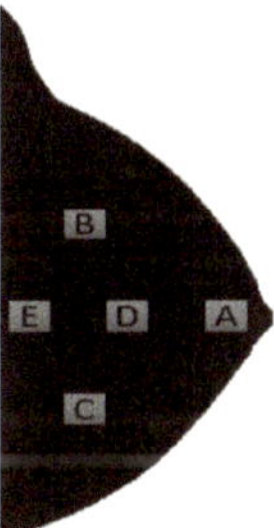


**Figure 2.** Different locations (**A–E**) where features were extracted for ROI evaluation.

For size analysis, 2 ROI's, one with a medium and other with a small size, were directly defined in the center of the larger, pre-defined, ROI, at locations A, B and C. The extracted features were the same that Huo extracted with the addition of a fractal dimension measure [28]. Stepwise feature selection with LDA was employed as a feature selection method, after feature extraction from the different ROI's was performed. Each feature was individually evaluated for its discriminative capacity between the considered groups, through ROC analysis, and the same was done to the linear discriminant analysis approach. Descriptors from Fourier analysis, co-occurrence, intensity-based and fractal dimension were chosen by the feature selection methodology, a process that is fully described in the paper and complemented in [29].

In what concerns to ROI size, for location A, the AUC for each individual feature ranged from 0.68 to 0.83 with the lower results being associated to smaller regions of interest. The performance of the LDA approach was of 0.92 in the original ROI, with AUC's of 0.87 and 0.89 in the medium and smaller ROI's, respectively. In the size analysis, significance was only achieved for one feature and for the LDA approach between the large and medium ROI and for the fractal dimension between the large and small ROI. The values of the AUC in the LDA approach for region B were substantially lower than the ones observed for region A and, besides that, no statistical significance was observed for this region neither in individual features assessment nor in the LDA approach. Finally, for region C, the only statistical significance was achieved for feature contrast, between large and medium, nonetheless, no statistical significance was obtained in the discriminative capacity between different sized ROI's.

Analyzing ROI location effects, for comparison purposes, only the features selected by the LDA approach for region A were considered. Most of the individual features were statistically significant different between different locations, and the same can be said about the LDA approach that presented AUC's of 0.92, 0.78, 0.69, 0.84, and 0.79 respectively for each region. A statistically significant decrease was observed in the LDA performance if the ROI was moved away from region A, that is located immediately behind the nipple, which probably explains why most of the approaches in this area of research use this location for feature extraction. The authors point out that the fact that the region immediately behind nipples has the best discriminative capacity may be due to the existence of a dense component in that breast location. The group still states that, in the future, research should extract features from the entire breast and compare the results with the ones obtained with a single-ROI approach that are, besides its limitations, positive.

In 2005, Li et al. [30] aimed to prove the usefulness of breast parenchymal patterns present on mammograms in the field of breast cancer risk assessment. As it had happened in the previous analyzed researches, the authors aimed to extract texture feature from mammograms to differentiate high-risk women, mutation carriers, from those who have a low-risk of developing the disease. In order to be considered for the low-risk group, besides the two conditions presented in the research done by Huo in 2000, women could not had been diagnosed with breast cancer in the past and, if they had done a biopsy, they were not considered. Once again, since age is an important risk factor, an age-matched group between low-risk women and mutation carriers was created. At each mammogram, features were extracted from the pre-defined ROI and, although ranging the common groups, they slightly differ from previous researches. Besides intensity-based, co-occurrence, Fourier and fractal dimension features, mean gradient, an edge frequency feature that measures the coarseness of a surface, was part of the studied features and is described in the paper. ROC analysis was used to evaluate the individual performance of each feature in the task of differentiating high-risk and low-risk women. The results ranged from $0.66 \pm 0.05$ (Entropy) to $0.86 \pm 0.03$ (co-occurrence contrast) in the entire dataset, and from $0.67 \pm 0.05$ to $0.86 \pm 0.05$ in the age-matched group, with statistical significance ($p$-value $< 0.001$) being achieved for all features. The authors proceed to present a figure where a distribution of skewness measure in the population can be observed. From that, it is drawn the conclusion that high-risk women present negative values of skewness, as advanced by Huo in 2000, since these women normally have denser breasts, when compared to lower-risk women, and, in general, high-risk women present lower skewness values than women at the low-risk group. Contrast helps to describe the local tissue variation, and higher values of these features were observed in low-risk women, which leads to the assumption that mutation carriers tend to present texture patterns low in contrast. Besides that, results analysis in terms of feature values lead the authors to state that mutation carriers tend to have coarser textures than low-risk women. Although these conclusions from contrast and coarseness can be made in general, not all women follow this trend. Nonetheless, this study presents itself as another proof that mammographic texture patterns can be successfully used in the field of breast cancer risk assessment.

### *3.2. A Disruption from the Classical One-ROI Approach for Risk Assessment*

Another interesting research is the one done by Tan et al. [31], where the authors aimed to evaluate the viability of predicting BC risk in women after they had a negative mammogram. Given a sample of screened women, each woman was considered for the study if had had two consecutive mammograms acquired in the authors' facilities and if the first mammography was negative. A dataset was then created with the accepted women, where each case was composed by two mammograms—defined as "prior" and "current" evaluations—and based on the current evaluation, the dataset was divided into three subgroups. The first was composed of women who had positive results, confirmed with other evaluation methods, and women who had pre-cancerous masses that were removed. The second subgroup consisted of women who had abnormalities in their mammograms, were recalled, but then the lesions proved to be benign. Finally, the third subgroup included women with negative mammograms and that were not recalled. In the study, the researchers used all the "prior" evaluations, that were negative, to assess breast cancer risk in the "current" evaluation. It is important to mention that, for each dataset case, age, family history of BC, and the density rating by the BI-RADS scale were the epidemiologic risk factors considered. For feature analysis purposes, the authors segmented the breast in different regions and extracted features in the segmented areas, therefore considering the whole breast for feature extraction. The extracted features could be divided, once again, in different subgroups: Intensity-based and co-occurrence features—in the horizontal direction—were extracted. Run-length (RL) features [32], that describe runs of same intensity pixels in a given image were also considered in both the vertical and horizontal direction. Besides that, another group of features, that the authors called "x-axis/y-axis histogram cumulative projection" was considered, a brief explanation of this group of features is given in the paper. Features were computed in the entire breast and also in dense breast regions, that are defined as regions that are composed of pixels with intensity above the median value of the whole breast.

Although the previously referred features were computed, they will not be directly used for risk assessment purposes. What is done is that each feature is calculated from the CC view of each breast and then, features that describe the bilateral asymmetry of each individual feature will be computed through the following equations:

$$F_{Assymetry\ 1-60} = \frac{|f_i - g_i|}{max(f_i, g_i)} \quad (1)$$

$$F_{Assymetry\ 61-120} = |f_i - g_i| \quad (2)$$

$$F_{Assymetry\ 121-180} = |f_i - g_i|^3 \quad (3)$$

A set of 180 asymmetry features were calculated and, adding the epidemiologic data, a final set of 183 features was considered. To choose the best features, a forward floating selection method was applied- proposed and described by [33]. Once the selected features were retrieved, a support vector machine (SVM) classifier, with a radial basis function kernel, was trained and tested with the referred dataset.

Classifier's performance validation was done using a 10-fold cross-validation methodology and, at each testing step the algorithm outputted a score ranging from 0 to 1. The higher the score, the higher the probability of having an "image-detectable" cancer in the next screening. Feature selection methodology, besides age, selected features from Run-length, Intensity-based, and cumulative projection groups. For classification purposes, using only the first and third subgroups the classifier had an AUC of 0.716 $\pm$ 0.020. Considering all the cases, the SVM model correctly predicted 71.3% using a 0.5 score as a decision threshold for a classification between negative/benign cases and positive cases. Some limitations of the developed work must be considered: The fact that the dataset used was produced in laboratory does not reflect the ratio between positive and negative cases in common BC screening programs; the methods used for validation may have resulted in bias and, the fact that the same portion of the dataset was used both for features selection and to evaluate the classifier accuracy may also have resulted in some bias in the process of optimizing the algorithm. Besides that, only asymmetry features were computed, which could lead to some masking effects of the effective texture of the parenchyma.

Zheng et al. [34] advocated, in 2015, that approaches that use a single ROI for risk assessment are insufficient since they cannot properly define all breast parenchyma, since it does not consider its heterogeneity. The authors stated that texture characterization should be done across all breasts, using structuring elements for feature extraction. The idea that these descriptors, calculated across all breasts recurring to structural elements, could improve texture description was advanced by these researchers and that resulted in the development of a software, that they call lattice-based approach, to extract features from structural elements across the entire breast. A comparison of the association of texture features with cancer between the lattice-based approach and the single-ROI methodology was performed. To use this methodology, breast area definition and pixel value normalization was performed. The next step was PD% computing, achieved by using a clustering algorithm that subdivided the breast into different regions with each region having approximately the same composition; then, a SVM algorithm would classify each subregion as being "fatty" or "dense". PD% was simply computed by dividing the number of dense areas by the total number of subdivisions defined.

The clustering algorithm used here is a variation of the fuzzy c-means (FCM), which works by giving each pixel/data point a membership degree to each cluster. This degree is related to distance metrics taken between the point and the cluster: the lower the distance, the higher the membership degree. Fuzzy c-means will then, through various iterations, try to minimize the intra-cluster variance while maximizing the inter-cluster variance. Besides breast segmentation for % PD calculation, this algorithm has been used for other purposes. A group of researchers used a variation of the FCM, where the influence of spatial neighbor pixels and similar super-pixels is incorporated in the model, for lesion segmentation on brain and breast MRI as also in mammograms [35]. The idea of modifying this algorithm was held by the fact of FCM being highly sensitive to noise because spatial information was not considered. The experimental results were evaluated with different metrics—specificity, accuracy and false alarm Rate (FAR)—and compared to other commonly used segmentation algorithms. Breast MRIs and mammograms were used to assess lesion segmentation, while brain MRI is used to evaluate algorithm's performance in noisy image enhancement. For the brain MRI image, the highest results for both accuracy and specificity were obtained for the methodology proposed by these authors, and the same can be said about the lowest FAR results, proving that best results are obtained by this methodology. Moreover, it also shows that the noise problem can be countered with this algorithm. Different types of breast tissue, breast size and tumor size were considered when studying segmentation of breast MRI images. Their results show that the standard FCM methodology achieved poor results, due to noise, while their methodology provide the best results, with tumor edge being as clear as possible (and not blurred as it happens with other algorithms). In terms of accuracy, specificity and FAR, the proposed method has the best results across all cases. Finally, for tumor segmentation in mammograms, four cases were analyzed, with different characteristics, and the results show that the methodology adopted by the authors was the one that was closer to the standard results obtained by clinicians/experts. Once again, accuracy, specificity and FAR achieved their best results for the authors' methodology. This study provided clues that the proposed methodology can outperform commonly used algorithms in the task of lesion or organ segmentation, even in the presence of noise.

Still concerning brain MRI, FCM has the potential to be used in a pipeline related to neuro-radiosurgery [36]. The authors that propose this approach relate that assessing necrotic tissue that occurs within the tumor might add knowledge about tumor development. The goal of the methodology proposed is then to use FCM for necrosis extraction, after a gross tumor volume segmentation(GTV). This pipeline might allow, for example, to selectively choose the given dose accordingly to zone resistance to radiation. The use of FCM after GTV will make the tumor characterization more precise, with necrotic and enhancement areas being distinguished—by clustering them. This brain tumor necrosis extraction will provide an increased clinically valuable insight about cancer characteristics, while playing an important role in neuro-radiosurgery, in terms of dose redistribution. Several metrics, ones related to spatial overlap, and others concerned with distance were calculated. The first (sensitivity, specificity, etc.) compared the regions that were achieved with this methodology, against the segmented areas obtained by an expert. The latter, contrary to overlap-based metrics, considers the boundary's voxel position in the space, which should be used, since boundary delineation is very important in radiosurgery or treatment planning. Considering overlap metrics, the proposed method provides higher results than conventional methodologies, providing clues that this is in fact an accurate and reliable method. These positive results are corroborated by the spatial metrics, which indicates that this pipeline serves its initial purposes. Given that, FCM presents itself once again as a good clustering algorithm for different goals.

Getting back to Zheng's research, once % PD calculation was performed, feature extraction could be conducted, and for that, the lattice-based approach needed to be considered. The authors display a grid over the entire breast tissue, where different values for the distance between each intersection point, $D$, and for structural element size, $W$, can be considered. The structural elements are centered in the intersection points and will serve as different ROIs for feature extraction, so, each computed feature will have different values across the breast. Although the optimal values for $D$ and $W$ might be different for different regions across the breast, authors considered a fixed and equal value for these components, resulting in a breast that is coated with structural elements. Intensity-based; co-occurrence; run-length; local binary pattern; fractal dimension, and structural features that describe "flow-like structures within the breast" were also considered—the authors provide references for these novel features. In order to look for the optimal W value, approaches with three increasing sizes—small, medium and large—were tested. Each "final value" of the features was defined as the mean value of the said feature across all structuring elements and the association between the computed features and breast cancer was evaluated with a logistic regression classifier with leave-one-out cross validation. Univariate and multivariate analysis were conducted, with feature selection being done in the later one through a forward feature selection methodology applied at each cross-validation loop. Considering univariate analysis and taking all window sizes into account, the average AUC over the GLCM features was of $0.58 \pm 0.03$, which is better than the one presented by the intensity-based features, being of $0.56 \pm 0.05$, the same value that was presented by the run-length features. Structural features presented a worst AUC than GLCM features, with a value of $0.57 \pm 0.06$. Comparing window sizes, the performance seems to be better for small $W$ values, with an average AUC of $0.58 \pm 0.07$ for a small window against AUCs of $0.57 \pm 0.05$ and $0.54 \pm 0.03$ for sizes medium and large, respectively. The feature that presented a higher discriminative capacity was fractal dimension for sizes small and medium, presenting an AUC of $0.69 \pm 0.03$. In where it comes to multivariate analysis, using a logistic regression, the performance was also better with smaller $W$ sizes and the AUCs values obtained were of $0.85 \pm 0.02$, $0.81 \pm 0.02$ and $0.76 \pm 0.03$ for sizes small, medium and large respectively. All the features outperformed PD% performance and no significance was obtained in the model when PD% was added to the set of features. The lattice-based approach significantly outperformed the single-ROI approach either from the retroaerolar area (AUC = $0.60 \pm 0.03$) or the central breast region (AUC = $0.74 \pm 0.03$), despite the $W$ size considered. The results may cause some surprise once, contrary to what was proven by Li in 2004, the central breast region ROI performed better than the retroaerolar area. Given what was discussed about this topic, some conclusion must be drawn: the extraction of features by itself does not result in better discriminative capacity but it is the combination of those features that gives positive results; and $W$ size is important for a better discriminative capacity, with an approach that considers smaller $W$'s providing better outcomes. Nonetheless, some problems related to the work done here must also be considered: the use of equal values for $W$ and $D$ might be a limitation, since much more combinations could be tested if that condition was not present, what could result in a better discriminative capacity; and PD% calculation was done by considering one of the many possible options available to perform that computation, what could also bias the results.

Changes in mammography texture features for breast cancer risk assessment were studied by Tan et al. [37] in a study where, as done in 2013, the authors conjectured that features that describe bilateral asymmetry might be important markers to predict near-term breast cancer risk. What is done differently here is that the authors aim to found features that allow the prediction models to have a better performance, and they compare the risk scores generated by their model with a time-lapse between a negative and positive mammogram of a patient with a number of sequential mammograms. For this study, women with at least four sequential mammograms were considered, with the cancer/risk cases being the ones that were diagnosed with breast cancer in the most recent mammogram, and the remaining being considered as control. The most recent mammograms were considered as "current" evaluations and all the previous mammograms were considered "prior" evaluations, classified as negative in an evaluation done by radiologists, with no recalls happening in the "prior" group. The authors provide an extensive description of two used groups of features—structural similarity features and Weber descriptors—that will not be replicated here. Besides those groups, GLCM, RL and intensity-based features were likewise computed. For GLCM and RL features, only the mean value and the maximum value of each feature across computation directions (0°, 45°, 90°, and 135°) were considered. The breast was segmented and the ratios of the area, within the segmentation, with intensity values above mean pixel intensity values for the whole segmented breast, were considered for PD% measures. Concretely, the authors used three thresholds to compute this ratio: (a) values above the mean; (b) values above the maximum; (c) values below the minimum. Once this was achieved, the study proceeded with the calculation of four features based in % region cutoff of the density function. Using a Sobel gradient operator, statistical measures driven from gradients were considered. Finally, the difference between the number of pixels present in each breast for the same patient was calculated. Equation (3) was used, and the result represented the bilateral asymmetry features between the left and right breast. A SVM algorithm, with a linear kernel, was trained and tested using a leave-one-out cross-validation methodology and, at each training session, stepwise regression was used to select the most relevant features. This procedure was done three times, one for each of the "prior" mammograms. Besides evaluating risk score evolution across the three "prior" mammograms, the authors also aimed to study variations of individual feature values between groups, what was done by computing the mean and standard deviation of the features between the "negative" and "positive" group for each "prior image". After that, using a t-test, $p$-values that assessed the difference between groups at each "prior image" were generated. Given the already explored problem of significantly different ages between high and low risk women, the authors repeated the SVM procedure with two different age-matched groups using a criteria of $\pm 1$ year and a criteria of $\pm 3$ years. Apart from that, the authors trained and tested three different classifiers and, at each time, they used the features selected through one of the prior images sets. Concerning the results, the AUC increased as the time approached the current evaluation, with the values being $0.666 \pm 0.029$, $0.710 \pm 0.028$, and $0.730 \pm 0.027$. As for feature difference results, different trends can be observed, with features having higher discriminatory capacity across the three "prior" examinations (structural similarity), others having significant discriminatory capacity in one or two of the examinations (run-length), and others with no discriminatory capacity in any of the mammograms (contrast). In line with previous research, one can conclude that although individual features might have good discriminatory capacities, it is the use of a multi-feature ensemble, recurring to a machine learning algorithm, that allows a good breast cancer risk assessment. Considering the predicted risks and defining the midpoint as a threshold, the SVMs had an accuracy of 65.7%, presenting a sensitivity of 46.5% and a specificity of 83.0%. When considering the algorithms trained and tested with the age-matched group, no significant difference in AUCs was observed. This, and the 2013 study are approaches widely different from the common ones since they add time-dependent variables for risk prediction that can be used to develop novel techniques for risk assessment in a personalized fashion. The authors proved a decreasing trend in AUC values from the most recent "prior" evaluation to the oldest but got results that point to the fact that this decrease might not occur linearly.

In this research the authors aimed to avoid the 2013 limitation of the cases not representing a screened population by ensuring that the cases were randomly selected by people who were not involved in model construction, what made the average women age in the "positive" group to be higher than the average in the "negative group" mimicking what happens in screening programs. When analyzing this study, some limitations must be considered: (a) model reproducibility might be affected by different acquisition systems and noise and, therefore, methodologies to reduce acquisition impact must be developed; (b) image features related to local region bilateral asymmetry were not used and might improve the obtained AUC; (c) this model does not include epidemiological/ risk factors which is a flaw, when compared to existing models; (d) the low accuracy values for individual features might be an obstacle to clinical use; (e) an examination of how features varied across the prior examinations was not considered but might be an interesting line of research to pursue.

In 2019, a group of researchers [38] aimed for a novel approach for breast cancer risk assessment. In this study each cancer case had age, ethnicity, and BMI matched controls. In what comes to ethnicity, all cases were correctly matched, 83% of the cases were matched for age (±5 years) and 94% of the BMI cases were also correctly matched (±1.5 kg/m$^2$). Feature extraction and PD% (done with the Volpara software, Lynnwood, WA, USA) were extracted from the CC view and, for cancer cases, contralateral images were used to assess risk. For feature extraction, five different subgroups of features can be considered: Intensity-based; GLCM; run-length; structural patterns, like LBP and fractal dimension measures; Weber local descriptors; Sobel gradient approaches introduced in the previous articles; and a new set of features called MRELBP, that can describe macro and microstructure information, having low effort computation and that are robust to image noise; and finally, spectral features, related to Gabor, were also computed. Model validation was done with leave-one-out cross-validation. Stepwise regression was used for feature selection and, at each iteration, F-statistic was calculated in order to assess if each feature had a statistically significant contribute to the model. Spearman's rank correlation was also computed to check for correlation between the more commonly selected features at each leave-one-out loop. After feature selection was conducted, the selected features were merged using linear discriminant analysis with the LDA classifier producing a risk score of each case to have breast cancer, meaning that, at each leave-one-out step, 500 risk scores were generated. When comparing the mean risk scores between cancers and controls, the system output a risk of 0.55 for cancer cases and of 0.44 for controls and this difference was statistically significant ($p < 0.001$). The same cannot be said about PD%, that was of 16.7% for cancer cases and of 16.2% for controls, but with a $p$-value of 0.50. The LDA classifier provided an AUC of 0.68 (95% CI 0.64–0.73), while the Volpara methodology presented an AUC of only 0.52 (95% CI 0.47–0.57), this difference was tested and achieved statistical significancy, proving that the classifier is able to extract more useful information than the measures of PD% done by the software. Intensity-based, co-occurrence, gradient and MRELBP features were amongst the chosen ones by feature selection. The six selected features were not all correlated, and some were only correlated to another two selected features, which proves, by the relative positive results obtained in the discrimination between cases and controls, that the LDA classifier could combine information from both correlated and uncorrelated features. Nonetheless, it can be noted that the obtained AUC was relatively low when compared to other studies and the authors pointed that this might happen due to differences in age and ethnicity of the women used in this study. Nonetheless, this study provides a proof that the methodology used for risk stratification in Caucasian women can be used, here, in Asian population at the same time that also provides new features that have a great discriminative capacity in what concerns to breast cancer risk assessment.

Still referring to Asian population, Gandmokar et al. [39], in the Fifteenth International Workshop on Breast Imaging (Leuven, Belgium, 2020), presented a breast cancer risk prediction model based not only in mammographic texture feature but also in an enormous set of epidemiological features (or risk factors), categorizing women into high-risk and low-risk groups. For each woman in the study the following epidemiologic factors were obtained: height, weight, BMI, age at menarche, menopause status, age at menopause, age at first delivery, parity history, number of children, breastfeeding history, personal history of breast cancer, family history of breast cancer, and degree of consanguinity. For feature assessment, contralateral images from the cancer patients were used and considered the high-risk group and control women images were labeled as "low-risk", it should be noted than only CC views were used. Breast segmentation for density calculations was done using a software called AutoDensity [40], from which results two thresholds, the first that represents the bright area of the mammogram and the second, since that is computed based on the dense area, represents the brightest area. Features concerning intensity-based groups, GLCM manipulation, and Fractal Dimension were extracted (from the bright and the brightest area) and added to the epidemiologic set of features. Then, these features were fed to an ensemble of decision trees, acoplated with AdaBoost that was validated with a leave-one-out cross-validation methodology and presented an AUC of 0.884 (CI 0.838–0.913) in differentiating risk groups. Although the results are promising, study limitation must be assessed; (a) the model was validated in a small dataset; (b) contralateral images were used as high-risk but since the goal is to do a risk prediction the model should be constructed using prior mammograms; (c) study population was from women recruited from a single city which does not represent the usually found differences between women from different locations; (d) the control and cancer cases were driven from different datasets.

### 3.3. Deep Learning in Risk Analysis

Deep learning, a sub-field of machine learning that can learn directly from a raw input, is also used for breast cancer risk prediction. In 2016, Kallenberg et al. [41] aimed to use unsupervised deep learning to perform breast density segmentation and mammographic risk scoring. In order to overcome that problem, this research uses deep learning methods to learn features from mammograms, in an automated fashion. The DL model used is called convolutional sparse autoencoder. An autoencoder can be understood, in general, as a neural network that works towards the aim of learning the input so well that will also learn to replicate it as the output of the model; the process by which this occurs is based in the learning of how to correctly compress and encode the input that will ultimately be reconstructed. An autoencoder has an encoder, that maps the input layer to the hidden layer, and a decoder, that maps the hidden layer to the output layer. Once the features are learned and extracted, the resulting set of descriptors will be used to associate the data with previously defined labels. This model was applied in two distinct phases; first, it was asked to the model to make breast segmentation based on density values, and second to address mammographic parenchymal patterns, considering the goal of predicting future breast cancer development. The methods used here are based in a denoising autoencoder, an approach in where the hidden layers have a higher dimension than the input layer. The ground base idea is that the encoder will receive a corrupted version of the data and will then learn how to reconstruct a version of the data that is not corrupted [42]. What also happens in this methodology, is that various autoencoders can be assembled together so that the learned features increase progressively in level of abstraction. The process by which this occurs makes features to be learned by one encoder, with the respective decoder being removed but the features being kept, then, the processed data is passed through a new autoencoder, where data is reconstructed. This process occurs until the reconstruction of the last hidden layer occurs [42]. The goal here is not to extract specific features, but rather to learn features directly from the mammograms, hoping that this methodology will be highly generalized, in opposite to what happens, in general, to a manual extraction approach. The models are trained in a forward propagation model, with a constant update of the learned weights, in order to optimize the process. A way of optimizing the features is to look for a minimization of the difference (or loss) between the predictions of the "top most layer" and the real labels. A division into multiple layers is done for feature learning, before a classifier is trained to make prediction in the "top most layer". This results in a "multioptimization" problem, that the authors point to have some advantages, like the fact that features are learned faster and in a more secure way, since each layer is specifically optimized, or the fact that these methodologies can incorporate other units, like classifiers, that can be independently optimized. In this case, the authors use a sparse autoencoder, which is a regular autoencoder where a sparsity limitation was forced in the hidden layers [43], for learning features that represent information at multiple scopes. The goal is to predict a "label mask" to each image, and, not only the entire image cannot be computationally used to retrieve label masks, as downsampling the data is also not possible since important information could be lost. What is suggested, instead, is that the algorithm should learn local neighborhood regions in an image-patches. Concerning the notation used in the paper, the goal is to map a patch, $x \subset X = \mathbb{R}^{c \times m \times m}$, with $m \times m$ being the size of the patch and $c$ being the number of channels in the patch, to a label patch, $y \in Y = \mathbb{R}^{C \times M \times M}$, with $M \times M$ being the size of the patch and having one channel per label. Although the image and the label patches may have different sizes, they are centered in the same location. Then, for training purposes, training data will be used to map X to Y. The training data consists in (x,y) pairs extracted from random locations across the images that are concerned with this part of the work. The mapping from X to Y does not occur directly, what happens is that abstract feature representations are learned across multiple layers. In other words, the input enters the algorithm and crosses multiple layers, with the output of one layer being the input of the next, where multiple transformations are made and learning is performed, until the last layer is reached, and a final feature representation is obtained. Finally, a classifier will be used to map the final feature representation to Y. For testing purposes, the hypothesis that was trained will be applied to a new image in all possible patches within the said image—using a sliding window. When doing this, a problem can arise: if the predicted output region is bigger than a pixel, there are predictions that may overlap. The problem is solved by calculating the average probability for each class. Mammogram analysis, conducted in a multi-scale fashion, is done by applying the discrete scale space theory, through a Fourier implementation. Algorithm unsupervised architecture consists of four layers: a convolutional layer, a pooling layer, and two final convolutional layers. Going deeper in convolutional architecture, what happens is that the convolutional layer will receive the input data, convolve it, do some transformations, and then send the results as a non-linear activation function to create an activation/features map. The output of the layer can be fed to another convolutional layer or to a pooling layer. The pooling layer was defined based on the goal of the study, once it is invariant to small distortions, but it is highly sensitive to small-scale details. Features are learned for each scale alone and only merged after the learning process.

The approach proposed by the authors aims for an overcomplete feature representation, which means that this representation is larger than the input, and resorts to the concept of sparsity. Sparsity can be used in feature representation by: (1) forcing most of the entries to be zero and leaving few non-zero entries to represent the input signal; (2) narrowing the number of examples that activate each unit. In this work, both approaches are combined, leading to, respectively, "a compact encoding per example" and to "example specific features". Sparse overcomplete approach is robust to noise and, since each example is going to be represent by specialized features, this methodology is designed to unscramble hidden aspects in the data. As it was said, the algorithm will be used in two different tasks and applied in three different datasets: the density dataset contained both MLO and CC views for both left and right breast but, for each woman, only one view was available; the texture dataset contained cancer cases and controls, that were matched both for age and time of the first image available; finally, the Dutch breast screening dataset was composed of cancer cases and healthy controls and the same matching as before was made. As for the classification part, a two-layer neural network was used, with one layer being a previously used and trained convolutional layer, and the other being a SoftMax classifier, meaning that the previously learned parameters will be tuned through a supervised methodology. Broyden–Fletcher–Goldfarb–Shanno algorithm was used as optimizer, and 5-fold cross-validation was performed for a classification task that considers: "pectorales muscle and background", "fatty tissue", and "dense tissue" as labels for the density scoring; in what concerns texture scoring, "cancer" and "normal" are the considered labels. Regarding results, for density scoring, the output is a score, from 0 to 1, that represents the probability of a given pixel to belong to the "dense class". Classification was done by choosing a value of probability to be a threshold, and the best results were obtained with a threshold of 0.75. The results explored the correlation between mammographic % PD done by the authors and by the radiologist, and a performance measure called Dice, that is given by:

$$D = \frac{|A \cap B|}{|A| + |B|} \tag{4}$$

where $A$ is the automated segmented region and B is the segmentation done by the radiologist. For this dataset, the correlation coefficient had a value of 0.85 (95% CI: 0.83–0.88), the Dice scores for fat and dense tissue were, respectively, of $0.95 \pm 0.05$ and $0.63 \pm 0.19$. The algorithm trained for this dataset was used to estimate % PD in the Dutch dataset and the cases had a value of $0.19 \pm 0.11$, the controls had a slightly smaller value—$0.15 \pm 0.11$. The correlation of % PD between both breasts was of 0.93 (95% CI: 0.92–0.95) and the obtained AUC for differentiating cases and controls was of 0.59 (95% CI: 0.57–0.62). On the other hand, the texture scoring represents the probability of a given pixel to be a part of the cancer class. In order to get one texture score per image, the scores from 500 patches randomly selected across the breast area were averaged. Besides the developed algorithm, two other methods were used and evaluated in the performance of this task; one that is based in multiscale local jet features, and other that uses static histograms. To avoid bias, the algorithm was tested multiple times and outperformed the two well established mammographic texture scores, with an AUC of 0.61 (95% CI: 0.57–0.66) vs. AUCs of 0.60 and 0.56 (95% CI: 0.51–0.61) for local jet and static histogram approaches, respectively. Applying the algorithm to the Dutch dataset resulted in an AUC of 0.57 (95% CI: 0.54–0.61) in the differentiation between cases and controls and produced a correlation for the mammographic texture of 0.91 (95% CI: 0.90–0.92), between left and right breast. Based on their results for correlation and % PD based classification, the authors advoke that this methodology is close to the ones that are present in the scientific community and, based in the outperformance of their methodology in the texture scoring task, they proceed to state that this could be a better alternative to the handcrafted texture extraction that is the current state-of-the-art. One of the downsides of this approach is that the authors assumed that changes in mammography due to cancer occur in a generalized way across the breast, but the opposite can also be true, with texture changes being visible only in restricted areas, and an algorithm that could take this hypothesis into account should be considered as a future development of the considered work.

In 2014, Petersen et al. [44] sought to do breast segmentation and risk scoring using deep learning methodologies. Patients were considered as cancer cases for this study if they had had a screen-detected or an interval cancer; in the first case, mammograms four years prior from the diagnosis were considered; for interval cancer cases, mammograms from 2–4 years before cancer appearance were examined. Patient cases, as happens in other research, were age-matched with controls. An experienced radiologist rated the mammograms in BI-RADS scale and computed mammographic PD%. The model used, which is once again a convolutional sparse autoencoder, will learn features in an increasing abstraction fashion, in order to associate the computed set of features to the considered labels. The tasks that are going to be considered are: segmentation, with labels being "background", "pectoral muscle", "breast tissue"; % PD scoring, with the labels being "fatty tissue" and "dense tissue"; and texture scoring, with labels being "diseased" and "healthy". As mentioned before, this methodology considers patches from different scales retrieved from the original image. The methodology for testing is the same as explained in the previous paper, but a special reference must be made to the fact that when the sliding window reaches the image border, the images is padded with constant values. The architecture used in this study is the same as used by Kallenberg, and the authors make a reference that, usually, convolutional and pooling layers are displayed in an alternate fashion, but in both studies, one of the pooling layers is replaced by another convolutional layer in order to grant the conditions of noise invariance and small-scale details sensibility. As for results, for segmentation, a dice metric was computed for an automated segmentation and a segmentation done by an expert and the results are: 0.99 $\pm$ 0.01, for background; 0.95 $\pm$ 0.08, for pectorales muscle; and 0.98 $\pm$ 0.01 to breast tissue. The correlation between automated and manual mammographic % PD scores was of 0.87, and the AUC for differentiating cases and controls, using the autoencoder was of 0.56 (95% CI: 0.51–0.61). For texture analysis, the performance of the algorithm for both left and right view were compared to a state-of-the-art texture scoring method and outperformed it (AUC = 0.65 [95% CI: 0.60–0.70] vs. 0.62 [95% CI: 0.57–0.67]). This research, although widely similar to the last one, proves that the used methodology generalizes relatively well to other datasets.

Back to 2016, when Qiu et al. [45] tried to create an algorithm that could, in an unsupervised fashion, estimate bilateral mammographic tissue density asymmetry, an important risk factor for the development of breast cancer. The authors aim to verify if this deep learning approach provides better results than the conventional machine learning methodology. For this study, each case had a "prior" and a "current" evaluation, all the prior mammograms were negative and the division between cancer and control cases was done based on the "current evaluation". The algorithm developed aims to predict, based in the "prior" exam, the likelihood of a case (women) to have an image-detectable cancer in the "current" mammogram. The deep learning network proposed by the authors has 8 layers and can be divided in two subsets: feature learning set, and classification set. The first is composed by alternate convolutional and pooling layers, actually creating three convolutional-pooling pairs. Convolutional layers apply convolutional kernels to the input and, then, the pooling layers are responsible for granting that the bilateral asymmetry is size and rotation independent. After passing through the first pair, a 20 $\times$ 48 $\times$ 48 feature map is created, and, when passing through the following two pairs, the final result will be a 5 $\times$ 6 $\times$ 6 feature map, that is directly linked to the classifier—multiple layer perceptron—that will generate the probability of having an image detectable cancer in the next mammogram. During the training, with the 200 cases, a method called mini-batch statistic gradient descend was used to optimize the algorithm, which the authors say that provides better optimized parameters with lower computational effort. The algorithm was tested with the test set and evaluated through a confusion matrix and ROC curve analysis. This metrics allowed the authors to state that the specificity of the classifier was of 0.60, and the sensitivity achieved a value of 0.703. The AUC value was of 0.697 $\pm$ 0.063 and the overall accuracy, based on the confusion matrix, was of 71.4%. This methodology allowed to overcome the problem of manually choosing features to describe bilateral asymmetry once the features are optimal and directly learned from the input. The authors proceed to state that, even though the metrics to evaluate the algorithm provide confidence, this is yet an early study, with a small dataset that does not incorporate inter-women variations and that, having only 8 layers, is not deep enough, which are limitation that need to be overcome in order for this type of approaches to be considered in clinical practice.

Tables 1–3 present a summary of the works assessed in this review. The first addresses questions concerning dataset description and host Institutions, while Table 2 is more related to the methodology used. As for the final table, results and main conclusions are addressed. In this table, for studies with more than one AUC result, only the highest value is considered.

**Table 1.** Studies Data Summary.

| Study | Institution | Mammogram View | Group-Matched? | Full Dataset Size |
|---|---|---|---|---|
| Hou et al., 2000 [18] | University of Chicago | CC | Yes. Age-matched | 158 women = 15 high/143 low risk |
| Li et al., 2004 [27] | University of Chicago | CC | Yes. Age-matched | 90 women = 30 high/60 low risk |
| Li et al., 2005 [30] | University of Chicago | CC | Yes. Age-matched | 172 women = 30 high/142 low risk |
| Tan et al., 2013 [31] | University of Pittsburgh | CC | No. | 645 women = 283 high/362 low risk |
| Zheng et al., 2015 [34] | University of Pennsylvania | MLO | Yes. Age-matched | 424 women = 106 high/318 low risk |
| Tan et al., 2016 [37] | University of Pittsburgh | CC and MLO | Yes. Age-matched | 335 women = 159 high/175 low risk |
| Tan et al., 2019 [38] | Subang Jaya Medical Center | CC | Yes. Age, Ethnicity, BMI-matched | 500 women = 250 high/250 low risk |
| Gandomkar et al., 2020 [39] | Fudan University Shanghai | CC | No. | 1079 women = 85 high/993 low risk |
| Kallenberg et al., 2016 [41] | University of Copenhagen | CC and MLO | Yes. Age and Acquisition time | Density: 493 healthy<br>Texture: 226 cancer and 442 controls<br>Dutch: 384 cancer and 1182 controls |
| Petersen et al., 2014 [44] | University of Copenhagen | MLO | Yes. Age-matched | 495 women = 245 cases/250 controls |
| Qiu et al., 2016 [45] | University of Oklahoma | CC | No. | 270 women = 135 cases/135 controls |

**Table 2.** Methods Summary.

| Study | ROI Analyzed | Intensity-Based | GLCM | RL | Other Features | Classifier/Algorithm |
|---|---|---|---|---|---|---|
| Hou et al., 2000 [18] | 256 × 256, manually placed behind the nipple. | x | - | - | NGTDM, Spectral | LDA. |
| Li et al., 2004 [27] | 256 × 256, 128 × 128 and 64 × 64 in referred locations | x | - | - | NGTDM, Spectral | ROCA. |
| Li et al., 2005 [30] | 256 × 256, manually placed behind the nipple | x | x | - | Fractal, Spectral, Edge | ROCA. |
| Tan et al., 2013 [31] | Entire breast considered—segmented into regions. | x | x | x | Cumulative Projection | SVM. |
| Zheng et al., 2015 [34] | Lattice-based approach. D = W = 63, 127 and 255. | x | x | x | LBP, Fractal, Edge | Logistic Reg. |
| Tan et al., 2016 [37] | Entire breast considered | x | x | x | Weber, Structural sim. | SVM. |
| Tan et al., 2019 [38] | Entire breast considered | x | x | x | Structural, Spectral | LDA. |
| Gandomkar et al., 2020 [39] | Two segmented areas using AutoDensity | x | x | - | Fractal | Decision Tree |
| Kallenberg et al., 2016 [41] | Patches with the smaller scale being 4.8 mm × 4.8 mm and the biggest 3.7 cm × 3.7 cm. | - | - | - | - | Sparse autoencoder. |
| Petersen et al., 2014 [44] | Patches. | - | - | - | - | Sparse autoencoder |
| Qiu et al., 2016 [45] | 256 × 256, manually placed behind the nipple | - | - | - | - | Multiple Layer Perception. |

**Table 3.** Results Summary.

| Study | AUC Results | Main Conclusion |
|---|---|---|
| Hou et al., 2000 [18] | AUC = 0.91 | Mammographic features were found to be associated with breast cancer risk. High-risk women tend to have dense breasts and the patterns present e mammograms tend to have low contrast and to be coarse. |
| Li et al., 2004 [27] | AUC = 0.93 (highest value) | Features extracted immediately behind the nipple tend to have the best performance. Concerning size, results were not statistically significant. |
| Li et al., 2005 [30] | AUC = 0.66 ± 0.05 − 0.86 ± 0.03 (only assessed individual features) | High-risk women tend to have dense breasts and their pattern tend to be coarser, to have a lower fractal dimension, to be lower in contrast and to have a small edge gradient measure. |
| Tan et al., 2013 [31] | AUC = 0.716 ± 0.020 (first and third subgroup) AUC = 0.725 ± 0.018 (all groups) | Risk calculation based on texture features of mammographic asymmetry through a SVM classifier has a good potential to predict the near-term risk of breast cancer in women. |
| Zheng et al., 2015 [34] | AUC = 0.85 ± 0.02 (highest value) | Lattice-based approach allows parenchyma characterization across the entire breast, meaning that the extracted features are provide better information than the ones extracted from classic approaches. |
| Tan et al., 2016 [37] | AUC = 0.730 ± 0.027 (highest value) | Proved a relationship between the risk scores generated by the proposed model and the near-term risk of having breast cancer. |
| Tan et al., 2019 [38] | AUC = 0.68 (95% CI: 0.64–0.73) | Breast texture analysis has a great potential as an independent risk factor. The study used an Asian population and confirmed previous studies performed in Caucasian women about the relationship between texture patterns and breast cancer risk. |
| Gandomkar et al., 2020 [39] | AUC = 0.884 (CI 0.838–0.913) | A model that combines texture information and epidemiological factors might lead to an increased discriminatory capacity of risk prediction. |
| Kallenberg et al., 2016 [41] | Density: AUC = 0.59 (95% CI: 0.57–0.62) Texture: AUC = 0.61 (95% CI: 0.57–0.66) and 0.57 (Dutch) (95% CI: 0.54–0.61) | Obtained breast density scores are positively related to manual density scores, and texture scores have a predictive value in what concerns to breast cancer. |
| Petersen et al., 2014 [44] | AUC = 0.65 (95% CI: 0.60–0.70) | PMD scores correlate positively to manual scores and mammographic texture are more related to future breast cancer risk than scores related to mammographic density. |
| Qiu et al., 2016 [45] | AUC = 0.697 ± 0.063 | This study concluded that deep learning technologies may have the potential to develop new risk predicting methods, that help to achieve an early detection of breast cancer through negative mammograms. |

## 4. Conclusions

Mainly, the reviewed articles, in terms of extracted features, had in common three major groups (intensity-based, GLCM and RL), and then present many feature-group variations, with spectral analysis being also vastly considered. In what comes to the feature extraction procedure, older papers used a manually single-ROI approach, while more recent ML studies opted to diversify the region analyzed. Some authors used several ROIs across the breast, others segmented the breast in different regions and extracted features from them, and yet, some research consider the entire breast for feature extraction. In papers that compared their approach with the classical single-ROI methodology, authors usually find that their procedure outperformed the use of a unique ROI. This may happen because, considering breast tissue heterogeneity, a single region does not account for this diversity, and therefore a tissue characterization that takes into account the entire breast (or more than one region) appears to be more robust. In terms of classifiers, the papers varied widely, from LDA to SVM, passing through decisions trees and logistic regression. More studies should be performed to assess if there is a classifier that is clearly superior to others. Nonetheless, the LDA approach proposed in the first analyzed paper achieved higher results than the other three classifiers, which would point that this classifier, for this type of tasks might outperformed the others. However, the work proposed by Tan in 2019, that used an LDA, was outperformed by a work conducted in 2013, that used an SVM. As it can be perceived, there is not a clear conclusion to be made in terms of what classifier is the best. Nevertheless, the results obtained by the reviewed papers allowed to conclude that texture analysis along with machine learning algorithms can be correctly employed in risk analysis, either by differentiating risk groups, or by giving a risk score to each patient. Besides understanding that this type of methodology can be used, the research also points out that procedures that consider the entire breast for feature extraction might provide more useful information. While many of these studies were conducted in Caucasian population, the study presented by Tan in 2019 allowed to understand that ML algorithms and texture analysis can also be used, with good outcomes, in Asian populations. The results of the deep learning approaches, although lower than the ones presented by the classical ML approach, appear to be very promising, especially because dismisses the laborious work of extracting handcrafted features, and allows the possibility of automatically finding predictors that better serve the purpose of the study.

Given the articles discussed in this paper, excluding the ones that use deep learning, two great future endeavors should be examined: first, considering the substantial differences in age and other risk factors between high-risk and low-risk groups, studies should start using larger matched-groups and consider other risk factors than age, in an approach analogous to what was done in 2019 [38], but with more dataset cases; secondly, most of the papers did the validation of their model through cross-validation, meaning that training and testing samples came from the same dataset, so, novel studies should try to validate their models in an independent dataset. Machine learning methodologies are widely used in this area, which is demonstrated by the given publications' date range considered here but should be interesting for new studies in the field of breast cancer risk assessment to consider deep learning, as it happens to the last three papers that were analyzed. Machine learning approaches proved to be substantially good in differentiating risk groups, but what might be more valuable in terms of medical application is the generation of risk scores, as done by Tan in 2013. The restriction to a high-risk/low-risk classification seems very limitative and the focus in giving a risk score specific to each woman should be considered.

While the development of new methodologies in both machine and deep learning, that suppress the weaknesses discussed in this section, might result in better outcomes, authors should start looking for breast cancer risk assessment in the perspective of transforming these algorithms and methods into real clinical applications.

The extensive review performed here allowed to have a general idea of what has been done for breast cancer risk prediction using textural analysis, that is sometimes combined with important risk factors. Although there are some downsides that can be pointed out to research's methodologies, they serve as a proof of concept that parenchymal texture patterns provide important information about breast cancer risk and should, once methodology's flaws are overcome, be used in clinical practice, and have a positive effect in millions of women that are diagnosed with breast cancer each year, worldwide.

**Author Contributions:** Conceptualization, J.M. and N.M.; methodology, J.M. and N.M.; writing—original draft preparation, J.M. and N.M.; writing—review and editing, J.M. and N.M. All authors have read and agreed to the published version of the manuscript.

**Funding:** This work was supported by Fundação para a Ciência e Tecnologia–Portugal (FCT-IBEB Strategic Project UIDB/00645/2020).

**Institutional Review Board Statement:** Not applicable.

**Informed Consent Statement:** Not applicable.

**Data Availability Statement:** Not applicable.

**Conflicts of Interest:** The authors declare no conflict of interest.

## References


1. American Cancer Society. *Breast Cancer Facts & Figures 2019–2020*; American Cancer Society: Atlanta, GA, USA, 2019.
2. Anand, P.; Kunnumakara, A.B.; Sundaram, C.; Harikumar, K.B.; Tharakan, S.T.; Lai, O.S.; Sung, B.; Aggarwal, B.B. Cancer is a preventable disease that requires major lifestyle changes. *Pharm. Res.* **2008**, *25*, 2097–2116. [CrossRef]
3. Hammer, G.D.; McPhee, S.J. Breast carcinoma. In *Pathophyisiology of the Disease*, 8th ed.; McGraw-Hill Education: New York, NY, USA, 2019; pp. 267–273.
4. DeSantis, C.E.; Ma, J.; Gaudet, M.M.; Newman, L.A.; Miller, K.D.; Sauer, A.G.; Jemal, A.; Siegel, R.L. Breast cancer statistics, 2019. *CA Cancer J. Clin.* **2019**, *69*, 438–451. [CrossRef]
5. National Health Service. Overview: Breast Cancer Screening. Available online: https://www.nhs.uk/conditions/breast-cancer-screening/ (accessed on 23 October 2020).
6. American Cancer Society. American Cancer Society Guidelines for the Early Detection of Cancer. Available online: https://www.cancer.org/healthy/find-cancer-early/cancer-screening-guidelines/american-cancer-society-guidelines-for-the-early-detection-of-cancer.html#:~{}:text=Women%20ages%2040%20to%2044,or%20can%20continue%20yearly%20screening (accessed on 23 October 2020).
7. Singletary, S.E. Rating the risk factors for breast cancer. *Ann. Surg.* **2003**, *237*, 474–482. [CrossRef] [PubMed]
8. Tretli, S. Height and weight in relation to breast cancer morbidity and mortality. A prospective study of 570,000 women in Norway. *Int. J. Cancer* **1989**, *44*, 23–30. [CrossRef] [PubMed]
9. Pike, M.C.; Krailo, M.D.; Henderson, B.E.; Casagrande, J.T.; Hoel, D.G. 'Hormonal' risk factors, 'breast tissue age' and the age-incidence of breast cancer. *Nature* **1983**, *303*, 767–770. [CrossRef]
10. Barnard, M.E.; Boeke, C.E.; Tamimi, R.M. Established breast cancer risk factors and risk of intrinsic tumor subtypes. *Biochim. Biophys. Acta* **2015**, *1856*, 73–85. [CrossRef]
11. Li, H.; Giger, M.L.; Lan, L.; Brown, J.B.; MacMahon, A.; Mussman, M.; Olopade, O.I.; Sennett, C. Computerized analysis of mammographic parenchymal patterns on a large clinical dataset of full-field digital mammograms: Robustness study with two high-risk datasets. *J. Digit. Imaging* **2012**, *25*, 591–598. (In English) [CrossRef] [PubMed]
12. Wolfe, J.N. The prominent duct patter as indicator of cancer risk. *Oncology* **1969**, *23*, 140–158. [CrossRef]
13. Wolfe, J.N. A study of breast parenchyma by mammography in the normal woman and those with benign and malignant disease. *Radiology* **1967**, *89*, 201–205. [CrossRef]
14. Wolfe, J.N. Risk for breast cancer development determined by mammographic parenchymal pattern. *Cancer* **1976**, *37*, 2486–2492. [CrossRef]
15. Wolfe, J.N. Breast patterns as an index of risk for developing breast cancer. *Am. J. Roentgenol.* **1976**, *126*, 1130–1139. [CrossRef]
16. Boyd, N.F.; Martin, L.J.; Yaffe, M.J.; Minkin, S. Mammographic density and breast cancer risk: Current understanding and future prospects. *Breast Cancer Res.* **2011**, *13*, 223. [CrossRef]
17. Jakes, R.W.; Duffy, S.W.; Ng, F.C.; Gao, F.; Ng, E.H. Mammographic parenchymal patterns and risk of breast cancer at and after a prevalence screen in Singaporean women. *Int. J. Epidemiol.* **2000**, *29*, 11–19. [CrossRef] [PubMed]
18. Huo, Z.; Giger, M.L.; Wolverton, D.E.; Zhong, W.; Cumming, S.; Olopade, O.I. Computerized analysis of mammographic parenchymal patterns for breast cancer risk assessment: Feature selection. *Med. Phys.* **2000**, *27*, 4–12. [CrossRef]
19. Niell, B.L.; Freer, P.E.; Weinfurtner, R.J.; Arleo, E.K.; Drukteinis, J.S. Screening for breast cancer. *Radiol. Clin. N. Am.* **2017**, *55*, 1145–1162. [CrossRef] [PubMed]
20. Coleman, C. Early detection and screening for breast cancer. *Semin. Oncol. Nurs.* **2017**, *33*, 141–155. (In English) [CrossRef] [PubMed]
21. Loberg, M.; Lousdal, M.L.; Bretthauer, M.; Kalager, M. Benefits and harms of mammography screening. *Breast Cancer Res.* **2015**, *17*, 63. [CrossRef]
22. Lehman, C.D.; White, E.; Peacock, S.; Drucker, M.J.; Urban, N. Effect of age and breast density on screening mammograms with false-positive findings. *AJR Am. J. Roentgenol.* **1999**, *173*, 1651–1655. [CrossRef] [PubMed]
23. Biller-Andorno, N.; Juni, P. Abolishing mammography screening programs? A view from the Swiss Medical Board. *N. Engl. J. Med.* **2014**, *370*, 1965–1967. [CrossRef] [PubMed]
24. Le, E.; Wang, Y.; Huang, Y.; Hickman, S.; Gilbert, F.J.C.R. Artificial intelligence in breast imaging. *Clin. Radiol.* **2019**, *74*, 357–366. [CrossRef]
25. National Cancer Institute. The Breast Cancer Risk Assessment Tool. Available online: https://bcrisktool.cancer.gov/ (accessed on 15 December 2020).
26. Haralick, R.M.; Shanmugam, K.; Dinstein, I. Textural features for image classification. *IEEE Trans. Syst. Man Cybern.* **1973**, *3*, 610–621. [CrossRef]
27. Li, H.; Giger, M.L.; Huo, Z.; Olopade, O.I.; Lan, L.; Weber, B.L.; Bonta, I. Computerized analysis of mammographic parenchymal patterns for assessing breast cancer risk: Effect of ROI size and location. *Med. Phys.* **2004**, *31*, 549–555. [CrossRef]

28. Shanmugavadivu, P.; Sivakumar, V. Fractal dimension based texture analysis of digital images. *Procedia Eng.* **2012**, *38*, 2981–2986. [CrossRef]
29. Lachenbruch, P.A.; Mickey, R.M. Estimation of error rates in discriminant analysis. *Technometrics* **1968**, *10*, 1–11. [CrossRef]
30. Li, H.; Giger, M.L.; Olopade, O.I.; Margolis, A.; Lan, L.; Chinander, M.R. Computerized texture analysis of mammographic parenchymal patterns of digitized mammograms. *Acad. Radiol.* **2005**, *12*, 863–873. [CrossRef]
31. Tan, M.; Zheng, B.; Ramalingam, P.; Gur, D. Prediction of near-term breast cancer risk based on bilateral mammographic feature asymmetry. *Acad. Radiol.* **2013**, *20*, 1542–1550. (In English) [CrossRef]
32. Galloway, M.M. Texture analysis using gray level run lengths. *Comput. Graph. Image Process.* **1975**, *4*, 172–179. [CrossRef]
33. Pudil, P.; Novovicova, J.; Kittler, J. Floating search methods in feature-selection. *Pattern Recognit. Lett.* **1994**, *15*, 1119–1125. (In English) [CrossRef]
34. Zheng, Y.; Keller, B.M.; Ray, S.; Wang, Y.; Conant, E.F.; Gee, J.C.; Kontos, D. Parenchymal texture analysis in digital mammography: A fully automated pipeline for breast cancer risk assessment. *Med. Phys.* **2015**, *42*, 4149–4160. [CrossRef]
35. Kumar, S.N.; Fred, A.L.; Varghese, P.S. Suspicious lesion segmentation on brain, mammograms and breast MR images using new optimized spatial feature based super-pixel fuzzy c-means clustering. *J. Digit. Imaging* **2019**, *32*, 322–335. [CrossRef]
36. Rundo, L.; Militello, C.; Tangherloni, A.; Russo, G.; Vitabile, S.; Gilardi, M.C.; Mauri, G. NeXt for neuro-radiosurgery: A fully automatic approach for necrosis extraction in brain tumor MRI using an unsupervised machine learning technique. *Int. J. Imaging Syst. Technol.* **2018**, *28*, 21–37. [CrossRef]
37. Tan, M.; Zheng, B.; Leader, J.K.; Gur, D. Association between changes in mammographic image features and risk for near-term breast cancer development. *IEEE Trans. Med. Imaging* **2016**, *35*, 1719–1728. (In English) [CrossRef] [PubMed]
38. Tan, M.; Mariapun, S.; Yip, C.H.; Ng, K.H.; Teo, S.H. A novel method of determining breast cancer risk using parenchymal textural analysis of mammography images on an Asian cohort. *Phys. Med. Biol.* **2019**, *64*, 035016. [CrossRef] [PubMed]
39. Gandomkar, Z.; Li, T.; Shao, Z.; Tang, L.; Xiao, Q.; Gu, Y.; Di, G.; Lewis, S.; Mello-Thoms, C.; Brennan, P. Breast cancer risk prediction in Chinese women based on mammographic texture and a comprehensive set of epidemiologic factors. In Proceedings of the Fifteenth International Workshop on Breast Imaging, Leuven, Belgium, 25–27 May 2020; SPIE: Bellingham, WA, USA, 2020.
40. Nickson, C.; Arzhaeva, Y.; Aitken, Z.; Elgindy, T.; Buckley, M.; Li, M.; English, D.R.; Kavanagh, A.M. AutoDensity: An automated method to measure mammographic breast density that predicts breast cancer risk and screening outcomes. *Breast Cancer Res.* **2013**, *15*, R80. [CrossRef] [PubMed]
41. Kallenberg, M.; Petersen, K.; Nielsen, M.; Ng, A.Y.; Diao, P.; Igel, C.; Vachon, C.M.; Holland, K.; Winkel, R.R.; Karssemeijer, N.; et al. Unsupervised deep learning applied to breast density segmentation and mammographic risk scoring. *IEEE Trans. Med. Imaging* **2016**, *35*, 1322–1331. [CrossRef] [PubMed]
42. Peterson, K.; Chernoff, K.; Ng, A. Breast density scoring with multiscale denoising autoendoers. In *Sparsity Techniques in Medical Imaging 2012, Proceedings of the International Conference on Medical Image Computing and Computer-Assisted Intervention MICCAI, Nice, France, 1–5 October 2012*; Springer: Cham, Switzerland, 2012.
43. Ng, A. CS294A Lecture Notes—"Sparse Autoencoder". Available online: https://web.stanford.edu/class/cs294a/sparseAutoencoder.pdf (accessed on 2 February 2021).
44. Peterson, K.; Nielsen, M.; Diao, P.; Karssmeijer, M.; Lillhom, M. Breast Tissue segmentation and mammographic risk scoring using deep learning. In Proceedings of the International Workshop on Digital Mammography, Gifu City, Japan, 29 June–29 July 2014.
45. Qiu, Y.; Yunzhi, W.; Shiju, Y.; Maxine, T.; Samuel, C.; Hong, L.; Bin, Z. An initial investigation on developing a new method to predict short-term breast cancer risk based on deep learning technology. *Proc. SPIE Med Imaging 2016 Comput. Aided Diagn.* **2016**, *9785*, 978521.

Journal of *Imaging*

MDPI

*Article*

# Feature Selection Using Correlation Analysis and Principal Component Analysis for Accurate Breast Cancer Diagnosis

**Sara Ibrahim [1], Saima Nazir [2,*] and Sergio A. Velastin [3,4]**

1 Department of Computer Science, Capital University of Science and Technology, Islamabad 45730, Pakistan; sara.ibrahim@cust.edu.pk
2 Department of Software Engineering, National University of Modern Languages, Rawalpindi 46000, Pakistan
3 Applied Artificial Intelligence Research Group, Department of Computer Science and Engineering, University Carlos III de Madrid, 28270 Madrid, Spain; sergio.velastin@ieee.org
4 School of Electronic Engineering and Computer Science, Queen Mary University of London, London E1 4NS, UK
* Correspondence: saima.nazir@cust.edu.pk

**Abstract:** Breast cancer is one of the leading causes of death among women, more so than all other cancers. The accurate diagnosis of breast cancer is very difficult due to the complexity of the disease, changing treatment procedures and different patient population samples. Diagnostic techniques with better performance are very important for personalized care and treatment and to reduce and control the recurrence of cancer. The main objective of this research was to select feature selection techniques using correlation analysis and variance of input features before passing these significant features to a classification method. We used an ensemble method to improve the classification of breast cancer. The proposed approach was evaluated using the public WBCD dataset (Wisconsin Breast Cancer Dataset). Correlation analysis and principal component analysis were used for dimensionality reduction. Performance was evaluated for well-known machine learning classifiers, and the best seven classifiers were chosen for the next step. Hyper-parameter tuning was performed to improve the performances of the classifiers. The best performing classification algorithms were combined with two different voting techniques. Hard voting predicts the class that gets the majority vote, whereas soft voting predicts the class based on highest probability. The proposed approach performed better than state-of-the-art work, achieving an accuracy of 98.24%, high precision (99.29%) and a recall value of 95.89%.

**Keywords:** breast cancer diagnosis; Wisconsin Breast Cancer Dataset; feature selection; dimensionality reduction; principal component analysis; ensemble method



**Citation:** Ibrahim, S.; Nazir, S.; Velastin, S.A. Feature Selection Using Correlation Analysis and Principal Component Analysis for Accurate Breast Cancer Diagnosis. *J. Imaging* **2021**, *7*, 225. https://doi.org/10.3390/jimaging7110225

Academic Editors: Leonardo Rundo, Carmelo Militello, Vincenzo Conti, Fulvio Zaccagna and Changhee Han

Received: 6 July 2021
Accepted: 18 October 2021
Published: 26 October 2021



## 1. Introduction

Breast cancer is one of the leading causes of death among women [1]. Although cancer is largely preventable in its primary stages, there are still many women who are diagnosed with cancer at a late stage. Better performing diagnosis techniques are very important in personalized care and treatment, and the use of such techniques can also help to control and reduce the recurrence of cancer. In the medical field, clinicians generally use data from various sources, such as medical records, laboratory tests, and studies related to the disease for accurate diagnosis and prediction of breast cancer. The use of artificial intelligence (AI) techniques in the medical field is also increasing to automate disease diagnosis and to get better results in terms of performance.

Breast cancer occurs in breast cells of the fatty tissues or the fibrous connective tissues within the breast. Breast cancer is a type of tumor that tends to become gradually worse and that grows fast, which leads to death. Breast cancer is more common among females, but it can also occur among males, although rarely. Various factors, such as age and family history, can also contribute to breast cancer risk. Two main types of breast tumors can be identified.

Benign: If the cells are not cancerous, the tumor is benign (not dangerous to health). It will not invade nearby tissues or spread to other areas of the body (metastasize). A benign tumor is not worrisome unless it is pressing on nearby tissues, nerves, or blood vessels and causing damage.

Malignant: This means that the tumor is made of cancerous cells and it can invade nearby tissues and thus be potentially hazardous. Some cancer cells can move into the bloodstream or lymph nodes, where they can spread to other tissues within the body, which is known as metastasis. This is a tumor that is more dangerous and causes death. The main types or forms of breast cancer include:

1. Ductal carcinoma in situ (DCIS): It is the earliest stage of breast cancer and can be diagnosed and is curable. The vast majority of women diagnosed with it get cured. Although it is non-invasive, it might lead to invasive cancer.
2. Invasive ductal carcinoma (IDC): It begins in the milk duct and can spread to the surrounding breast tissues. It is the most common type of breast cancer.
3. Invasive lobular carcinoma (ILC): It starts in a lobule of the breast. It can spread fast to the lymph nodes and other areas of the body.

Approximately one million females are diagnosed with breast cancer approximately every year worldwide. As many as 81% of females with early-stage breast cancer survive for five years. However, only 35% of females with late or advanced-stage breast cancer survive for five years. The work proposed here highlights the significance of the use of the best performing machine learning classifiers with ensembles techniques for accurate diagnosis of breast cancer. The objective of the proposed research was to implement a feature reduction algorithm which can find a subset of features that can guarantee a highly accurate breast cancer classification as either benign or malignant. Principal component analysis (PCA) was used for dimensionality reduction and hyper-parameter tuning was performed to gain performance. We also compared different state-of-the-art machine learning classification algorithms. We used the publicly available Wisconsin Breast Cancer Dataset (WBCD) [2], and its features were computed from a digitized image of a fine needle aspirate (FNA) of a breast mass. They describe the characteristics of the cell nuclei present in the image. We evaluated the performances of logistic regression, support vector machine, k-nearest neighbors, stochastic gradient descent learning, naïve Bayes, random forest, and decision tree. These seven classifiers were used for further processing and ensembled with voting techniques that included hard voting and soft voting.

The rest of the paper is structured as follows: Section 2 describes related work and different state-of-the-art approaches used for breast cancer diagnosis. Section 3 describes the proposed framework and related details of the proposed work. Section 4 deals with experimentation and discussion. Section 5 presents a comparison with the state-of-the-art. Section 6 ends the paper with conclusions and proposed future work.

## 2. Literature Review

Many studies have used artificial intelligence (AI) techniques for breast cancer diagnosis to enhance the accuracy of classification and its speed. Here, we have reviewed some relevant work dealing with breast cancer diagnosis that has used machine and deep learning approaches.

Nguyen et al. [3] used the WBCD Dataset to evaluate the performance of supervised and unsupervised breast cancer classification models. Scaling and principal component analysis were used for the selection of features, and they split the data into a 70:30 ratio for training and testing. They argued that the ensemble voting method is suitable as a prediction model for predicting breast cancer. After feature selection techniques, various models were tested and trained on the data. Among all the models used for the prediction, they stated that only four models, i.e., ensemble-voting classifier, logistic regression, support vector machine, and adaboost, provided approximately 90% accuracy. They reported the performance of the proposed model using accuracy, recall tests, ROC-AUC (receiver operating characteristic curve- area under the curve), F1-measure, and computational time.

To compare the models, the data from the Iranian Center for Breast Cancer dataset were analyzed to explore risk factors in breast cancer prediction. Ahmed et al. [4] used decision trees (DTs), artificial neural networks (ANNs), and support-vector machines (SVMs). The results show that SVM outperformed both the decision tree and the MLP (multilayer perceptron) in all the parameters of sensitivity, specificity, and accuracy. There are some limitations to their study, as many cases were lost in the follow-up and there were records with missing values that were omitted. Apart from missing data, some important variables such as S-phase fraction and DNA index were not included in the study because of their unavailability, which may have decreased the performance of the models.

Omondiagbe et al. [5] discussed the classification of different types of breast cancer (benign and malignant) in the Wisconsin Diagnostic datasets using support vector machine (SVM), artificial neural network (ANN), and naive Bayes approaches. Their main goal was to propose the most suitable approach by integrating machine learning techniques with different feature selection/feature extraction methods. They proposed a hybrid approach for breast cancer diagnosis by reducing the high dimensionality of features using LDA (linear discriminant analysis) and then applying the new reduced feature dataset to a support vector machine. Their approach showed 98.82% accuracy, 98.41% sensitivity, 99.07% specificity, and 0.9994 area under the receiver operating characteristic curve (AUROC).

Yesuf et al. [6] used the CFS (correlation based feature selection) technique for feature selection in which a 0.7 correlation filter value was set and features with means above 0.7 were omitted from the training dataset. Another technique used for feature selection was recursive feature elimination (RFE) [7], which used the wrapper approach. In that approach, all the feature subsets were rated on the basis of accuracy score, and subsets were selected which had features having top ranking scores. Their research was on the basis of a technique that used PCA (principal component analysis) on neural networks. They used PCA and LDA for feature extraction and CFS and RFE for feature selection.

Jamal et al. [8] worked on two machine learning algorithms, a support vector machine (SVM) and extreme gradient boosting, and compared their performances. For classification they reduced the number of data attributes by extracting the features with the help of principal component analysis (PCA) and clustering with k-means. They reported the performances of four models using accuracy, sensitivity, and specificity from confusion matrices. Their results indicated that k-means was the best method, which was not generally used for dimensionality reduction, but can perform well compared to PCA. Four algorithms were employed—namely, PCA, factor analysis, linear discriminant analysis, and multidimensional scaling. The result of simulation on the WBCD showed that maximum accuracy was obtained by the use of PCA and a back-propagation neural network.

Subrata et al. [9] proposed the diagnosis of breast cancer by comparing naïve Bayes (NB), logistic regression (LR), and decision tree (DT) classifiers; the time complexity of each of the classifiers was also measured. It was concluded that the logistic regression classifier was the best classifier with the highest accuracy as compared to the other two classifiers. Kumar et al. [10] worked on the WBCD dataset and evaluated the performance of their proposed work on with adaboost, a decision table, J48, logistic regression, Lazy IBK, Lazy K-star, a multiclass classifier, a multilayer–perceptron, naïve Bayes, J-Rip random forest, and a random tree.

Lucas et al. [11] used Bayesian network and decision tree machine learning classifiers on the WBCD dataset. The Bayesian network gave the best accuracy of 97.80%. Bharat et al. [12] evaluated the performance of their proposed work with three popular machine learning classifiers: naïve Bayes, J48, and RBF networks. The models showed that naïve Bayes obtained the best accuracy of 97.3%, followed by RBF with 96.77%, and J48 came up with 93.41%.

Ravi et al. [13] worked on the Extensible Breast Cancer Prognosis Framework (XBPF) for breast cancer prognosis, which included susceptibility or risk assessment, recurrence, or redevelopment of cancer after the resolution, and survivability. A representative feature for subset selection (RFSS) algorithm was used along with SVM to improve efficiency in prognosis. SVM-RFSS showed a significant performance improvement over state-of-the-art prognosis methods. Chaurasia et al. [14] used three common machine learning classifiers: Bayes' theorem, a radial basis function network, and decision tree J48. They acquired the UCI dataset (683 instances). They further applied techniques on this dataset such as data selection, preprocessing, and transformation for the development of accurate diagnosis models. The results showed that the naive Bayes performed better, having classification accuracy of 97.36%; and the next two, RBF network and J48, showed 96.77% and 93.41%, respectively.

Haifeng and Won Yoon [15] presented a study on breast cancer diagnosis using different machine learning classifiers. They formulated an effective way to predict breast cancer based on patients' clinical records. They used four machine learning classifiers: support vector machine (SVM), artificial neural network (ANN), naive Bayes classifier, and adaboost tree. They used two datasets: Wisconsin Diagnostic Breast Cancer and WBCD. In their research work, they also discussed feature space reduction, proposed a hybrid network between various machine learning models and principal component analysis (PCA), and implemented the k-fold cross-validation for the estimation of test errors for each models to select the best method. They also suggested that there were some other models, such as k-means, which can be used for feature space reduction.

Abdollel et al. [16] used relative and absolute area density-based breast cancer measurements. They assessed cancer diagnosis through time of screening mammography and took 392 images from effected cases of breast cancer and 817 images from age matched controls. Multi-variable logistic regression and AUROC (area under the receiver-operating characteristic) were used to assess three risks models. The first model used clinical risk factors, the second model used measures of density-related images, and a third model used clinical risk factors and density-related measurements. They reported that the clinical risk factors model had an AUROC of 0.535, the second model got an AUROC of 0.622, and the third model gave the best result—0.632—outperforming the clinical risk model.

Shravya et al. [17] focused on improving predictive models aimed at high performance in diagnosis of disease outcomes with the help of supervised machine learning methods. They proposed and analyzed the implementations of different machine learning classifiers, logistic regression (LR), support vector machine (SVM), and k nearest neighbors, on the WBCD dataset. SVM performed best with an accuracy of 92.7%.

William et al. [18] focused on naïve Bayes and the J48 decision tree, two machine learning classifiers, to predict breast cancer risks in patients in Nigeria. The J48 decision tree proved to be the most efficient and effective method for predicting breast cancer with the help of highest accuracy level of 94.2% and low error rates as compared to naïve Bayes, having accuracy of 82.6%. Recently, several researchers proposed machine learning (ML) methods for classifying breast abnormality in mammogram images. Assiri et al. [19] proposed an ensemble classifier based on a majority voting mechanism. The performances of different state-of-the-art ML classification algorithms were evaluated for the WBCD dataset. Their classifier achieved an accuracy of 99.42%.

Darzi et al. [20] addressed feature selection for breast cancer diagnosis. They presented a process with a genetic algorithm (GA) and case-based reasoning (CBR). The genetic algorithm was used for searching the problem space to find all of the possible subsets of features, and case-based reasoning was employed to estimate the evaluation result of each subset. The results show that the proposed model performed comparably to the other models on the WBCD dataset. They achieved an accuracy of 97.37%, after feature selection.

When dealing with data that do not have a significant number of training samples, unsupervised machine learning techniques have also proven to be of significant importance in biomedical applications. Marrone et al. [21] have used the 2D fuzzy c-means (FCM) clustering along with geometrical breast anatomy characterization through well defined keypoints. They used FCM to shift the base mask extraction from a simple gray-level-based segmentation to a membership probability. They stated that key point characterization of breast anatomy can be effectively used to weight FCM membership probability, allowing one to accurately separate pectoral muscle from the chest wall. Rundo et al. [22] also used the fuzzy c-means algorithm for the automatic detection and delineation of the necrotic regions within the planned GTV for neuro-radiosurgery therapy.

Most of the published literature has evaluated the performances of classifiers based on accuracy, i.e., a value that is higher when the frequencies of true positives (TPs) and true negatives (TNs) are high compared to those of false positives (FPs) and false negatives (FNs). However, measuring performance in terms of false negatives (recall) and false positives (precision) and F-measures score is equally important, because missing a condition could have serious consequences for patients.

## 3. Methodology

In this paper, an ensemble method is proposed for accurate breast cancer classification, which was made by selecting the appropriate features for processing.

The public UCI breast disease dataset (WBCD) [2] was used as input data. The large size of the dataset and the multiple sources make the data highly useful. WBCD contains 569 instances and 32 attributes. We split the data into a ratio of 70:30 for training and testing. For splitting the dataset, we used the Scmap plot showing the correlaiKit-Learn library in Python—the train-test-split method. Details about the libraries used are mentioned in Appendix A. The training set contained a known output, and the model learned on this data in order to be generalized to other data later on. We used the test dataset (or subset) in order to test our model's prediction on this subset.

The pre-processing of the data was done via data cleaning, data transformation, and normalization. As shown in Figure 1, after pre-processing, we performed feature selection and dimensionality reduction by analyzing the correlation and variance of the input features. Later, the most significant features were used for classification using seven state-of-the-art classification algorithms: logistic regression, support vector machine, k-nearest neighbors, stochastic gradient descent learning, naïve Bayes, random forest, and decision tree. Later, these classifiers were ensembled using voting-based ensembled methods. Hard voting predicts the class that gets the majority vote and soft voting predicts the class based on highest probability. Details about each step of the proposed methodology are given in the sections below.

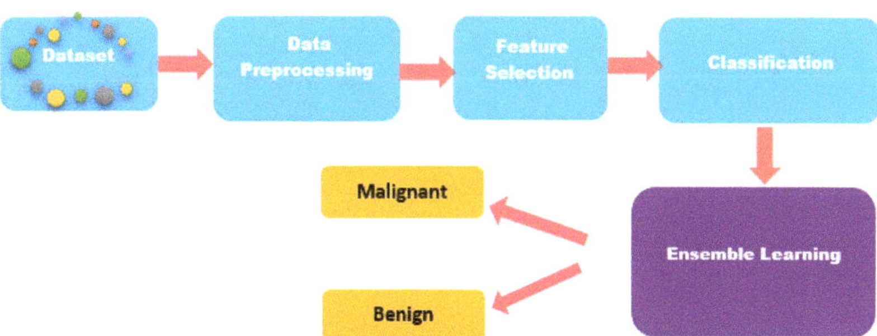


**Figure 1.** Proposed methodology for accurate breast cancer tumor classification.

### 3.1. Data Pre-Processing

Data pre-processing was performed to improve data quality and get a clean dataset which could be used for building the model. Without pre-processing, several challenges will occur—inconsistencies, error, noise, missing values, model over-fitting, etc. To evaluate the impacts of the pre-processing steps on the results of the classification algorithms, breast cancer diagnosis was evaluated separately with and without pre-processing. For pre-processing, we used two feature selection methods and chose the better performing one.

3.1.1. Dimensionality Reduction Using Correlation Analysis

Dimensionality reduction is a technique to remove features that are less significant for predicting the outcome(s). In this work, dimensionality reduction was performed by analyzing the correlations among the features of input data, dropping features that had high variance. As shown in Figure 2, a heat map was used to analyze the correlations between features of the dataset. A high correlation was observed among "radius-mean", "parametric-mean", and "area mean" features, as all these features contain information about the size of breast cancer cells. Therefore, only the "radius-mean" feature was selected to further represent the information about the size of breast cancer cell.

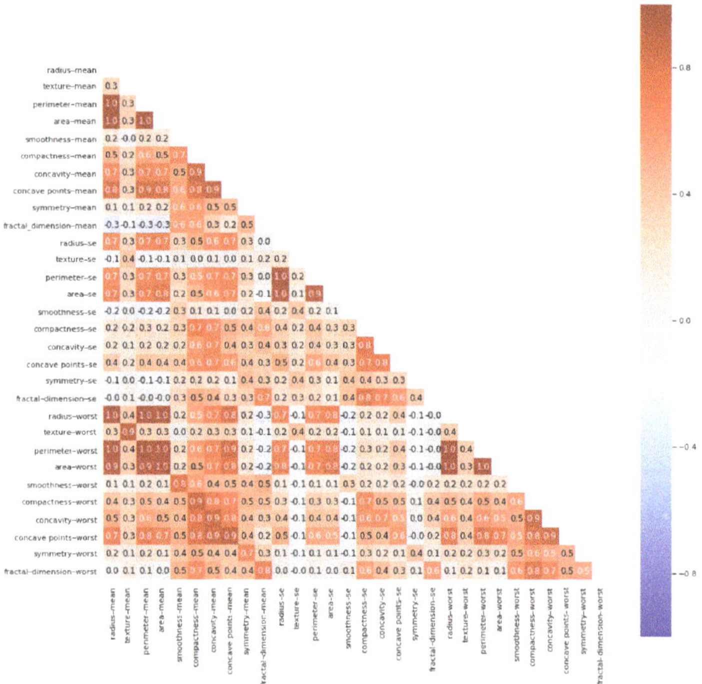


**Figure 2.** Heat map plot showing the correlations among input features of WBCD dataset.

High correlations were observed between the features representing the "mean" and "worst" values of different features. For instance, the "radius-mean" feature is highly correlated with the "radius-worst" feature. The feature representing the "worst" value of "radius" was dropped, as it is just a subset of the "mean" value feature. Similarly, high correlations were observed between the features containing information about the shape of breast cancer cell—i.e., compactness, concavity, and concave points. For better breast cancer cell shape representation, we decided to only consider the "compactness-mean" feature for further processing. We dropped a total of nine features: "area-mean, perimeter-mean, radius-worst, area-worst, perimeter-worst, texture-worst, concavity-mean, perimeter-se, area-se." This way, we had 22 features remaining for further processing. Figure 3 shows the correlations among the selected features.

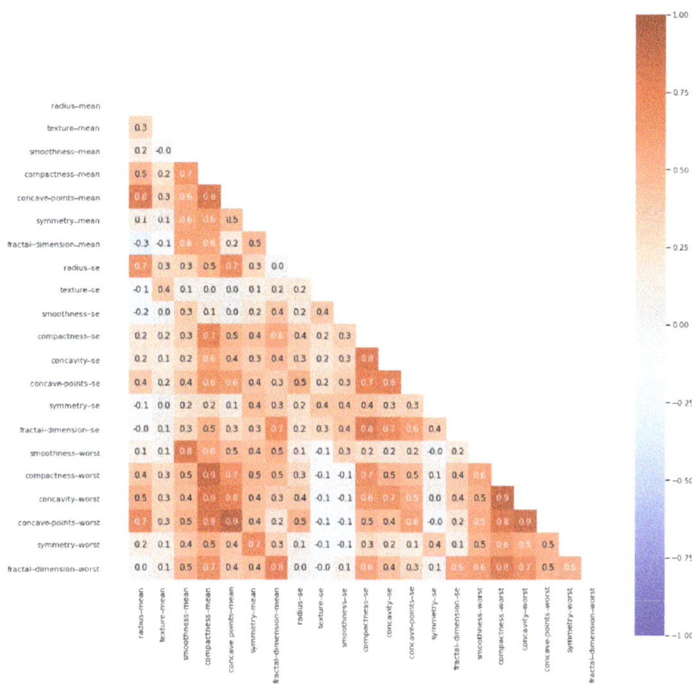


**Figure 3.** Heat map plot showing the correlations among selected features of WBCD dataset.

### 3.1.2. Dimensionality Reduction Using Principal Component Analysis

The selected features were further analyzed based on their variance. To perform dimensionality reduction based on their variance, we used the well-known principal component analysis (PCA) algorithm. We used the sklearn library, the sklearn.decomposition function, to import PCA (linear dimensionality reduction using singular value decomposition of the data to project it to a lower dimensional space) for feature selection. For PCA we had to ensure that all features were on the same scale; otherwise, the features that have high variance would have affected the outcomes of the PCA. "StandardScaler" was used to standardize features, followed by PCA for dimensionality reduction.

Figure 4, shows the variance of different features for the dataset. This graph shows that most of variance can be represented using 10 features only. These 10 features are "radius-mean, texture-mean, compactness-mean, concave points-mean, symmetry-mean, fractal-dimension-mean, smoothness-mean, radius-se, texture-se, and smoothness-se."

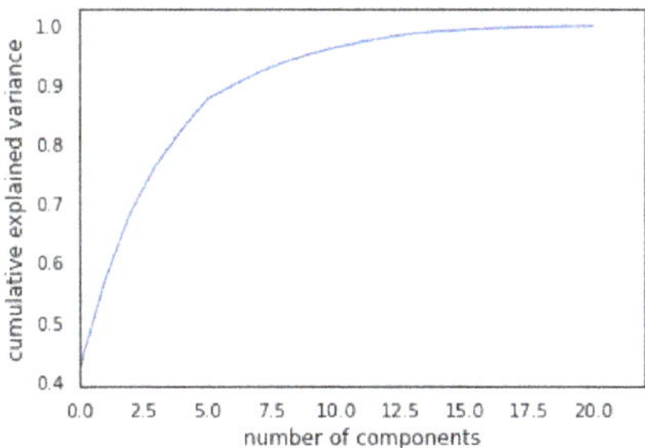


**Figure 4.** Number of features and their cumulative variance in the WBCD dataset.

#### 3.1.3. Feature Selection by Using a Wrapper Subset Selection Method

We used a wrapper subset selection method for feature selection. Wrapper methods work by evaluating a subset of features using a machine learning algorithm that employs a search strategy to look through the space of possible feature subsets, evaluating each subset based on the quality of the performance of a given algorithm. Wrapper methods generally result in better performance than filter methods because the feature selection process is optimized for the classification algorithm to be used. However, wrapper methods are too expensive for high dimensional data in terms of computational complexity and time, since each feature set considered must be evaluated with the classifier algorithm used. The working of wrapper methods is illustrated in Figure 5,

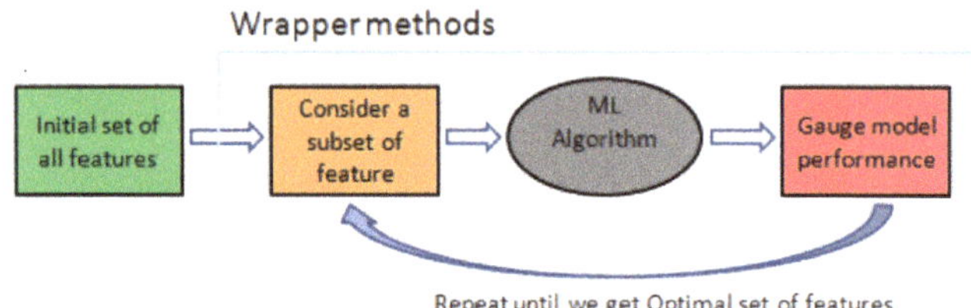


**Figure 5.** Wrapper method.

Summarizing, wrapper methods work in the following way.

- Search for a subset of features: Using a search method, we select a subset of features from the available ones.
- Build a machine learning model: In this step, a chosen ML algorithm is trained on the previously-selected subset of features.
- Evaluate model performance: Finally, the newly-trained ML model is evaluated with a chosen metric.
- Repeat: The whole process starts again with a new subset of features, a new trained ML model. The process stops when the desired condition is met, at which point the subset with the best result in the evaluation phase is chosen.

As a part of first step of feature selection, the search method used was BestFirst, and the chosen machine learning classifier was J48; we set the values of fold to 10, seed to 1, and threshold to $-1.0$. BestFirst selects the $n$ best features for modeling a given dataset, using a greedy algorithm. It starts by creating $N$ models, each of them using only one of the $N$ features of the dataset as input. The feature that yields the model with the best performance is selected. In the next iteration, it creates another set of $N-1$ models with two input features: the one selected in the previous iteration and another of the $N-1$ remaining features. Again, the combination of features that gives the best performance is selected. The script stops when it reaches the number of desired features. One improvement we made to this script was including k-fold cross-validation in the model evaluation process at each iteration. This ensured that the good or bad performance of one model was not produced by chance because of a single favorable train/test split.

The result provided nine attributes, concavity-mean, concave points-mean, perimeter-se, area-se, texture-worst, area-worst, smoothness-worst, symmetry-worst, and fractal-dimension-worst, as shown in Figure 6. The total number of subsets evaluated was 955, and the best subset figure of merit was 96.8%.

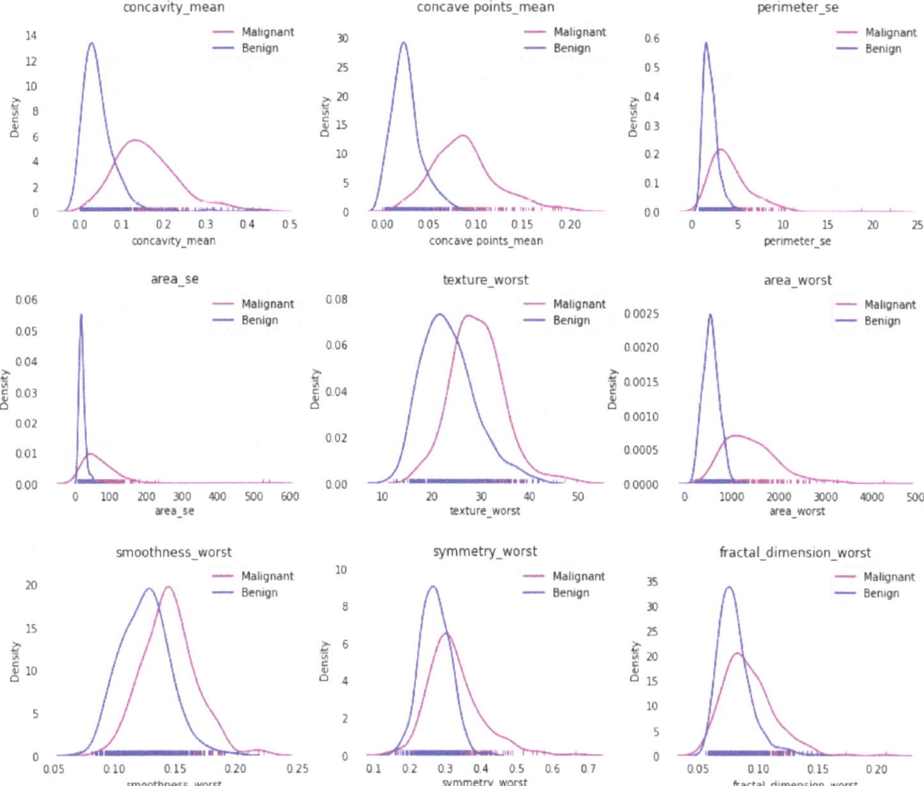


**Figure 6.** Distribution of malignant and benign cells for reduced features for WBCD.

### 3.2. Breast Cancer Tumor Classification

The following classification algorithms were evaluated for the task of breast cancer tumor classification, and hyper-parameter tuning was performed for classifiers using "GridSearchCv", which performs exhaustive searching over specified parameter values for an estimator. GridSearchCV tries all the combinations of the values passed in the dictionary and evaluates the model for each combination using the Cross-Validation method. From the exhaustive set of accuracies thus obtained, the best one is chosen.

#### 3.2.1. NaïVe Bayes Classification

The naïve Bayes model is very effective for large datasets because of its simplicity. It works on the probability basis p(c | x), where p(c | x) is the posterior probability of the class (c) and predictor (x).

#### 3.2.2. Support Vector Machine (SVM)

We performed hyper-parameter tuning for SVM with GridSearchCv. SVM-CV performance was then compared with default SVM performance. Both showed the same accuracy, precision, and recall score. The parameters values showing the best performances were C = 1 and degree = 1, where C is a SVM cost function used for SVM optimization and degrees is a value of polynomial used to find the hyper-plane to split the data. The default setting for SVM is C = 1, degree = 3.

3.2.3. Decision Tree

Decision trees use multiple algorithms to decide to split a node into two or more sub-nodes. The creation of sub-nodes increases the homogeneity of resultant sub-nodes. In other words, it can be said that the purity of the node increases with respect to the target variable. The decision tree splits the nodes on all available variables and then selects the split which results in the most homogeneous sub-nodes.

3.2.4. K-Nearest Neighbors (KNN)

KNN is non-parametric method, as it does not consider the dimensionality of dataset for diagnosis because it relies upon nearest training data points. The "GridSearchCv" was used to figure out the total number of neighbors for the KNN training needed to achieve superior performance.

3.2.5. The Random Decision Forest Method

A random forest is considered as a highly accurate and robust method because of the number of decision trees participating in the process. It tries to build k different decision trees by picking a random subset $S$ of training samples. It generates fully Iterative Dichotomiser 3 (ID3) trees with no pruning. It makes a final prediction based on the mean of each prediction, and it tends to be robust to overfitting, mainly because it takes the average of all the predictions, which cancels out biases.

3.2.6. Simple Logistic Regression

Logistic regression is a statistical method for evaluating a dataset in which a result is calculated by one or more independent variables. It is a supervised learning technique similar to linear regression.

3.2.7. Stochastic Gradient Descent Learning for Support Vector Machine

In stochastic gradient decent, in each interaction only a few samples are selected randomly instead of the entire dataset. The samples are shuffled at random and chosen to perform the interaction.

*3.3. Ensemble Classification*

Ensemble learning strategically brings together several machine learning models for achieving better performance. There are three different types of ensemble techniques: bagging based ensemble learning, boosting based ensemble learning, and voting-based ensemble learning. In our work, we used voting-based ensemble learning.

Voting-based ensemble learning is one of the basic or straightforward ensemble learning techniques in which diagnoses from multiple models are combined with either hard or soft voting.

3.3.1. The Majority-Based Voting Mechanism (Hard Voting)

In hard voting, we assign or predict the final class label as the class label that the classification models has most often predicted. Hard voting is the simplest case of majority voting. In majority voting, the class label $y$ is predicted via majority (plurality) vote the classifiers $C$:

$$y = mode\,\{C_1(x),\ C_2(x),\ ..,\ C_n(x)\} \tag{1}$$

3.3.2. The Probability-Based Voting Mechanism (Soft Voting)

In soft voting, we predict the class labels based on the predicted probabilities $p$ for the classifiers [23]. Soft voting attains the best results by averaging out the probabilities calculated by individual algorithms. Soft voting predicts the label as:

$$\hat{y} = \underset{i}{argmax} \sum_{j=1}^{m} w_j p_{ij} \tag{2}$$

where $w_j$ is the weight that is assigned to the j[th] classifier and $p_{ij}$ is the predicted membership probability of the i[th] classifier for class label $j$.

## 4. Experimentation and Discussion

### 4.1. The Wisconsin Breast Cancer Dataset (WBCD)

This public dataset [2] is based on microscopic examination of aspiration tests using fine needles on breast masses. The breast mass attribute is determined from a digital fine-needle aspirate (FNA) scan. Breast mass FNA is an important way of assessing malignancy. The WBCD was created by Dr. William H. Wolberg at the University of Wisconsin-Madison Hospital. There are 569 instances in this database, consisting of two cases: 357 benign instances and 212 malignant ones. These 569 instances are of human breast tissue from the FNA and were clinically evaluated based on 32 characteristics. All attributes can be considered as symptoms of a patient's breast cancer. Finally, 70:30 training:testing split was used for evaluation.

### 4.2. Results and Discussion

As can be seen in Table 1, results without pre-processing are unreliable and inaccurate. Some classifiers—support vector machine, naïve Bayes, etc.—did not perform well and produced low precision and recall scores.

**Table 1.** Breast cancer diagnosis without data pre-processing.

| Classification Algorithms | Accuracy (%) | Precision | Recall | F-Measures | F2-Measures |
|---|---|---|---|---|---|
| Naive Bayes Classification | 84.50% | 0.70% | 0.57% | 0.62% | 0.59% |
| Simple Logistic Regression | 87.94% | 0.88% | 0.87% | 0.87% | 0.87% |
| Random Decision Forest Method | 99.47% | 0.99% | 0.99% | 0.98% | 0.99% |
| Support Vector Machine | 62.00% | 0.62% | 0.40% | 0.48% | 0.43% |
| K-Nearest Neighbor Classification | 90.00% | 0.89% | 0.80% | 0.84% | 0.81% |
| Decision Tree | 88.00% | 0.88% | 0.86% | 0.86% | 0.86% |
| Stochastic Gradient Decent Learning | 90.30% | 0.83% | 0.88% | 0.85% | 0.86% |

After feature selection, we compared the performances of different machine learning classification methods for breast tumor classification. To find out best parameters, hyper-parameter tuning using GridSearchCv was used, and the performance of each classifier was improved after that. As shown in Table 2, logistic regression outperformed the other classifiers with an accuracy of 97.49% and high precision and recall of 97.89% and 95.21%, respectively.

**Table 2.** Comparison of different classification methods on WBCD after feature scaling, and hyper-parameter tuning of features using PCA and correlation analysis.

| Classification Algorithms | Accuracy | Precision | Recall | F-Measures | F2-Measures |
|---|---|---|---|---|---|
| Simple Logistic Regression Learning | 97.49% | 97.89% | 95.21% | 96.53% | 95.73% |
| K-Nearest Neighbor Classification | 97.49% | 98.48% | 89.70% | 93.88% | 91.32% |
| Support Vector Machine | 96.23% | 91.88% | 93.94% | 92.89% | 93.52% |
| Random Decision Forest | 94.22% | 93.86% | 82.88% | 88.02% | 84.86% |
| Stochastic Gradient Descent Learning | 92.11% | 84.38% | 89.20% | 86.72% | 88.19% |
| Decision Tree | 90.45% | 87.14% | 87.00% | 87.06% | 87.02% |
| Naïve Bayes Classification | 91.60% | 91.90% | 91.80% | 91.84% | 91.81% |

Table 3 shows that probability-based soft voting mechanism performed better than majority-based (hard voting) voting, because soft voting uses more information by using individual classifiers' uncertainties in the final diagnosis.

**Table 3.** Evaluation results of ensemble voting after pre-processing, using method 1.

| Voting Classifiers | Accuracy (%) | Precision (%) | Recall (%) | F-Measures (%) | F2-Measures (%) |
|---|---|---|---|---|---|
| Soft Voting | 99.00 | 99.29 | 96.00 | 97.61 | 96.64 |
| Hard Voting | 97.29 | 96.48 | 95.70 | 96.08 | 95.85 |

Results after applying wrapper feature selection methods:

The nine attributes: concavity-mean, concave points-mean, perimeter-se, area-se, texture-worst, area-worst, smoothness-worst, symmetry-worst, and fractal-dimension-worst were provided by a wrapper feature selection method. Performance results of machine learning classifiers with reduced numbers of features from the initial set are shown in Table 4. Kernel density estimation (KDE) plots were used to check the distribution of malignant and benign cases for selected features. The visualization of the above-mentioned features is shown in Figure 6

**Table 4.** Comparison of the performances of different classification methods for WBCD after applying the wrapper feature selection method.

| Classification Algorithms | Accuracy | Precision | Recall | F-Measures | F2-Measures |
|---|---|---|---|---|---|
| Simple Logistic Regression Learning | 98.10% | 98.10% | 98.10% | 96.90% | 98.10% |
| K-Nearest Neighbor Classification | 95.43% | 95.40% | 95.40% | 95.40% | 95.40% |
| Support Vector Machine | 95.80% | 96.00% | 95.80% | 95.70% | 95.83% |
| Random Decision Forest | 96.70% | 96.70% | 96.70% | 96.60% | 96.70% |
| Stochastic Gradient Descent Learning | 97.40% | 97.40% | 97.40% | 97.40% | 97.40% |
| Decision Tree | 96.83% | 96.80% | 96.80% | 96.80% | 96.80% |
| Naïve Bayes Classification | 92.80% | 92.80% | 92.80% | 92.80% | 92.80% |

After this, we also analyzed the performance of this reduced set of features from the wrapper method when using ensemble voting. Table 5 shows that the probability-based soft voting mechanism performed better than majority-based (hard voting) voting, because soft voting gets more information by using individual classifiers' uncertainties in the final diagnosis.

**Table 5.** Evaluation results of ensemble voting after pre-processing by using the wrapper features subset selection method.

| Voting Classifiers | Accuracy (%) | Precision (%) | Recall (%) | F-Measures (%) | F2-Measures (%) |
|---|---|---|---|---|---|
| Soft Voting | 97.70 | 97.70 | 97.70 | 97.70 | 97.70 |
| Hard Voting | 97.40 | 97.40 | 97.40 | 97.30 | 97.40 |

Comparing both methods for feature selection, it can be concluded that the performances of machine learning classifiers were improved at the individual level by using a wrapper method. As can be seen in Table 4, simple logistic regression learning provided 98.10% accuracy, random decision forest 96.70%, stochastic descent learning 97.40%, decision tree 96.83%, and naïve Bayes 92.80%. However, from the evaluation results of ensemble voting, there was only a small improvement for hard voting.

## 5. Comparison with Existing Work

Table 6, shows a comparison with existing work for breast cancer diagnosis using ensemble techniques. Nguyen et al. [3] analyzed the performances of different supervised and unsupervised breast cancer classification models on the WBCD dataset. They analyzed the performance of an ensemble voting method for breast cancer detection. They applied principal component analysis for feature analysis and reported an accuracy of 98.00%. Compared to this approach, our proposed feature selection and ensemble method classification shows an improvement of 1.00%.

Rodrigues et al. [11] achieved a performance of 97.80% on the WBCD dataset using a Bayesian network; however, they evaluated performance only using a machine learning classification algorithm and did not analyze the significance of important features needed for better performance. They have evaluated the performances of two different classification algorithms, i.e., a Bayesian network and a decision tree. The Bayesian network performed better than the decision tree.

To compare the performances of different classification models, Ahmed et al. [4] used data from the Iranian Center for Breast Cancer dataset and explored the risk factors for predicting breast cancer. There are some limitations in this study, as many cases were lost in the follow-up and the records with missing values were omitted. Some important variables, such as S-phase fraction and DNA index, were not included in the study because of their unavailability, which may have decreased the performances of the models.

Shravya et al. [17] used three well known classification algorithms for the detection of breast cancer. They used logistic regression, a support vector machine, and k-nearest neighbors. The SVM outperformed the other two classifiers and showed better performance with 92.70% accuracy. It is noted that there is a lot of room for improvement when the ensemble method classification is used instead of using individual classification algorithms. Darzi et al. [20] addressed feature selection for breast cancer diagnosis. Their process contains a wrapper approach based on a genetic algorithm (GA) and case-based reasoning (CBR), and reported an accuracy of 97.37% on WBCD. As compared to this approach, our proposed feature selection and ensemble method classification show an improvement of 2.00%.

Bharat et al. [12] achieved a performance of 97.3% on the WBCD dataset using naïve Bayes; however, they evaluated the performance only using a machine learning classification algorithm and did not analyze the significance of important features needed for better performance. They evaluated the performances of three different classification algorithms, i.e., naïve Bayes, the J48 network, and the RBF network. Naïve Bayes performed better than the other two. Assiri et al. [19] proposed an ensemble classifier based on a majority voting mechanism. The performances of different state-of-the-art ML classification algorithms were evaluated for the WBCD dataset, achieving an accuracy of 99.42%. However, they did not evaluate different feature selection algorithms that could help them to determine the smallest subset of features that can assist in accurate classification of breast cancer as either benign or malignant.

Lucas et al. [11] used Bayesian network and decision tree machine learning classifiers. The Bayesian network gave the best accuracy of 97.80% on WBCD. As compared to this approach, our proposed feature selection and ensemble method classification showed an improvement of 2.00%.

**Table 6.** Comparison with the existing work for breast cancer diagnosis.

| Authors | Classifiers | Accuracy (%) |
|---|---|---|
| Proposed Approach | Dimensionality Reduction and Ensemble based learning | 99.00 |
| Darzi et al. [20] | CBR-Genetic (case-based reasoning) | 97.37 |
| Nguyen et al. [3] | Ensemble Method | 98.00 |
| Rodrigues et al. [11] | Bayesian Network<br>Decision Tree | 97.80<br>92.00 |
| Subhani et al. [17] | Logistic Regression<br>Support Vector Machine<br>K Nearest Neighbor | 88.00<br>92.70<br>82.00 |
| Ahmed et al. [4] | Decision tree<br>Artificial neural network<br>Support Vector Machine | 93.60<br>94.70<br>95.70 |
| Lucas et al. [11] | Bayesian network<br>J48 Decision tree | 97.80<br>96.05 |
| Bharat et al. [12] | Decision tree C4.5<br>Support Vector Machine<br>Naive Bayes<br>K Nearest Neighbor | 95.00<br>96.20<br>97.00<br>91.00 |
| Assiri et al. [19] | Ensembled machine learning method | 99.42 |

## 6. Conclusions and Future Work

Early detection of breast cancer is important, as it is one of the leading causes of death among women, so its detection at early stages is very important. Early breast cancer tumor detection can be improved with the help of modern machine learning classifiers. In medical research, the false positive and false negative examples have great significance, but most existing work has evaluated performance based only accuracy evaluation measure. Therefore, we focused not only on accuracy but also evaluated performance based on precision and recall. In this work, feature selection and dimensionality reduction were performed using principal component analysis and by analyzing the correlations among different sets of features and their variance. The performances of different machine learning algorithms, including logistic regression, support vector machine, naïve Bayes, k-nearest neighbor, random forest, decision tree, and stochastic gradient decent learning, were evaluated. We reported the performances of different classifiers using different performance measures, including accuracy, precision, and recall. A voting ensemble method was used to improve the performances of the classifiers. The three best classifiers were then used for final classification using a voting ensemble method. We used hard voting (majority-based voting) and soft voting (probability-based voting) for ensemble classification. The average-probability-based voting (soft voting) showed better results as compared to hard voting. For big datasets, how these machine learning classifiers algorithms behave is one of the future scopes of this project. This work could be enhanced through the use of deep learning techniques for classification and identification of particular stage s of breast cancer.

**Author Contributions:** Conceptualization, S.I. and S.N.; methodology, S.N. and S.I. ; software, S.I.; validation, S.I., S.N. and S.A.V.; formal analysis, S.I.; investigation, S.I.; resources, S.N. and S.I.; data curation, S.I.; writing—original draft preparation,S.I.; writing—review and editing, S.N. and S.A.V.; visualization, S.I., S.N. and S.A.V.; supervision, S.N.; project administration, S.I. and S.N. All authors have read and agreed to the published version of the manuscript.

**Funding:** This research received no external funding.

**Institutional Review Board Statement:** Not applicable.

**Informed Consent Statement:** Not applicable.

**Data Availability Statement:** Dataset is publicly available.

**Conflicts of Interest:** The authors declare no conflict of interest

## Appendix A

This section contains the information about the libraries used for implementation [24].

- Sklearn.metrics: used to import confusion-matrix
- sklearn.model-selection: used to import cross-val-core (Evaluate a score by cross-validation).
- sklearn.metrics: Used to import precision-score, recall-score, f1-score, f2-score
- NumPy: It provides support for large multidimensional array objects and various tools to work with them.
- Pandas: Pandas allow importing data from various file formats. Pandas allows various data manipulation operations such as merging, reshaping, selecting, as well as data cleaning, and data wrangling features.
- Matplotlib: Matplotlib is a plotting library for the Python programming language and its numerical mathematics extension NumPy.
- Seaborn: Seaborn is a library in Python predominantly used for making statistical graphics. Seaborn is a data visualization library built on top of matplotlib and closely integrated with pandas data structures in Python.
- Scikit-learn: Scikit-learn is also known as sklearn. It is free and the most popular machine learning library for Python and used to build machine learning models.

## References


1. Coccia, M. The increasing risk of mortality in breast cancer: A socioeconomic analysis between countries. *J. Soc. Adm. Sci.* **2019**, *6* 218–230.
2. UCI. Breast Cancer Wisconsin Dataset. Available online: https://archive.ics.uci.edu/ml/datasets/Breast+Cancer+Wisconsin+(Diagnostic) (accessed on 9 June 2019).
3. Nguyen, Q.H.; Do, T.T.; Wang, Y.; Heng, S.S.; Chen, K.; Ang, W.H.M.; Philip, C.E.; Singh, M.; Pham, H.N.; Nguyen, B.P.; et al. Breast Cancer Prediction using Feature Selection and Ensemble Voting. In Proceedings of the 2019 International Conference on System Science and Engineering (ICSSE), Dong Hoi City, Vietnam, 19–21 July 2019; pp. 250–254.
4. Ahmad, L.G.; Eshlaghy, A.; Poorebrahimi, A.; Ebrahimi, M.; Razavi, A.; Using three machine learning techniques for predicting breast cancer recurrence. *J. Health Med. Inform.* **2013**, *4*, 3.
5. Omondiagbe, D.A.; Veeramani, S.; Sidhu, A.S. Machine Learning Classification Techniques for Breast Cancer Diagnosis. In Proceedings of the IOP Conference Series: Materials Science and Engineering, Kazimierz Dolny, Poland, 21–23 November 2019; Volume 495, p. 012033.
6. Yesuf, S.H. Breast cancer detection using machine learning techniques. *Int. J. Adv. Res. Comput. Sci.* **2019**, *10*, 27–33. [CrossRef]
7. Chen, X.W.; Jeong, J.C. Enhanced recursive feature elimination. In Proceedings of the Sixth International Conference on Machine Learning and Applications (ICMLA 2007), Cincinnati, OH, USA, 13–15 December 2007; pp. 429–435.
8. Jamal, A.; Handayani, A.; Septiandri, A.; Ripmiatin, E.; Effendi, Y. Dimensionality Reduction using PCA and K-Means Clustering for Breast Cancer Prediction. *Lontar Komput. J. Ilm. Teknol. Inf.* **2018**, *9*, 192–201. [CrossRef]
9. Mandal, S.K. Performance analysis of data mining algorithms for breast cancer cell detection using Naïve Bayes, logistic regression and decision tree. *Int. J. Eng. Comput. Sci.* **2017**, *6*.
10. Kumar, V.; Mishra, B.K.; Mazzara, M.; Thanh, D.N.; Verma, A. Prediction of Malignant and Benign Breast Cancer: A Data Mining Approach in Healthcare Applications. In *Advances in Data Science and Management*; Springer: Berlin/Heidelberg, Germany, 2020; pp. 435–442.
11. Borges, L.R. Analysis of the wisconsin breast cancer dataset and machine learning for breast cancer detection. *Group* **1989**, *1*, 15–19.
12. Bharat, A.; Pooja, N.; Reddy, R.A. Using Machine Learning algorithms for breast cancer risk prediction and diagnosis. In Proceedings of the 2018 3rd International Conference on Circuits, Control, Communication and Computing (I4C), Bengaluru, India, 3–5 October 2018; pp. 1–4.
13. Aavula, R.; Bhramaramba, R. XBPF: An Extensible Breast Cancer Prognosis Framework for Predicting Susceptibility, Recurrence and Survivability. *Int. J. Eng. Adv. Technol.* **2019**, 2249–8958
14. Chicco, D.; Rovelli, C. Computational prediction of diagnosis and feature selection on mesothelioma patient health records. *PLoS ONE* **2019**, *14*, e0208737. [CrossRef]
15. Wang, H.; Yoon, S.W. Breast cancer prediction using data mining method. In Proceedings of the IIE Annual Conference, Institute of Industrial and Systems Engineers (IISE), Nashville, TN, USA, 31 May–3 June 2015; p. 818.

16. Abdolell, M.; Tsuruda, K.M.; Lightfoot, C.B.; Payne, J.I.; Caines, J.S.; Iles, S.E. Utility of relative and absolute measures of mammographic density vs clinical risk factors in evaluating breast cancer risk at time of screening mammography. *Br. J. Radiol.* **2016**, *89*, 20150522. [CrossRef]
17. Shravya, C.; Pravalika, K.; Subhani, S. Prediction of Breast Cancer Using Supervised Machine Learning Techniques. *Int. J. Innov. Technol. Explor. Eng. (IJITEE)* **2019**, *8*, 1106–1110.
18. Williams, K.; Idowu, P.A.; Balogun, J.A.; Oluwaranti, A.I. Breast cancer risk prediction using data mining classification techniques. *Trans. Networks Commun.* **2015**, *3*, 01. [CrossRef]
19. Assiri, A.S.; Nazir, S.; Velastin, S.A. Breast tumor classification using an ensemble machine learning method. *J. Imaging* **2020**, *6*, 39. [CrossRef]
20. Darzi, M.; AsgharLiaei, A.; Hosseini, M.; others. Feature selection for breast cancer diagnosis: A case-based wrapper approach. *Int. J. Biomed. Biol. Eng.* **2011**, *5*, 220–223.
21. Marrone, S.; Piantadosi, G.; Fusco, R.; Petrillo, A.; Sansone, M.; Sansone, C. Breast segmentation using Fuzzy C-Means and anatomical priors in DCE-MRI. In Proceedings of the 2016 23rd International Conference on Pattern Recognition (ICPR), Cancun, Mexico, 4–8 December 2016; pp. 1472–1477.
22. Rundo, L.; Militello, C.; Tangherloni, A.; Russo, G.; Vitabile, S.; Gilardi, M.C.; Mauri, G. NeXt for neuro-radiosurgery: A fully automatic approach for necrosis extraction in brain tumor MRI using an unsupervised machine learning technique. *Int. J. Imaging Syst. Technol.* **2018**, *28*, 21–37. [CrossRef]
23. Wang, H.; Yang, Y.; Wang, H.; Chen, D. Soft-voting clustering ensemble. In *International Workshop on Multiple Classifier Systems*; Springer: Berlin/Heidelberg, Germany, 2013; pp. 307–318.
24. Raschka, S. *Python Machine Learning*; Packt Publishing Ltd: Birmingham, UK, 2015.

Journal of *Imaging*

MDPI

*Article*

# Study on Data Partition for Delimitation of Masses in Mammography

**Luís Viegas [1,†], Inês Domingues [2,†] and Mateus Mendes [1,3,*,†]**

1 Polytechnic of Coimbra—ISEC, Rua Pedro Nunes, Quinta da Nora, 3030-199 Coimbra, Portugal; a21250789@isec.pt
2 Medical Physics, Radiobiology and Radiation Protection Group, IPO Porto Research Centre (CI-IPOP), 4200-072 Porto, Portugal; inesdomingues@gmail.com
3 ISR (Instituto de Sistemas e Robótica), Departamento de Engenharia Electrotécnica e de Computadores da UC, University of Coimbra, 3004-531 Coimbra, Portugal
* Correspondence: mmendes@isec.pt
† These authors contributed equally to this work.

**Abstract:** Mammography is the primary medical imaging method used for routine screening and early detection of breast cancer in women. However, the process of manually inspecting, detecting, and delimiting the tumoral massess in 2D images is a very time-consuming task, subject to human errors due to fatigue. Therefore, integrated computer-aided detection systems have been proposed, based on modern computer vision and machine learning methods. In the present work, mammogram images from the publicly available Inbreast dataset are first converted to pseudo-color and then used to train and test a Mask R-CNN deep neural network. The most common approach is to start with a dataset and split the images into train and test set randomly. However, since there are often two or more images of the same case in the dataset, the way the dataset is split may have an impact on the results. Our experiments show that random partition of the data can produce unreliable training, so the dataset must be split using case-wise partition for more stable results. In experimental results, the method achieves an average true positive rate of 0.936 with 0.063 standard deviation using random partition and 0.908 with 0.002 standard deviation using case-wise partition, showing that case-wise partition must be used for more reliable results.

**Keywords:** mammography; computer-aided detection; breast mass; mass detection; mass segmentation; Mask R-CNN; dataset partition

**Citation:** Viegas, L.; Domingues, I.; Mendes, M. Study on Data Partition for Delimitation of Masses in Mammography. *J. Imaging* **2021**, *7*, 174. https://doi.org/10.3390/jimaging7090174

Academic Editors: Leonardo Rundo, Carmelo Militello, Vincenzo Conti, Fulvio Zaccagna and Changhee Han

Received: 11 July 2021
Accepted: 26 August 2021
Published: 2 September 2021

**Publisher's Note:** MDPI stays neutral with regard to jurisdictional claims in published maps and institutional affiliations.



## 1. Introduction

In 2020, there were 2.3 million new cases of brest cancer in the world [1]. That makes it the most common malignant tumor affecting women, accounting for a total of 11.7% of all cancer cases diagnosed. It is also the fifth leading cause of cancer mortality, with 685,000 deaths worldwide [1]. Among women, breast cancer is responsible for 1 in 4 cancer cases and 1 in 6 cancer-related deaths [1].

Despite these worrying figures, mortality from breast cancer is relatively low. In general, the disease has a good prognosis if the tumours are diagnosed in the early stages. About 90% of women with breast cancer are well five years after the original diagnosis [2]. However, due to the high incidence, this illness ranks first among all causes of cancer-related deaths in the female population. Mortality due to breast cancer has been decreasing continuously and consistently for several years. Early screening, that allows for the diagnosis of carcinomas at increasingly earlier stages, is one of the most important factors for the success of treatment and consequent reduction of mortality [3].

The present paper describes a method to detect and segment breast masses, based on a popular deep learning model known as Mask R-CNN. This model has been used before, with good results, by researchers such as Min et al. [4]. However, the focus of the present paper is a comparison to determine the importance of splitting the dataset properly, in order to avoid overfitting of the data. Experiments were performed splitting the images, to create the test set, randomly and by case. While this seems to be a small detail, in data preparation, it may have a significant impact on the results. The dataset used is the publicly available INbreast [5]. Experiments show that the method has competitive results compared to state-of-the-art methods. Additionally, division of the dataset by case instead of by image leads to more stable training procedures.

The paper is organized as follows: Section 2 explains in more detail what a mammogram image is and how computer aided detection can facilitate the diagnosis process. Section 3 presents a short survey of the state of the art related to detection and segmentation of masses from mammograms using deep learning. Section 4 describes methods to detect and segment tumoral masses. Section 5 contains a summary of the experiments and the results. Section 6 gives a brief discussion with comparison of results. Section 7 draws conclusions and highlights possible future research directions.

## 2. Mammography Images

Mammography has long been considered the most effective diagnostic imaging test for the early detection of breast cancer. The exam is simple and non-invasive. It must be performed routinely, in asymptomatic women (screening), or for diagnosis, being a fundamental tool in the detection of lesions in early stages, allowing a favorable prognosis and an increase in the success rate of treatments [6].

The imaging technique most used in the screening and diagnosis of breast cancer is X-ray mammography. It is a fast, low-cost technique with high spatial resolution. The basic views performed in a mammography exam are the Craniocaudal view (CC) and the Mediolateral Oblique view (MLO). Both are performed for each breast, up to a total of four images per patient. The main signs of breast cancer are the masses and clusters of microcalcifications, so the analysis of a mammographic image begins with the search for these types of formations.

There are different types of breast abnormalities. The abnormalities that can be seen in mammograms include masses, calcifications, asymmetry, or breast distortion. However, the breast masses, which are areas of thicker tissue that show in the mammography, are the most important sign of the illness. The analysis of mammogram images is a difficult task, even for trained radiologists. The main challenges are due to the different breast patterns, variations of color and shape of the tumoral masses, their possible locations, and different sizes possible. This variability often makes the abnormalities difficult to detect, segment, and classify.

The huge number of mammograms that can be generated and need to be analyzed during breast cancer screening programs require a significant workload, which often leads to fatigue and consequently errors of the radiologists that have to process and analyze hundreds or thousands of medical images over several days in a row. Therefore, Computer-Aided Detection (CAD) systems have been proposed, with the aim of assisting technicians and radiologists in the task, facilitating the process and contributing to lowering the probability of generating false negatives and false positives. CAD systems are used as a second opinion in the interpretation of mammograms, by the radiologists, contributing for more confidence in the diagnosis. However, such CAD systems need to operate at high levels of precision and accuracy. They must be robust, both to false positives and false negatives. A false positive can lead to unnecessary further testing, while a false negative can lead to further complications which might have been avoided.

The tumoral masses are volumes of abnormal density. Mammogram images are only an incomplete description of the 3D structure of the mass. The masses show in 2D mammography images with a high variability of shapes, sizes and locations. Most of the times they are difficult to distinguish from the background, even for experienced technicians. Existing CAD systems and modern detection and segmentation models have shown promising results, but the problem is still subject to heavy research. Training machine learning algorithms is also a challenge per se, for there are not many large datasets, containing Full Field Digital Mammograms (FFDM), annotated by experts and available for general use. This poses additional difficulties for developing modern CAD systems.

Recent developments in methods based on Deep Learning (DL) can contribute to develop robust solutions to undertake these problems. Particularly, the methods that use Convolutional Neural Networks (CNNs) to automatically learn a relevant hierarchy of features directly from inputting images. The topic has been subject to heavy research and there have been important developments. However, most developments are just in the specific area of detection, where the result is a bounding box [7], or in the specific area of region segmentation, to tell the region of interest from the background [8,9]. Nonetheless, there are also a number of important developments proposing a completely integrated system, able to detect and segment tumoral masses in the pipeline with minimal human intervention. The most common approaches still deal with two-dimensional images. Three-dimensional approaches have already been studied [10–12], and even stereoscopic approaches [13]. However, the state-of-the-art CAD systems are mostly based on 2D methods and trained on datasets consisting of 2D images. This makes the methods of pre-processing the images and partitioning the datasets a very important and still open issue.

## 3. Related Work

Tumor mass detection and segmentation in mammogram images have been subject to heavy research in recent years. One of the latest techniques to be applied is DL machine models, namely CNNs. CNNs have been applied in different medical image analysis with success. The review focuses on research papers that use the publicly available database INbreast, or other databases, for training and testing, having the focus on implementation of CNNs to address the issues of detection and/or segmentation of breast masses in mammograms.

### 3.1. Detection of Tumoral Masses

Many modern object detection models have achieved good performance in object detection and segmentation tasks. Nonetheless, those tasks still remain a challenge when detecting breast tumor masses in medical images, due to the low signal-to-noise ratio and the variability of size and shape of masses.

Dhungel et al. [14] presented an architecture that contains a cascade of DL and Random Forest (RF) classifiers for breast mass detection. Particularly, the system comprises a cascade of multi-scale Deep Belief Network (m-DBN) and a Gaussian Mixture Model (GMM) to provide mass candidates, followed by cascades of Region-based Convolutional Neural Network (R-CNNs) and RF to reduce false positives.

Wichakam et al. [15] proposed a combination between CNNs for feature extraction and Support Vector Machines (SVM) as the classifiers to detect a mass in mammograms. Choukroun et al. [16] presented a patch based CNN for detection and classification of tumor masses where the mammogram images are tagged only on a global level, without local annotations. The method classifies mammograms by detecting discriminative local information from the patches, through a deep CNN. The local information is then used to localize the tumoral masses.

### 3.2. Segmentation of Tumoral Masses

A fundamental stage in typical CAD systems is the segmentation of masses. Most popular segmentation approaches are based on pre-delimited Regions Of Interest (ROI) of the images.

Dhungel et al. [17] proposed the use of structured learning and deep networks to segment mammograms—specifically, using a Structured Support Vector Machine (SSVM) with a DBN as a potential function. In a first stage, the masses are manually extracted; then, a DBN is used to detect the candidates and a Gaussian Mixture Model classifier performed the segmentation step.

In [18,19], two types of structured prediction models are used, combined with DL based models as potential functions, for the segmentation of masses. Specifically, SSVM and Conditional Random Field (CRF) models were combined with CNNs and DBNs. The CRF model uses Tree Re-Weighted Belief Propagation (TRW) for label inference, and learning with truncated fitting. The SSVM model uses graph cuts for inference and cutting plane for training.

However, these methods [17–19] have some limitations due to their dependence on prior knowledge of the mass contour. Zhu et al. [20] proposed an end-to-end trained adversarial network to perform mass segmentation. The network integrates a Fully Convolutional Network (FCN), followed by a CRF to perform structured learning.

Zhang et al. [21] proposed a framework for mammogram segmentation and classification, integrating the two tasks into one model by using a Deep Supervision scheme U-Net model with residual connections.

Liang et al. [22] proposed a Conditional Generative Adversarial Network (CGAN) for segmentation of the tumoral masses in a very small dataset using only images with masses. The CGAN consists of two networks, the Mask-Generator and the Discriminator. The Mask-Generator network uses a modified U-Net, where the feature channels between low level feature layers are discarded, and the ones between high level feature layers are preserved. For the Discriminator network, a convolutional PatchGAN classifier is used. As a condition to achieve CGANs, an image sample with its ground truth is added into the Mask-Generator.

### 3.3. Detection and Segmentation of Masses

The approaches described above focus either on detection or on segmentation of the masses. However, there are also approaches that address both problems in a pipeline system. Pipeline techniques have recently received increasing attention in machine learning. A pipeline is created, so that successive transformations are applied on the data, the last being either a model training or prediction operation. The pipeline model is regarded as a block, connecting each task in the sequence to the successor and delivering the result at the end [23].

Sawyer Lee et al. [24] compare the performance of segmentation-free and segmentation-based machine learning methods applied to detection of breast masses. Rundo et al. [25] use genetic algorithms in order to improve the performance of segmentation methods in medical magnetic resonance images. Tripathy et al. [26] perform segmentation using a threshold method on mammogram images, after enhancing contrast using the CLAHE algorithm.

Some systems that integrate both detection and segmentation stage still require manual rejection of false positives before the segmentation stage, as happens in [27,28]. Dhungel et al. [27], presented a two-stage pipeline system for mass detection and segmentation. Specifically, they adopted a cascade of m-DBNs and GMM classifier to provide mass candidates. The mass candidates are then delivered to cascades of deep neural nets and random forest classifiers, for refinement of the detection results. Afterwards, segmentation is performed through a deep structured learning CRF model followed by a contour detection model.

Al-antari et al. [28] presented a serial pipeline system designed for detection, segmentation, and classification, also based on DL models. A YOLO CNN detector is implemented for mass detection. The results of the YOLO detector are then fed to an FCN to perform segmentation. The result is then fed to a basic deep CNN for classification of the mass as benign or malign.

In [29], the authors address detection, segmentation, and classification in a multi-task CNN model enabled by cross-view feature transferring. With an architecture built upon Mask R-CNN, the model enables feature transfer from the segmentation to the classification task to improve the classification accuracy.

Min et al. [4] presented a method for sequential mass detection and segmentation using pseudo-color mammogram images as inputs to a Mask R-CNN DL framework. During the training phase, the pseudo-color mammograms are used to enhance contrast of the lesions, compared to the background. That boosts the signal-to-noise ratio and contributes to improving the performance of the model in both tasks. The model comprises a Faster R-CNN object detector and an FCN for mask prediction. The method used for the experiments performed in the present work was based on the same framework. However, Min uses 5 fold cross validation, and this is not used in the present work.

## 4. Materials and Methods

The experiments were performed using an implementation of a Mask R-CNN to detect and segment tumoral masses in the INbreast dataset.

### 4.1. Database

The dataset used in the present study is obtained from INbreast, a publicly available full-field digital mammographic database with precise ground truth annotations [5]. The resolution of each image is 2560 × 3328 or 3328 × 4084 pixels, and they are in Digital Imaging and Communications in Medicine (DICOM) format. The confidential information was removed from the DICOM file but a randomly generated patient identification keeps the correspondence between images of the same patient. The database includes examples of normal mammograms, mammograms with masses, calcifications, architectural distortions, asymmetries, and images with multiple findings. For each breast, both CC and MLO views were provided. Among the 410 mammograms from 115 cases in INbreast, 107 contain one or more masses. There is a total of 116 benign or malignant masses. The average mass size is 479 mm$^2$. The smallest mass has an area of 15 mm$^2$, and the biggest one has an area of 3689 mm$^2$.

The dataset is very small for training modern deep learning models, which require a large number of samples for proper training. However, large datasets are rare because of the difficulty in obtaining good quality medical images. Medical images require highly qualified people to provide the ground truth. There are also many privacy concerns because of the sensitive information they carry. Therefore, such images are rare and very important. Sometimes, the datasets are also imbalanced, with just a small number of samples showing a particular but important condition. Bria et al. [30] address the problem of class imbalance in medical images. A common technique is to use data augmentation, adding copies of some images with a transformation such as mirroring or rotation [31]. The present approach applies data augmentation through a random transformation, as described in Section 4.3.

### 4.2. Data Pre-Processing

One important step to start image processing is to tell the region of interest from the background. This can be done based on threshold methods [32]. Militello et al. use a different approach, based on quartile information [33], to distinguish epicardial adipose tissue from the background in medical cardiac CT scans. In the present work, the same procedure as in [4] was adopted. To prepare the images, the breast region is extracted using a threshold value to crop away the redundant background area. Specifically, and since the intensity of the background pixels of the INbreast mammograms is zero, the region where the pixels have a non-zero intensity value is extracted as the breast region [4,34]. The mammogram image is then resized to one fourth of the original image size. Afterwards, it is normalized to 16-bit. The normalized image is finally padded into a square matrix.

After cropping and normalization, the mammogram is converted to pseudo-color mammogram (PCM), in order to enhance the areas of thicker masses. The gray images were also changed to colour RGB images, which have the ability to convey colour information. In this way, the red, green and blue channels are filled respectively with the grayscale mammogram (GM), and two images generated by the Multi-scale Morphological Sifting

(MMS) algorithm [4]. The images generated by MMS and the GM are linearly scaled to 8-bit. Therefore, a PCM RGB image comprises a GM in the first (R) channel, the output image of the MMS transform scale 1 into the second (G) channel and the MMS transform scale 2 in the third (B) channel.

The MMS makes use of morphological filters with oriented linear structuring elements to extract lesion-like patterns. The MMS can enhance lesion-like patterns within a specified size range. To deal with the size variation of breast masses, the sifting process is applied in two scales.

The result is that a relatively smaller mass in the size range of scale 1 will have higher intensity in the second channel. Therefore, this is interpreted as a higher amount of green, and it tends to yellow on the PCM image. Figure 1a shows an example of a yellowish mass. A relatively larger size mass will have higher components in the range of scale 2, and therefore that will be interpreted as more of the blue component. The result is that it will tend to purple on the PCM image. This result is exemplified in Figure 1c. This transformation enhances the masses, which are then easier to differentiate from the background. As in [4], better results were achieved using PCM rather than using GM, so PCM was used for this work.

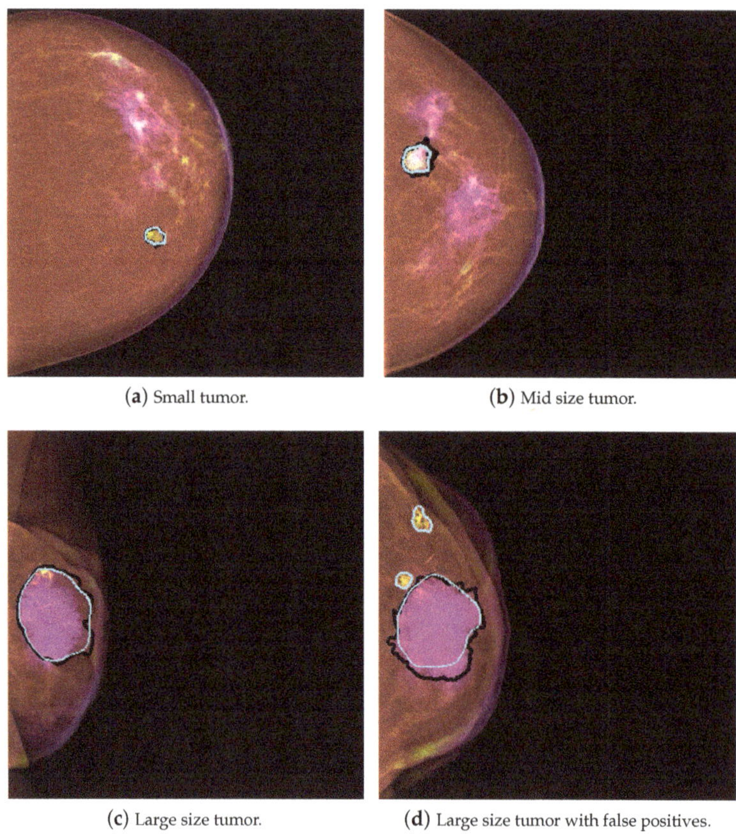

(**a**) Small tumor. (**b**) Mid size tumor. (**c**) Large size tumor. (**d**) Large size tumor with false positives.

**Figure 1.** Some visual results of automatic detection and segmentation of breast masses. Black and cyan lines respectively stand for ground truth of the masses and segmentation of the detected regions.

### 4.3. The Mask R-CNN Model

The present work applied transfer learning technique. Transfer learning is a common machine learning procedure where a pre-trained model is used as the basis to create a new model. In the present work, a pre-trained Mask R-CNN model was used, in order to speed up the training process. The dataset used is limited in size, thus starting with a pre-trained model not only speeds up the training process but also increases the chances of success. The Mask R-CNN is a framework that allows sequential mass detection and segmentation in mammograms. It integrates a Faster R-CNN object detector with an FCN for mask prediction. The Faster R-CNN utilizes the Region Proposal Network (RPN) to generate ROI candidates and then, for each candidate, performs classification and bounding-box regression. The FCN performs segmentation on the ROI candidates, generating the masks. During training, a multitasking loss function given by Equation (1) was used:

$$L = L_{cls} + L_{bbox} + L_{msk} \tag{1}$$

where $L_{cls}$ is the classification loss, $L_{bbox}$ is the bounding box regression loss, and $L_{msk}$ is the mask loss, defined as the binary cross-entropy loss [35].

To make use of the transfer learning technique, the Mask R-CNN model training was initialized starting with the pre-trained "mask_rcnn_balloon" model. It consists of a network that was previously trained for a detection and binary classification problem of separation of balloons from the background [36].

A deep residual neural network, the ResNet101, was used as the model backbone. The images are resized into 1024 × 1024 pixels. To expand the number of images, data augmentation is implemented. Specifically, images are augmented by randomly selecting one of the available operations, namely, flipping up, down, left, right, and rotations in 90, 180 and 270 degrees. The network is then trained through 10 epochs, with a batch size of 1. The parameters settings mentioned above are the same as those utilized in [4]. For all the parameters which were not specified above, the default values in [36] were adopted.

For the experiments, we used Python 3.6 (available at http://www.python.org (accessed on 1 September 2021)) and ran on an Asus laptop with Intel(R) Core(TM) i7-7500U CPU @ 2.90 GHz, 16 GB RAM (Coimbra, Portugal).The generation of the pseudo-color image was implemented in MATLAB 2019b (available at https://www.mathworks.com/products/matlab.html (accessed on 1 September 2021)) using the same machine.

### 4.4. Evaluation Method

Experiments on the INbreast dataset were performed using all the 410 images available. Those 410 images must be split into at least the train and test set. Most of the experiments in the literature divide the data randomly, for example setting 70% for training, 15% for validation, and 15% for testing. However, as stated above, there are multiple images of the same patient and also of the same tumor. Therefore, some authors mention that data must be split case-wise to avoid contamination of the test and validation sets with images of patients or cases contained in the training set [16]. To the best of our knowledge, however, the impact of this possible contamination has not been tested before.

In the present work, different experiments were performed, with case-wise partition of the dataset and with random split partition. In all cases, the dataset was split into 280 images for training, 65 images for validation and 65 images for testing. Data augmentation doubles the number of images. In the case-wise partition, when performing the division, it was guaranteed that images of the same patient were in the same subset. The division is based on cases, ensuring that there were no case overlaps between the splits.

For the images with masses, segmentation masks are used as the ground truth, while, for the images without any masses, their ground truths are the black background.

For the evaluation of the performance of the method, the metrics used were Sensitivity (S) or True Positive Rate (TPR ) and False Positive Per Image (FPPI) for the mass detection task, and the Dice Similarity Index (Dice) for the mass segmentation task. The criteria for these metrics are defined as follows:

$$\mathrm{TPR} = \frac{\mathrm{TP}}{\mathrm{TP} + \mathrm{FN}} \tag{2}$$

$$\mathrm{FPPI} = \frac{\mathrm{FP}}{\mathrm{TP} + \mathrm{FP}} \tag{3}$$

$$\mathrm{Dice} = \frac{2 \times \mathrm{TP}}{2 \times \mathrm{TP} + \mathrm{FP} + \mathrm{FN}} \tag{4}$$

where TP, FP, and FN represent the number of true positive, false positive and false negative detections, respectively. The mass is considered to be detected (TP) if the Intersection over Union (IoU) between the predicted bounding box and ground truth is greater than or equal to 0.2 [4].

## 5. Experiments and Results

Six experiments were performed. Three of them use random split partition of the images. The other three use case-wise partition. Mass detection and segmentation performance comparison between experiments are shown in Table 1. Experiments R1, R2 and R3 use random split of the images. Experiments C1, C2 and C3 use case-wise partition. The hyperparameters and other settings of the model were all the same, so that the results of the experiments could be compared.

More experiments could be performed for more confidence in the results. However, the results clearly show that case wise partition of the data seems to provide more stable results. In C1, C2, and C3, the TPR is very similar and the Dice only differed about 1%. Using randomly split data, however, the results for TPR varied between 0.875 and an overoptimistic 1.000 and the Dice varied between 0.857 and 0.885. In addition, R2 and R4 show a larger Dice variance than C1, C2 and C3.

**Table 1.** Comparison of TPR and Dice metrics between experiments. Experiments R1, R2 and R3 use a random split of the images. Experiments C1, C2 and C3 use case-wise partition of the images.

| Experiment | TPR @ FPPI | Dice |
|---|---|---|
| R1 | 0.875 @ 1.47 | 0.885 ± 0.044 |
| R2 | 0.933 @ 1.35 | 0.857 ± 0.118 |
| R3 | 1.000 @ 1.09 | 0.874 ± 0.097 |
| Average | 0.936 @ 1.30 | 0.872 ± 0.086 |
| STD | 0.063 @ 0.19 | 0.014 ± 0.038 |
| C1 | 0.909 @ 0.77 | 0.891 ± 0.050 |
| C2 | 0.909 @ 1.32 | 0.880 ± 0.061 |
| C3 | 0.906 @ 1.33 | 0.897 ± 0.036 |
| Average | 0.908 @ 1.14 | 0.889 ± 0.049 |
| STD | 0.002 @ 0.32 | 0.009 ± 0.012 |

The results show that using Mask R-CNN with PCMs, with case-wise dataset partition, achieves an average TPR of 0.909 at 0.77 FPPI and a Dice of 0.89 with some confidence on the results as shown in Table 1. The average TPR is 0.936 @ table. 1.30, with a standard deviation of 0.063 @ 0.19 using a random split of the samples. For case wise partition, the average is a bit lower, but the standard deviation is also lower: the average TPR is 0.908 @ 1.14 and the standard deviation is 0.002 @ 0.32. Thus, there is much less variation in the results obtained using case wise partition. As for Dice, using random split, the average Dice is 0.872 ± 0.086, with a standard deviation of 0.014 ± 0.038. The average Dice for mass segmentation using case wise partition is 0.889 ± 0.049, with a standard deviation of 0.009 ± 0.012. Therefore, in the case wise experiments, the standard deviation is always considerably lower than in the random split partition. Some visual results of detection and segmentation of breast masses are shown in Figure 1.

## 6. Discussion

Most medical image analysis applications require object detection, segmentation and classification. Modern DL models contribute to automation of all the tasks in a pipeline. Therefore, they are a useful technical solution to address the different tasks in a row.

The Mask R-CNN integrates mass detection and segmentation stages in one pipeline. Since a very small data set was used and training was initialized with pre-trained weights, there was no need to train for too long.

A public available dataset, INbreast, was used for evaluating the method. For quantitative analysis, three evaluation metrics, TPR or Recall, FPPI and Dice were utilized.

Case-wise partition was performed on dataset division to prevent images of the same case from appearing in more than one subset. Otherwise, contamination of the validation set and test set with images of the same patient could impact the results. This division by case seemed to have a small positive impact on the results obtained in the test set, compared to the results obtained when random split was used.

The global performance comparison between this method and several others methods are shown in Table 2. As the table shows, the results are competitive with the best published in the literature for the same dataset, and slightly better than other methods that perform detection and segmentation. Using case-wise partition, the results are also stable.

From Table 2, it can be seen that the PCMs + Mask R-CNN model, when compared to single task models, achieves a higher detection performance to [14,16], and outperforms [17,21] in segmentation. In addition, the model underperforms to a certain degree compared to [18–20] in segmentation. The reason may be that, in these [18–20], the input training samples were manually detected ROI masses, and this helped to improve the performance of segmentation results.

In comparison to Liang et al. [22], the method underperformed in segmentation. One of the reasons may be that Liang et al. used a very small and imbalanced dataset, consisting of only images with tumoral masses. In comparison with models which tackle both detection and segmentation, the model outperformed [27] in both tasks, achieving a similar sensibility and a higher segmentation performance than [29], and underperformed [28] in segmentation. For the lower result in comparison to [28], the reason may be that, like as in [27], they manually excluded all the false positive detections before segmentation. On the other hand, the PCMs + Mask R-CNN model is a fully automatic model, which can operate without human input.

**Table 2.** Performance comparison between PCMs + Mask R-CNN and several other state-of-the-art methods. The PCMs + Mask R-CNN is marked in bold.

| Method | Database | TPR @ FPPI | Dice |
| --- | --- | --- | --- |
| Dhungel et al. [14] | INbreast | 0.87 ± 0.14 @ 0.8 | n.a. |
| Wichakam et al. [15] | INbreast | n.a. | n.a. |
| Choukroun et al. [16] | INbreast | 0.76 @ 0.48 | n.a. |
| Dhungel et al. [17] | INbreast | n.a. | 0.88 |
| Dhungel et al. [18] | INbreast | n.a. | 0.90 ± 0.06 |
| Zhu et al. [20] | INbreast | n.a. | 0.9097 |
| Dhungel et al. [19] | INbreast | n.a. | 0.90 |
| Zhang et al. [21] | INbreast | n.a. | 0.85 |
| Liang et al. [22] | INbreast | n.a. | 0.91 |
| Dhungel et al. [27] | INbreast | 0.90 ± 0.02 @ 1.3 | 0.85 ± 0.02 |
| Al-antari et al. [28] | INbreast | n.a. | 0.9269 |
| Gao et al. [29] | INbreast | 0.91 ± 0.05 @ 1.5 | 0.76 ± 0.03 |
| Min et al. [4] | INbreast | 0.90 ± 0.05 @ 0.9 | 0.88 ± 0.10 |
| **PCMs+Mask R-CNN** | **INbreast** | **0.909 @ 0.769** | **0.891 ± 0.05** |

## 7. Conclusions

An integrated mammographic CAD system based on deep learning is described. It is capable of simultaneous detection and segmentation of the masses, from mammograms based on Mask R-CNN. It does not require human intervention to operate.

Experimental results show that the system achieves state-of-the-art competitive performance in detection and segmentation. The results obtained from our experiments show that data preparation may have a small impact on the performance of the system. Namely, case-wide partition seems to have a small positive impact on the performance, preventing the system from overfitting compared to when the dataset is randomly split.

Future work includes tests with other datasets, as well as a study of the application of the methodology to other similar problems, such as other types of tumors. The method can also be tested with other medical imaging types and modalities, such as MRI and PET.

**Author Contributions:** Conceptualization and methodology, L.V. and I.D.; software, L.V.; validation and formal analysis, I.D. and M.M.; formal analysis; writing—original draft preparation, L.V. and M.M.; writing—review and editing, and supervision, M.M. and I.D. All authors have read and agreed to the published version of the manuscript.

**Funding:** The authors acknowledge Fundacao para a Ciencia e a Tecnologia (FCT) for the financial support to the project UIDB/00048/2020. FCT had no interference in the development of the research.

**Institutional Review Board Statement:** Not applicable.

**Informed Consent Statement:** Not applicable.

**Data Availability Statement:** Not applicable.

**Conflicts of Interest:** The authors declare no conflict of interest.

## Abbreviations

The following abbreviations are used in this manuscript:

| | |
|---|---|
| CAD | Computer-Aided Detection system |
| CC | Craniocaudal |
| CGAN | Conditional Generative Adversarial Network |
| CNN | Convolution Neural Network |
| CRF | Conditional Random Fields |
| DBN | Deep Belief Network |
| DL | Deep Learning |
| DICOM | Digital Imaging and Communications in Medicine |
| GM | Grayscale Mammogram |
| GMM | Gaussian Mixture Model |
| Faster R-CNN | Faster Region based Convolutional Neural Network |
| FCN | Fully Convolutional Network |
| FFDM | Full Field Digital Mammogram |
| FPPI | False Positive Per Image |
| FrCN | Full Resolution Convolutional Network |
| IoU | Intersection over Union |
| m-DBN | multi-scale Deep Belief Network |
| MG | Mammogram |
| MLO | Mediolateral Oblique |
| MMS | Multi-scale Morphological Sifting |
| MRI | Magnetic Resonance Imaging |
| MTL | Multi-Task Learning |
| PCM | Pseudo-Color Mammogram |
| PET | Positron Emission Tomography |
| R-CNN | Region based Convolutional Neural Network |
| RF | Random Forest |

| | |
|---|---|
| RGB | Red, Green, Blue color model |
| RPN | Region Proposal Network |
| ROI | Region Of Interest |
| SSVM | Structured Support Vector Machine |
| SVM | Support Vector Machine |
| TPR | True Positive Rate |
| TRW | Tree Re-Weighted Belief Propagation |
| YOLO | You-Only-Look-Once |

## References


1. Sung, H.; Ferlay, J.; Siegel, R.L.; Laversanne, M.; Soerjomataram, I.; Jemal, A.; Bray, F. Global Cancer Statistics 2020: GLOBOCAN Estimates of Incidence and Mortality Worldwide for 36 Cancers in 185 Countries. *CA A Cancer J. Clin.* **2021**, *71*, 209–249. [CrossRef]
2. SEER National Cancer Institute. Female Breast Cancer—Cancer Stat Facts. Available online: https://seer.cancer.gov/statfacts/html/breast.html (accessed on 21 April 2021).
3. Bessa, S.; Domingues, I.; Cardosos, J.S.; Passarinho, P.; Cardoso, P.; Rodrigues, V.; Lage, F. Normal breast identification in screening mammography: A study on 18,000 images. In Proceedings of the IEEE International Conference on Bioinformatics and Biomedicine (BIBM), Belfast, UK, 2–5 November 2014; pp. 325–330. [CrossRef]
4. Min, H.; Wilson, D.; Huang, Y.; Liu, S.; Crozier, S.; Bradley, A.P.; Chandra, S.S. Fully Automatic Computer-aided Mass Detection and Segmentation via Pseudo-color Mammograms and Mask R-CNN. In Proceedings of the 2020 IEEE 17th International Symposium on Biomedical Imaging (ISBI), Iowa City, IA, USA, 3–7 April 2020; pp. 1111–1115. [CrossRef]
5. Moreira, I.C.; Amaral, I.; Domingues, I.; Cardoso, A.; Cardoso, M.J.; Cardoso, J.S. INbreast: Toward a Full-field Digital Mammographic Database. *Acad. Radiol.* **2012**, *19*, 236–248. [CrossRef] [PubMed]
6. PDQ® Screening and Prevention Editorial Board. *PDQ Breast Cancer Screening*; National Cancer Institute: Bethesda, MD, USA, 2020. Available online: https://www.cancer.gov/types/breast/hp/breast-screening-pdq (accessed on 21 April 2021).
7. Malta, A.; Mendes, M.; Farinha, T. Augmented Reality Maintenance Assistant Using YOLOv5. *Appl. Sci.* **2021**, *11*, 4758. [CrossRef]
8. Coelho, J.; Fidalgo, B.; Crisóstomo, M.M.; Salas-González, R.; Coimbra, A.P.; Mendes, M. Non-Destructive Fast Estimation of Tree Stem Height and Volume Using Image Processing. *Symmetry* **2021**, *13*, 374. [CrossRef]
9. Domingues, I.; Cardoso, J.S. Mass detection on mammogram images: A first assessment of deep learning techniques. In Proceedings of the 19th Portuguese Conference on Pattern Recognition, Lisbon, Portugal, 1 November 2013; p. 2.
10. Ciatto, S.; Houssami, N.; Bernardi, D.; Caumo, F.; Pellegrini, M.; Brunelli, S.; Tuttobene, P.; Bricolo, P.; Fantò, C.; Valentini, M.; et al. Integration of 3D digital mammography with tomosynthesis for population breast-cancer screening (STORM): A prospective comparison study. *Lancet Oncol.* **2013**, *14*, 583–589. [CrossRef]
11. Zhang, X.; Zhang, Y.; Han, E.Y.; Jacobs, N.; Han, Q.; Wang, X.; Liu, J. Classification of Whole Mammogram and Tomosynthesis Images Using Deep Convolutional Neural Networks. *IEEE Trans. NanoBiosci.* **2018**, *17*, 237–242. [CrossRef]
12. Liang, G.; Wang, X.; Zhang, Y.; Xing, X.; Blanton, H.; Salem, T.; Jacobs, N. Joint 2D-3D Breast Cancer Classification. In Proceedings of the 2019 IEEE International Conference on Bioinformatics and Biomedicine (BIBM), San Diego, CA, USA, 18–21 November 2019; pp. 692–696. [CrossRef]
13. Ferre, R.; Goumot, P.A.; Mesurolle, B. Stereoscopic digital mammogram: Usefulness in daily practice. *J. Gynecol. Obstet. Hum. Reprod.* **2018**, *47*, 231–236. [CrossRef]
14. Dhungel, N.; Carneiro, G.; Bradley, A.P. Automated Mass Detection in Mammograms Using Cascaded Deep Learning and Random Forests. In Proceedings of the International Conference on Digital Image Computing: Techniques and Applications (DICTA), Adelaide, Australia, 23–25 November 2015; pp. 1–8. [CrossRef]
15. Wichakam, I.; Vateekul, P. Combining deep convolutional networks and SVMs for mass detection on digital mammograms. In Proceedings of the 8th International Conference on Knowledge and Smart Technology (KST), Chiang Mai, Thailand, 3–6 February 2016; pp. 239–244. [CrossRef]
16. Choukroun, Y.; Bakalo, R.; Ben-ari, R.; Askelrod-ballin, A.; Barkan, E.; Kisilev, P. Mammogram Classification and Abnormality Detection from Nonlocal Labels using Deep Multiple Instance Neural Network. *Eurographics Proc.* **2017**. [CrossRef]
17. Dhungel, N.; Carneiro, G.; Bradley, A.P. Deep structured learning for mass segmentation from mammograms. In Proceedings of the IEEE International Conference on Image Processing (ICIP), Quebec City, QC, Canada, 27–30 September 2015; pp. 2950–2954. [CrossRef]
18. Dhungel, N.; Carneiro, G.; Bradley, A.P. Deep Learning and Structured Prediction for the Segmentation of Mass in Mammograms. In *Medical Image Computing and Computer-Assisted Intervention (MICCAI)*; Navab, N., Hornegger, J., Wells, W.M., Frangi, A., Eds.; Springer International Publishing: Cham, Switzerland, 2015; pp. 605–612. [CrossRef]
19. Dhungel, N.; Carneiro, G.; Bradley, A.P. Combining Deep Learning and Structured Prediction for Segmenting Masses in Mammograms. In *Deep Learning and Convolutional Neural Networks for Medical Image Computing*; Springer: Berlin/Heidelberg, Germany, 2017; pp. 225–240. [CrossRef]

20. Zhu, W.; Xiang, X.; Tran, T.D.; Xie, X. Adversarial Deep Structural Networks for Mammographic Mass Segmentation. *arXiv* **2016**, arXiv:1612.05970. [CrossRef]
21. Zhang, R.; Zhang, H.; Chung, A.C.S. A Unified Mammogram Analysis Method via Hybrid Deep Supervision. In *Image Analysis for Moving Organ, Breast, and Thoracic Images*; Springer: Berlin/Heidelberg, Germany, 2018; pp. 107–115. [CrossRef]
22. Liang, D.; Pan, J.; Yu, Y.; Zhou, H. Concealed object segmentation in terahertz imaging via adversarial learning. *Optik* **2019**, *185*, 1104–1114. [CrossRef]
23. Litjens, G.; Kooi, T.; Bejnordi, B.E.; Setio, A.A.A.; Ciompi, F.; Ghafoorian, M.; van der Laak, J.A.; van Ginneken, B.; Sánchez, C.I. A survey on deep learning in medical image analysis. *Med. Image Anal.* **2017**, *42*, 60–88. [CrossRef] [PubMed]
24. Sawyer Lee, R.; Dunnmon, J.A.; He, A.; Tang, S.; Ré, C.; Rubin, D.L. Comparison of segmentation-free and segmentation-dependent computer-aided diagnosis of breast masses on a public mammography dataset. *J. Biomed. Inform.* **2021**, *113*, 103656. [CrossRef] [PubMed]
25. Rundo, L.; Tangherloni, A.; Cazzaniga, P.; Nobile, M.S.; Russo, G.; Gilardi, M.C.; Vitabile, S.; Mauri, G.; Besozzi, D.; Militello, C. A novel framework for MR image segmentation and quantification by using MedGA. *Comput. Methods Programs Biomed.* **2019**, *176*, 159–172. [CrossRef]
26. Tripathy, S.; Swarnkar, T. Unified Preprocessing and Enhancement Technique for Mammogram Images. *Procedia Comput. Sci.* **2020**, *167*, 285–292. [CrossRef]
27. Dhungel, N.; Carneiro, G.; Bradley, A.P. A deep learning approach for the analysis of masses in mammograms with minimal user intervention. *Med. Image Anal.* **2017**, *37*, 114–128. [CrossRef] [PubMed]
28. Al-antari, M.A.; Al-masni, M.A.; Choi, M.T.; Han, S.M.; Kim, T.S. A fully integrated computer-aided diagnosis system for digital X-ray mammograms via deep learning detection, segmentation, and classification. *Int. J. Med. Inform.* **2018**, *117*, 44–54. [CrossRef]
29. Gao, F.; Yoon, H.; Wu, T.; Chu, X. A feature transfer enabled multi-task deep learning model on medical imaging. *Expert Syst. Appl.* **2020**, *143*, 112957. [CrossRef]
30. Bria, A.; Marrocco, C.; Tortorella, F. Addressing class imbalance in deep learning for small lesion detection on medical images. *Comput. Biol. Med.* **2020**, *120*, 103735. [CrossRef] [PubMed]
31. Porcu, S.; Floris, A.; Atzori, L. Evaluation of Data Augmentation Techniques for Facial Expression Recognition Systems. *Electronics* **2020**, *9*, 1892. [CrossRef]
32. Shahedi, M.B.K.; Amirfattahi, R.; Azar, F.T.; Sadri, S. Accurate Breast Region Detection in Digital Mammograms Using a Local Adaptive Thresholding Method. In Proceedings of the Eighth International Workshop on Image Analysis for Multimedia Interactive Services (WIAMIS '07), Santorini, Greece, 6–8 June 2007; p. 26. [CrossRef]
33. Militello, C.; Rundo, L.; Toia, P.; Conti, V.; Russo, G.; Filorizzo, C.; Maffei, E.; Cademartiri, F.; La Grutta, L.; Midiri, M.; et al. A semi-automatic approach for epicardial adipose tissue segmentation and quantification on cardiac CT scans. *Comput. Biol. Med.* **2019**, *114*, 103424. [CrossRef]
34. Min, H.; Chandra, S.S.; Crozier, S.; Bradley, A.P. Multi-scale sifting for mammographic mass detection and segmentation. *Biomed. Phys. Eng. Express* **2019**, *5*, 025022. [CrossRef]
35. He, K.; Gkioxari, G.; Dollar, P.; Girshick, R. Mask R-CNN. In Proceedings of the IEEE International Conference on Computer Vision (ICCV), Venice, Italy, 22–29 October 2017; pp. 2980–2988. [CrossRef]
36. Abdulla, W. Mask R-CNN for Object Detection and Instance Segmentation on Keras and TensorFlow. 2017. Available online: https://github.com/matterport/Mask_RCNN (accessed on 17 April 2021).

Journal of *Imaging*

MDPI

*Article*

# Incoherent Radar Imaging for Breast Cancer Detection and Experimental Validation against 3D Multimodal Breast Phantoms

**Antonio Cuccaro [1], Angela Dell'Aversano [1], Giuseppe Ruvio [2,3], Jacinta Browne [4,5] and Raffaele Solimene [1,6,7,*]**

1 Dipartimento di Ingegneria, Università degli Studi della Campania Luigi Vanvitelli, 81031 Aversa, Italy; antoniocuccarox@gmail.com (A.C.); angela.dellaversano@gmail.com (A.D.)
2 School of Medicine, National University of Ireland Galway, Galway 8, Ireland; giuseppe@endowave.ie
3 Endowave Ltd., Galway 8, Ireland
4 Department of Radiology, Mayo Clinic, Rochester, MN 55905, USA; Browne.Jacinta@mayo.edu
5 Medical Ultrasound Physics and Technology Group, School of Physics and Clinical & Optometric Sciences, IEO, FOCAS, Technical University Dublin, Dublin 8, Ireland
6 Consorzio Nazionale Interuniversitario per le Telecomunicazioni-CNIT, 43124 Parma, Italy
7 Indian Institute of Technology, Madras, Chennai 600036, India
* Correspondence: raffaele.solimene@unicampania.it



**Citation:** Cuccaro, A.; Dell'Aversano, A.; Ruvio, G.; Browne, J.; Solimene, R. Incoherent Radar Imaging for Breast Cancer Detection and Experimental Validation against 3D Multimodal Breast Phantoms. *J. Imaging* **2021**, *7*, 23. https://doi.org/10.3390/jimaging7020023

Academic Editors: Leonardo Rundo, Carmelo Militello, Vincenzo Conti, Fulvio Zaccagna and Changhee Han
Received: 30 December 2020
Accepted: 25 January 2021
Published: 1 February 2021



**Abstract:** In this paper we consider radar approaches for breast cancer detection. The aim is to give a brief review of the main features of incoherent methods, based on beam-forming and Multiple SIgnal Classification (MUSIC) algorithms, that we have recently developed, and to compare them with classical coherent beam-forming. Those methods have the remarkable advantage of not requiring antenna characterization/compensation, which can be problematic in view of the close (to the breast) proximity set-up usually employed in breast imaging. Moreover, we proceed to an experimental validation of one of the incoherent methods, i.e., the I-MUSIC, using the multimodal breast phantom we have previously developed. While in a previous paper we focused on the phantom manufacture and characterization, here we are mainly concerned with providing the detail of the reconstruction algorithm, in particular for a new multi-step clutter rejection method that was employed and only barely described. In this regard, this contribution can be considered as a completion of our previous study. The experiments against the phantom show promising results and highlight the crucial role played by the clutter rejection procedure.

**Keywords:** microwave imaging; incoherent imaging; clutter rejection; breast cancer detection

## 1. Introduction

Global statistics have demonstrated that breast cancer is the most frequently diagnosed invasive cancer and the leading cause of death due to cancer among female patients [1]. In recent years the incidence of breast cancer in developed countries has continued to rise; but at same time, the rate of mortality has undergone a substantial decline [2]. This is due to the improvements in medical cancer treatment and in the implementation of screening programs as well as to improved imaging techniques [3].

As shown in [4], and as can be naturally expected, the survival of a patient is strongly determined by the stage of the disease at the time the treatment starts. Therefore, early diagnostics is crucial. This requires further improvements of the capabilities of current diagnostic modalities. In addition, over the last few years, this has steered efforts towards the development of new imaging modalities with the aim of supplementing the ones currently employed in the clinical practice.

Current conventional imaging modalities are X-ray mammography, digital breast tomo-synthesis, ultrasound and magnetic resonance imaging (MRI), with mammography actually being the golden standard in breast cancer imaging [5]. Among new imaging modalities, in this paper we focus on microwave breast imaging (MBI). Microwave imaging has triggered a great deal of research over the last decades because it offers a number of potential advantages related to the use of non-ionizing radiation, it does not require to compress the breast and requires a relatively cheap technology [6]. All these features along with the progress achieved in this field [7], show that MBI is actually a "promising imaging modality" [8,9].

Many algorithms for microwave imaging have been tailored for breast diagnostics [10]. Some of them reconstruct a 3D volume; others are based on a sliced approach and, for example, they reconstruct repeated coronal slices of the breast and thus reduce the imaging algorithm complexity and accelerate image reformatting [11]. In any case, microwave breast imaging entails solving a non-linear ill-posed inverse scattering problem since diffraction effects cannot be ignored as in X-ray tomography.

Microwave imaging algorithms can be coarsely grouped in two broad categories, depending on the way non-linearity is dealt with.

When the aim is to reconstruct the dielectric/conductivity profile of the breast tissue under examination, "quantitative" algorithms must be adopted. In these cases, the non-linearity of the problem must be taken into account and the reconstructions are basically achieved by iterative optimization procedures that try to minimize some cost function of the misfit between the available data and the model ones [12]. As such, non-linear inversions are generally computationally very intensive [13] and can suffer from convergence and reliability problems due to false solutions [14]. However, we need to mention that some hybrid approaches, that exploit a priory information provided by other modality, can help mitigate these issues.

The imaging problem is drastically simplified if the imaging method is based on linearized scattering models [15]. In this case, imaging results into linear procedures that are robust and computationally effective. However, only qualitative information can be obtained. Indeed, the corresponding images are more like hot maps where strong inhomogeneities are highlighted. Therefore, linear methods can be conveniently employed if the main objective is to detect and localize targets with a significant dielectric contrast as compared to the surrounding background tissues.

Linear imaging methods are commonly addressed as radar approaches and are the ones we are concerned with in this contribution.

Among the linear methods, beam-forming (BF) is probably the most popular in MBI. Basically, it consists in time-shifting the signals received over the measurement aperture in order to isolate signals scattered from (and hence to focus at) a particular synthetic focal point belonging to the spatial area to be imaged [16]. The BF approach is attractive for the excellent compromise between the achievable performance and the procedure complexity. In [17] the classical delay and sum (DAS) beam-former is used for breast cancer imaging, but many different versions of DAS beam-former have been employed and proposed in literature. For example the delay multiply and sum (DMAS) beam-former is proposed in [18] and the enhanced DAS (EDAS) beam-former in [19]. Besides BF, many other linear inversion methods can be found in literature. For example, a number of linear inversion methods that rely on the spatial spectral representation of the solutions of the wave equation have been developed in different applicative contexts [20]. Among them we mention the range migration [21], the Stolt migration [22], the wave-interference migration [23] and the Holographic Imaging (HI) [24]. These methods are very appealing since their implementation requires computing a Fourier transformation [25,26] which can be effectively achieved via a Fast Fourier Transform (FFT) algorithm. While the latter typically requires the scattered field data to be collected over a planar (rectilinear for 2D cases) measurement aperture, since the Cartesian spatial coordinates naturally match the spatial Fourier transform setting, the extension to deal with circular configurations (more suitable for breast imaging) was previously pursued, for example, in [27].

A detailed analytical comparison between beam-forming and holographic methods has been carried out (for a scattering scenario pertinent to breast imaging) in [28], where the role of critical parameters, such as the operating frequency range, the number of scatterers and data discretization, was considered. Instead, in [29] it is shown that all these methods are variants of the so-called generalized holography.

As remarked above, linear methods restrict the imaging stage to a mere detection and localization of strong in-homogeneities. However, even in light of this reduced task and under the simplified linear framework, the imaging problem is still extremely difficult for a number of reasons. One of the issue, is that all the previous methods "coherently" combine data collected at different frequencies. Therefore, the achievable performance is negatively affected by frequency dispersion of breast tissues (which are unknown or known with a considerable degree of uncertainty and vary from patient to patient) as well as by the antenna frequency response, which is hard to predict because it is in close proximity to breast. As shown in [30,31], this drawback can be mitigated by employing non-coherent imaging strategies. In particular in those papers we introduced and compared incoherent versions of beam-forming and MUSIC [32] (I-MUSIC) and showed that the performance remains stable by using different types of antennas although they were non-characterized, i.e., their frequency responses were not estimated nor enclosed in the model upon which the algorithms were based.

Another crucial aspect is the clutter that generally overwhelms the relatively weak signal coming from the cancer targets. Accordingly, before obtaining the image, data must be first processed in order to reduce the clutter due to the antenna's internal reflection, the skin interface and other non-tumor breast tissues. To this end, we employed a hybrid clutter removal method [33].

In this contribution we will give a quick review of the incoherent methods and detail the clutter mitigation procedure that was used, but actually not described, in [33]. In particular, the achievable cancer detection is checked by using experimental data collected by employing the multi-modal breast phantom developed in [33]. Accordingly, this paper focuses more on the image process and can be considered as a companion paper of [33] which, instead, mainly considered the development of the breast phantom and its tissue characterization for different imaging modalities.

## 2. Ideal Scattering Configuration and Beam-Forming

In order to introduce the notation and to more easily describe the incoherent imaging methods we consider an idealized scattering scenario. More specifically, the scattering problem is considered for a two-dimensional scalar configuration (see Figure 1). Here, invariance is assumed along the $z$-axis and the electromagnetic incident field has a transverse magnetic TM polarization.

According to the measurement set-up commonly used for breast imaging, sensors are assumed to be located over a circle which is in close proximity and embodies the scattering region $D$. The position of each sensor is identified by the vector $\underline{r}_o \in \Gamma_o$, $\Gamma_o$ being the circular measurement curve. The scattered signals are assumed to be collected only at the same position as the transmitter, while the latter can assume different positions over the circle in order to synthesize the measurement aperture. Hence, a multimonostatic configuration is considered. Note that, while the multimonostatic setting is by far the most common, more complex multiview/multistatic sensors' arrangement are often employed: these configurations are not considered herein.

As far as the background medium is concerned, it is assumed to be homogeneous and lossless. Of course, this does not match with realistic breast structures which consist of many layers of different materials that can have even articulated boundaries. Nonetheless, because the breast structure is actually unknown, and for the sake of simplicity, during the image stage an equivalent homogeneous background medium is usually considered.

According to the previous assumptions and under Born approximation [15], the scattered field in the frequency domain is given as:

$$E_S(\omega, \underline{r}_o) = \left(\frac{\omega}{2v}\right)^2 P(\omega) \int_D G^2(\omega/v, \underline{r}_o, \underline{r}) \chi(\underline{r}) d\underline{r} \tag{1}$$

where $D$ is the spatial region under investigation, $\omega$ is the angular frequency and $v$ the background propagation speed. Moreover, $E_S(\omega, \underline{r}_o)$ is the scattered field data, $P(\omega)$ is the temporal Fourier spectrum of the transmitted pulse, $G(.) = 1/4jH_0^{(2)}(\cdot)$ is the two-dimensional scalar background Green function, with $H_0^{(2)}(\cdot)$ being the Hankel function of second kind and zero order. Finally, $\chi(\underline{r})$ is the so-called contrast function which describes the scatterers in terms of their shape and electromagnetic parameters. Note that in general $\chi(\underline{r})$ is also frequency dependent but here such a dependence has been neglected. In particular, exploiting the asymptotic expansion of the Hankel function, i.e., $H_0^{(2)}(x) \simeq \sqrt{2/\pi x}\exp[-j(x-\pi/4)]$), Equation (1) can be recast as:

$$E_S(\omega, \underline{r}_o) = (j\omega/2\pi v)P(\omega)\int_D \frac{\exp\left(\frac{-j2\omega}{v}|\underline{r}_o - \underline{r}|\right)}{|\underline{r}_o - \underline{r}|}\chi(\underline{r})d\underline{r} \tag{2}$$

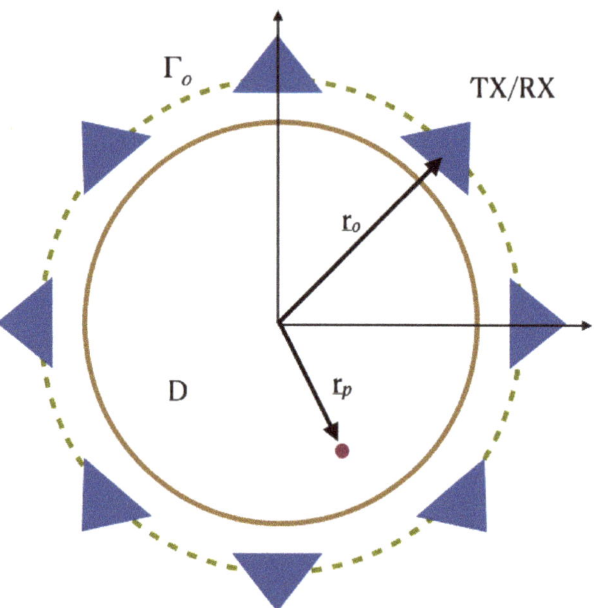


**Figure 1.** Pictorial view of the scattering scene. Invariance is assumed along the $z$-axis.

In practice, the quantity that is actually measured is not the scattered field but rather the system scattering parameters. This basically entails taking in account the antenna frequency response. Accordingly, (2) modifies as:

$$S(\omega, \underline{r}_o) = (j\omega/2\pi v)\tilde{P}(\omega)\int_D \frac{\exp\left(\frac{-j2\omega}{v}|\underline{r}_o - \underline{r}|\right)}{|\underline{r}_o - \underline{r}|}\chi(\underline{r})d\underline{r} \tag{3}$$

where $\tilde{P}(\omega) = H^2(\omega)P(\omega)$ now takes into account the squared (because the antenna acts as TX and RX) antenna response assumed to be solely dependent on the frequency $\omega$ and $S(\omega, \underline{r}_o)$ are the scattering measurements.

In order to introduce the beam-forming method, it is convenient to consider the time domain version of Equation (3). Hence, by Fourier transforming with respect to $\omega$, we obtain the time domain scattering measurements as:

$$s(t, \underline{r}_o) = \int_D \tilde{p}[t - \tau(\underline{r}_o, \underline{r})]\chi(\underline{r})d\underline{r} \tag{4}$$

where $\tilde{p}(.)$ is related to the transmitted pulse and is the Fourier transform of $(j\omega/2)\tilde{P}(\omega)/|\underline{r}_0 - \underline{r}|$ and $\tau(\underline{r}_o, \underline{r}) = 2/v|\underline{r}_0 - \underline{r}|$ is the round-trip delay.

Generally, the image obtained by the DAS beam-forming is given by:

$$I_{BF}(\underline{r}) = \int W(t) \left[ \int_{\Gamma_o} s[t - T(\underline{r}_o, \underline{r}), \underline{r}_o] d\underline{r}_o \right]^2 dt \tag{5}$$

where $W(.)$ is a suitable time window and $T(r_o, r) = T_W - \tau(r_o, r)$, with $T_W = max_{\underline{r}_o, \underline{r}}\{\tau(\underline{r}_o, \underline{r})\}$. Accordingly, the received signals are "aligned" at the time instant $T_W$ and then summed. In particular, by setting $W(t) = \delta(t - T_W)$ the reconstruction $I_{BF}$ becomes:

$$I_{BF}(\underline{r}) = \left[ \int_{\Gamma_o} s[T_W - T(\underline{r}_o, \underline{r}), \underline{r}_o] d\underline{r}_o \right]^2 \tag{6}$$

and returning back to the frequency domain (details can be found in [28]):

$$I_{BF}(\underline{r}) = \left| \int_{\Omega} \int_{\underline{\Gamma}_o} S(\omega, \underline{r}_o) \exp\left[j\omega\tau(\underline{r}_o, \underline{r})\right] d\underline{r}_o d\omega \right|^2 \tag{7}$$

Equation (7) is functional to appreciate the difference with the incoherent approach to be shown in the sequel. In addition, it allowed a closed-form derivation of the point-spread function, that is the reconstruction of a point-like target $\chi(\underline{r}_p) = \delta(\underline{r} - \underline{r}_p)$, that permits to evaluate the achievable resolution in terms of the configuration parameters, including the frequency range and the data discretization [28]. In particular, it was shown that the common belief that in order to achieve a finer resolution a wider frequency band is required does not necessarily hold true. Indeed, while this statement is correct for aspect-limited configurations, for the case at hand, where measurements can be taken all around the scattering region (i.e., non-aspect limited configuration, see Figure 1), finer resolution can be obtained by moving a fixed frequency band towards high frequencies. This is an important result which has practical implications since it promotes the use of cheaper hardware and simplifies the antenna design, which does not necessarily have to work on an ultra-wide band. All details can be found in [28].

## 3. Incoherent Image Procedures

As highlighted in (5), the measured scattering parameters depend on the antenna response. Indeed, this enters in shaping the frequency behaviour of $\tilde{P}(\omega)$ and in general introduces a frequency dependent propagation delay. The latter must be considered while setting the time window $W(t)$ and the alignment time $T_w$. This requires near-field antenna characterization/equalization that can be pursued by a suitable set of measurements or numerical simulations. However, as the breast properties change from patient to patient, residual errors still remain. With uncertainty levels as high as the magnitude of the tumor scattered field, the imaging procedure's robustness is dramatically endangered. This is particularly true for dense breasts as they present lower tumor/healthy-tissue contrast. It can be noted that this drawback arises because frequency data are coherently summed. Therefore, a viable way to mitigate this problem is to devise imaging schemes which do rely on such a coherence and process each frequency data separately. This is the topic addressed in this section.

### 3.1. Incoherent Beam-Forming

Basically, incoherent beam-forming (IBF) is achieved as follows:

$$I_{IBF}(\underline{r}) = \int_{\Omega} \left| \int_{\Gamma_o} S(\omega, \underline{r}_o) \exp\left[j\omega\tau(\underline{r}_o, \underline{r})\right] d\underline{r}_o \right|^2 d\omega \tag{8}$$

where the basic difference, with respect to (5), is clearly that data are summed in amplitude along the frequency domain. Of course, it is interesting to elucidate how (8) relates physically (meaning) and in terms of the achievable performance to (5). As shown in [28], the time domain counterpart of (8) is

$$I_{IBF}(\underline{r}) = \int \left[ \int_{\Gamma_o} s[t - T(\underline{r}_o, \underline{r}), \underline{r}_o] d\underline{r}_o \right]^2 dt \tag{9}$$

where basically the window function has been removed. Form the achievable performance point of view, in [28] the point-spread function was also analytically derived for the incoherent case and it was found that the main difference with respect to the coherent case is that side-lobes are slightly higher. However, the point-spread function main beams (and hence the resolution) are practically the same. Therefore, the cost to pay while using (8) in place of (5) is that side-lobe reconstruction increases a little bit (of course, the actual increase depends on the configuration parameters, especially by the frequency band) but this is largely rewarded since the need to estimate/compensate the antenna response is avoided.

### 3.2. Discrete Data Setting

In the previous section we implicitly considered the situation where data are collected continuously all around the scattering scene. In practice, the number of data samples must be finite. Accordingly, in this section we recast the previous argument within a discrete data setting which, in turn, is also necessary to introduce the I-MUSIC, as shown below.

Therefore, say $\underline{r}_{o1}, \underline{r}_{o2}, \cdots, \underline{r}_{oN_o}$ $N_o$ measurement points taken uniformly over the measurement circle $\Gamma_o$ and $\omega_1, \omega_2, \cdots, \omega_{Nf}$ the employed frequencies. In addition, denote as $\underline{r}_1, \underline{r}_2, \cdots, \underline{r}_{N_s}$ the coordinates of the pixels that divide the spatial region under test $D$. The finite dimensional (discrete) counterpart of (1) can then be written as:

$$\mathbf{S}^i(\omega_i) = (\omega_i/2v)^2 \tilde{P}(\omega_i) \mathbf{A}(\omega_i) \mathbf{b}(\omega_i) \tag{10}$$

where

$$\mathbf{S}^i(\omega_i) = [S(\underline{r}_{o1}, \omega_i), S(\underline{r}_{o2}, \omega_i), \cdots, S(\underline{r}_{oN_o}, \omega_i)]^T \in C^{N_o} \tag{11}$$

is the column vector of the scattering data collected at frequency $\omega_i$,

$$\mathbf{b}^i(\omega_i) = [b_1(\omega_i), b_2(\omega_i), \cdots, b_{N_s}(\omega_i)]^T \in C^{N_s} \tag{12}$$

is the vector of the pixel scattering coefficients), $(\cdot)^T$ denoting the transpose, and $\mathbf{A}(\omega_i) \in C^{N_o \times N_s}$ is the $N_o \times N_s$ matrix propagator (indeed a discrete version of Equation (1)) whose $n$-th column has the form:

$$\mathbf{A}^n(\underline{r}_n, \omega_i) = [G^2(\omega_i, \underline{r}_{o1}, \underline{r}_n), G^2(\omega_i, \underline{r}_{o2}, \underline{r}_n), \cdots, G^2(\omega_i, \underline{r}_{oN_o}, \underline{r}_n)]^T \tag{13}$$

where the Green function is the same as in Equation (2). Accordingly, the overall data scattering matrix is:

$$\mathbf{S} = [\mathbf{S}^1(\omega_1) \ \mathbf{S}^2(\omega_2) \ \cdots \ \mathbf{S}^{N_f}(\omega_{N_f})] \in C^{N_o \times N_f} \tag{14}$$

Due to this discrete setting, Equation (8) can be particularized as:

$$I_{IBF}(\underline{r}) = \sum_{m=1}^{N_f} \left| \sum_{l=1}^{N_o} S(\omega_m, \underline{r}_{ol}) \exp\left[j\omega_m \tau(\underline{r}_{on}, \underline{r})\right] \right|^2 \tag{15}$$

where $\underline{r} \in \underline{r}_1, \underline{r}_2, \cdots, \underline{r}_{N_s}$. A crucial question to be addressed within the discrete setting is the choice of the minimum number of sensor positions that should be deployed around the scattering scene in order to obtain the same results as the ideal case (i.e., data collected continuously) or at least to avoid aliasing effects that can result in reconstruction crowed by spurious artifacts that can be mistaken for actual targets. In particular, it is shown that to avoid aliasing a sufficient condition is that the number of measurement points be:

$$N_o \geq 4k_{max}R_c \tag{16}$$

where $k_{max}$ is the wave number corresponding to the highest adopted frequencies and $R_c < R$ the radius of the circular investigation domain. Basically, Equation (16) guarantees that data are properly "spatially" sampled for each adopted frequency. However, because of the multifrequency data, and the related mutifrequency reconstruction process, some degree of under-sampling can be tolerable for part of the frequency band. This is because aliasing spurious artifacts are frequency dependent. Thus, their positions change with the frequency. By contrast, the main contribution of the reconstruction always peaks at the actual scatterer's location. Therefore, even if condition (16) is not satisfied, while summing up different frequency contributions in (15), artifacts tend to be averaged out whereas the main beam (due to scatterer) is not.

### 3.3. Incoherent MUSIC

The starting point is the construction of the correlation matrix for each frequency, that is:

$$\mathbf{R}(\omega_i) = \mathbf{S}^i(\omega_i)\mathbf{S}^{iH}(\omega_i) = \mathbf{A}(\omega_i)\mathbf{B}(\omega_i)\mathbf{A}^H(\omega_i) \tag{17}$$

where $\mathbf{b}^H(\omega_i)$ and $\mathbf{A}^H(\omega_i)$ are the Hermitian vector and matrix of $\mathbf{b}(\omega_i)$ and $\mathbf{A}(\omega_i)$, respectively, and $\mathbf{B}(\omega_i) = \mathbf{b}(\omega_i)\mathbf{b}^H(\omega_i)$. According to [32], scatterers can be localized by finding the steering vectors which are orthogonal to the so called noise subspace. This requires computing the eigenspectrum of $\mathbf{R}(\omega_i)$ and the steering vectors which basically consists of the normalized columns of the propagator $\mathbf{A}(\omega_i)$, that is $\mathbf{Sv}(\underline{r}_n, \omega_i) = \mathbf{A}^n(\underline{r}_n, \omega_i)/\|\mathbf{A}^n(\underline{r}_n, \omega_i)\|$ being evaluated in correspondence to the trial position $\underline{r}_n$ within the spatial domain $D$. Hence, scatterers' positions are identified where the pseudospectrum

$$\phi(\underline{r}_n, \omega_i) = \frac{1}{\|\mathcal{P}_\mathcal{N}[\mathbf{Sv}^n(\omega_i)]\|^2} \tag{18}$$

peaks, with $\mathcal{P}_\mathcal{N}[\cdot]$ being the projection operator onto the noise subspace. However, as shown in [31,34], the correlation matrix is rank deficiency with rank one. Therefore, the scheme to identify the scatterers' location can be modified defining $\mathcal{P}_\mathcal{N} = (\mathcal{I} - \mathcal{P}_\mathcal{S})$, with $\mathcal{P}_\mathcal{S}[\cdot]$ being the projector onto the signal space. Hence, the detection is achievable by adopting the only significant singular vector associated to the signal subspace.

Note that Equation (18) refers to single frequency data. Multiple frequencies can be incoherently combined [35] giving rise to

$$\Phi_{I-MUSIC}(\underline{r}_n) = \prod_{i=1}^{N_f} \phi(\underline{r}_n, \omega_i) \tag{19}$$

Eventually, (19) is the proposed as algorithm for cancer detection.

### 3.4. Numerical Comparison

In this section a numerical comparison between the I-MUSIC and the beam-forming strategies is shown. Initially, single frequency ($f = 3$ GHz) data are considered and a background medium with $\epsilon_r = 9$. The investigation domain $D$ is assumed to be a circle of radius $R_c = 6$ cm whereas measurements are taken over a concentric circle of radius $R$ slightly greater than $R_c$. A point-like target is located in the centre of $D$. For the case at hand, Equation (16) suggests $N_o > 45$ to avoid artefacts. The reconstructions corresponding to this case are shown in panels (a) and (b) of Figure 2. Note that at single frequency there is no difference between BF and IBF. Accordingly, in Figure 2 we just refer to beam-forming. When the number of points is lowered, spurious artefacts actually corrupt the reconstructions. In particular, the bottom panels of the same figure depict the reconstructions when the number of sensors is reduced by seven times. These results are perfectly consistent with the theory developed in [28,34].

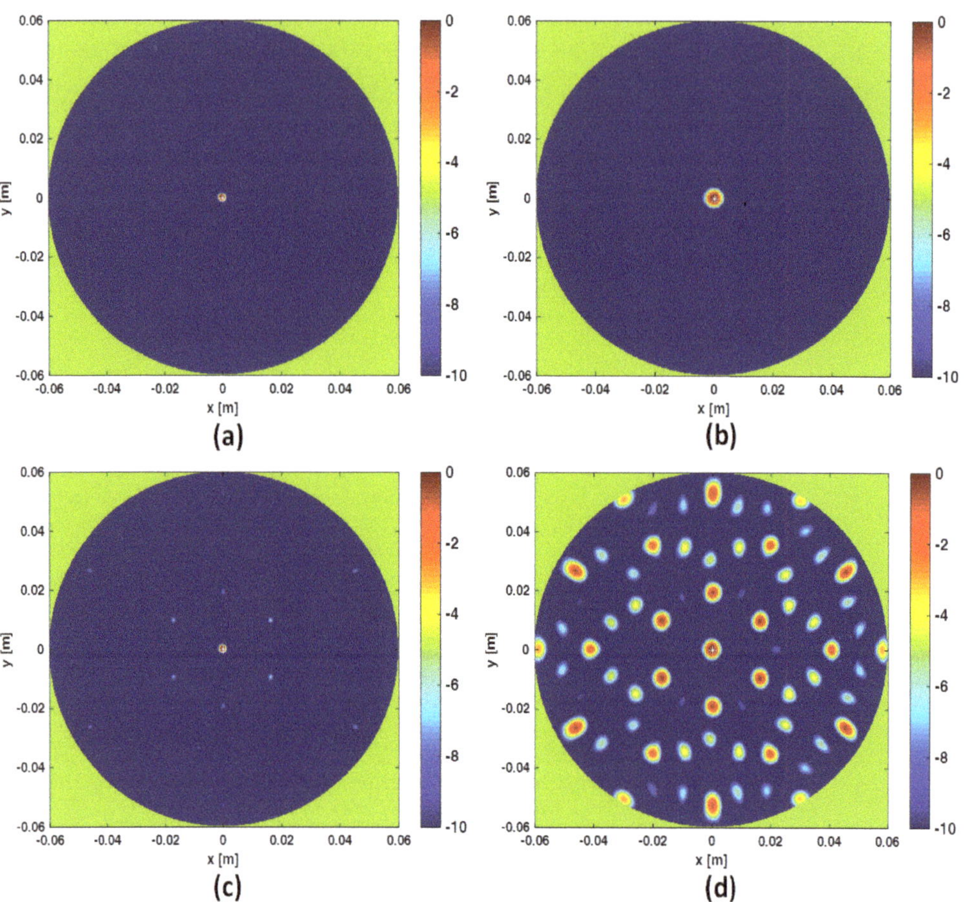


**Figure 2.** Comparing I-MUSIC and beam-forming (BF) for single frequency data. The left column refers to I-MUSIC; the right one to BF. In panels (**a**,**b**) $N_o = 49$ whereas in panels (**c**,**d**) $N_o = 7$.

As mentioned above, frequencies are a good ally to mitigate artifacts when the data are under-sampled. This can be observed in Figure 3 where the three reconstruction schemes, i.e., I-MUSIC, BF and IBF, are compared for the same cases as in panels (c) and (d) of Figure 2 but by considering two different frequency bands. As can be seen, the frequency band greatly helps in reducing artefacts. In addition, as expected and according to previous discussion, IBF presents higher side-lobe levels than BF. Furthermore, I-MUSIC outperforms BF schemes since it allows for a more sharper target localization and better resilience to aliasing. Therefore, I-MUSIC is the method that has been selected for undergoing the experimental validation reported in the sequel.

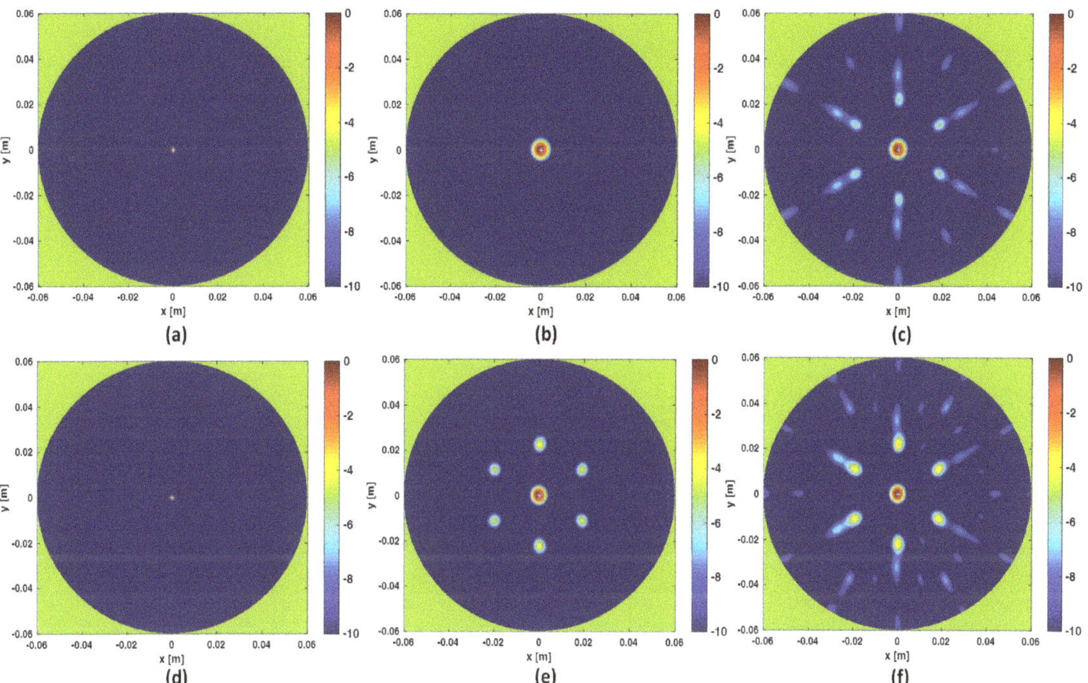


**Figure 3.** Illustrating the role of the frequency band. In all the reconstructions only $N_o = 7$ sensors are considered. In the top panels the frequency band is $[1, 3]$ GHz, in the bottom panels the frequency band is reduced to $[2, 3]$ GHz. Finally, (**a**,**d**) refer to I-MUSIC, (**b**,**e**) to BF and (**c**,**f**) to IBF.

## 4. Experimental Analysis

As mentioned in the introduction, this paper can be regarded as a companion paper of [33]. In that paper, we mainly focused on the design, construction and characterization of the breast phantom; microwave imaging algorithms were not described at all. While the detection algorithm was actually the I-MUSIC that we have already described in previous contributions (and whose main ingredients have been briefly recalled above in conjunction to the comparison with more classical BF methods) the clutter rejection algorithm deserves a more in-depth description. Therefore, the program for this section is to first briefly report about the measurement set-up and the breast phantom and then to move on to a detailed description of the clutter rejection method. Finally, a few experimental reconstructions are used to show the effectiveness of the I-MUSIC + de-cluttering procedure.

### 4.1. Measurement Set-Up

The pictorial view of the measurement set-up is shown in Figure 4 and basically coincides with the measurement scheme adopted in [36]. A breast phantom was scanned by an antipodal Vivaldi antenna in the frequency range [0.5–5] GHz connected to a VNA. In particular, at a given height the antenna rotated around the phantom (with a 5° angular step) in order to synthesize a multimonostatic configuration (i.e., $T_X$ and $R_X$ were co-located) for a total 72 scanning positions. In general, data collected at different heights can be simultaneously employed to get a 3D reconstruction. However, here we exploited the sliced approach. The phantom and the antenna were immersed within a coupling medium with relative dielectric permittivity equal to 12. This was done for antenna miniaturization purposes and to reduce the dielectric discontinuity from the antenna side to the breast, which can hinder microwave energy penetration [35]. Accordingly, such a value of the dielectric permittivity was used to define the equivalent homogeneous reference background medium which was used to build up the scattering model upon which the detection algorithm was based. No information concerning the phantom nor the antenna response (which was not estimated or compensated) was exploited in the following image stage.

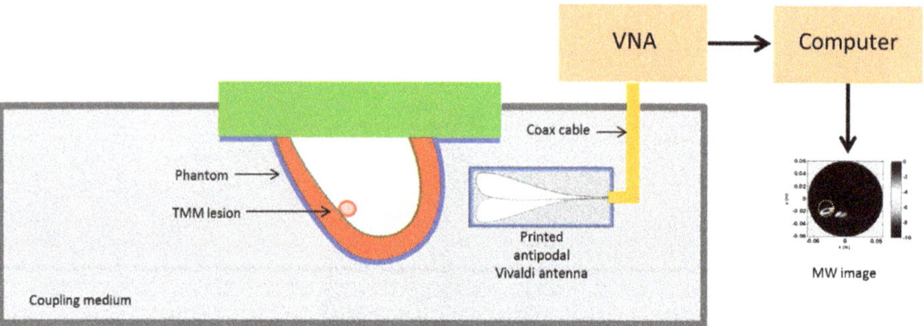


**Figure 4.** Schematic diagram showing the MBI scanning setup. The system antenna + phantom is immersed in a coupling medium. The antenna is connected to a Vector Network Analyzer (VNA) scanning the phantom at a fixed height in multimonostatic configuration. This allows collecting data for a single coronal slice.

### 4.2. Breast Phantom

In [33] we developed multimodal anthropomorphic breast phantoms suitable for evaluating the imaging performance of microwave imaging in comparison to the established diagnostic imaging modalities of Magnetic Resonance Imaging, Ultrasound, Mammography and Computed Tomography. In that study, the aim was to build a bridge between the numerical simulation environment and a more realistic diagnostic scenario. To this end, the constructed anthropomorphic phantoms mimic breast tissues in terms of their heterogeneity, anatomy, morphology, and mechanical and dielectric characteristics and reproduce different healthy and pathologic tissue types for each of the modalities, taking into consideration the differing imaging and contrast mechanisms for each modality. In that study, two phantoms were developed: the phantom (named as 'Phantom A') had a simple and less morphologically accurate interface between mammary fat and fibroglandular tissue; the second ('Phantom B') had a more relevant complex fat and fibroglandular interface. Both were extracted from real patient MRI datasets. Apart from the different morphological structure, the phantoms had the same five different tissue-mimicking materials: skin, subcutaneous fat, fibroglandular tissue, pectoral muscle and tumor. The phantoms' construction used non-toxic materials, and they were inexpensive and relatively easy to manufacture. Both phantoms were characterized and scanned using conventional modalities (MRI, US, mammography and CT). The details concerning all the steps required for their manufacturing, characterization and imaging can be found in [33]. Their MRI coronal slices are reported below and highlight the different tissue mimicking layers as well as the tumor. In particular, it can be seen that in both cases the tumor was located inside the fibroglandural structure. The tissue dielectric permittivity and the conductivity of the different tissues are reported in Figure 1 of [33] which shows that the dielectric contrast between tumor and fibroglandural tissue was at best 1.2:1 (hence extremely low) within the frequency band $[0.5, 4]$ GHz.

### 4.3. Clutter Rejection Algorithm

In order to obtain the reconstruction, in our case a single coronal slice, before the imaging stage, data had to be properly processed in order to reduce clutter disturbances that arose from the antenna internal reflections, the skin interface, and other breast tissues. As the clutter tends to mask the informative signal, it needed to be reduced before the image construction procedure was run. Different clutter rejection methods have been proposed in the literature. For example, in [36] a hybrid artefact removal algorithm for microwave imaging is used, while in [37] some of the most common algorithms used in Through Wall Imaging (TWI) applications were compared, including the simplest average trace subtraction strategy. In this paper, a new method for "extracting" the useful signal is proposed. It was based on a two-step entropy computation and a subspace projection stage. The first entropy step was used to set a time-gating procedure in order to remove the strong antenna's internal reflections and skin contribution; the second one was instead used to select the subset of sensors' positions where tumor contribution is stronger. Finally, a subspace projection procedure [38] was aimed at mitigating contributions due to breast inhomogeneities. After mitigating the clutter, I-MUSIC was employed to obtain the image.

For convenience we rewrote the scattering data matrix as follows:

$$\mathbf{S} = \begin{bmatrix} \mathbf{S}_1 \\ \mathbf{S}_2 \\ \vdots \\ \mathbf{S}_{N_o} \end{bmatrix} \tag{20}$$

where $\mathbf{S}_i$ are the rows of $\mathbf{S}$ and hence vectors whose indexes range over the frequencies, i.e., $\mathbf{S}_i \in C^{N_f}$. In order to compute the time gate, the first step is to transform the rows of $\mathbf{S}$ in time domain. Accordingly, upon applying an IDFT routine, we get:

$$\mathbf{s} = \begin{bmatrix} \mathbf{s}_1 \\ \mathbf{s}_2 \\ \vdots \\ \mathbf{s}_{N_o} \end{bmatrix} \tag{21}$$

with $\mathbf{s} \in C^{N_o \times N_t}$ being the time domain version of $\mathbf{S}$ and $N_t$ is the number of retained time domain samples. Hence, the rows of $\mathbf{s}$ are basically the time-traces (A-scan in usual radar literature) collected over the different antenna positions. Note that if data had already been collected in time domain, this step would have not be necessary.

In Reference [39], an entropy-based metric was used to discriminate between clutter and target signals. The same idea was adopted here to seek a suitable time-gating. Accordingly, normalized time traces, $\tilde{\mathbf{s}}_n$, were constructed whose entries are given by:

$$\tilde{s}_n(t_m) = \frac{|s_n(t_m)|}{\sum_{l=1}^{N_o} |s_l(t_m)|} \quad \forall t_m = t_1, \cdots, t_{N_t} \tag{22}$$

where $s_n(t_m)$ is just the m-th entry of $\mathbf{s}_n$, i.e., the n-th time trace. Now, $\tilde{s}_n(t_m) \geq 0$ and $\sum_{n=1}^{N_o} \tilde{s}_n(t_m) = 1$, $\forall t_m$. Therefore, for each instant of time, the vector of the normalized data could be assimilated to a probability density function. This observation suggested introducing the entropy measure as

$$\epsilon_s(t_m) = -\sum_{n=1}^{N_o} \tilde{s}_n(t_m)\log[\tilde{s}_n(t_m)] \tag{23}$$

It was expected that $\epsilon_s$ was high for those instants of time for which the received signals were similar across the different spatial acquisitions. Of course, this occurred when the antenna was receiving its internal reflections or the skin contribution. Figure 5 (left panel) shows a typical entropy behavior obtained for the collected data. As can be observed, $\epsilon_s$ was nearly constant and high until the time $t_{m_{min}}$ (marked by the dashed red circle), where the entropy attained its first abrupt change compared to its maximum value $\log(N_o)$. According to previous discussion, signals coming from the phantom should start to be received only beyond $t_{m_{min}}$. Signals before such an instant must be discarded. This can be enforced by adopting a time-gating with a time-windowing that removes signals for $t_m < t_{m_{min}}$, that is:

$$s_{Wn}(t_m) = W(t_m)s_n(t_m) \tag{24}$$

with

$$W(t_m) = \begin{cases} 0 & \text{if } t_m \leq t_{m_{min}} \\ 1 & \text{elsewhere} \end{cases} \tag{25}$$

After time gating, the scattering data matrix was denoted as:

$$\mathbf{s_W} = \begin{bmatrix} \mathbf{s}_{W1} \\ \mathbf{s}_{W2} \\ \vdots \\ \mathbf{s}_{WN_o} \end{bmatrix} \tag{26}$$

and a further entropy-based windowing was applied. More in detail, $\mathbf{s}_{Wn}$ were further normalized as follows:

$$\hat{s}_{Wn}(t_i) = \frac{|s_{Wn}(t_i)|}{\sum_{l=1}^{N_t}|s_{Wn}(t_l)|} \tag{27}$$

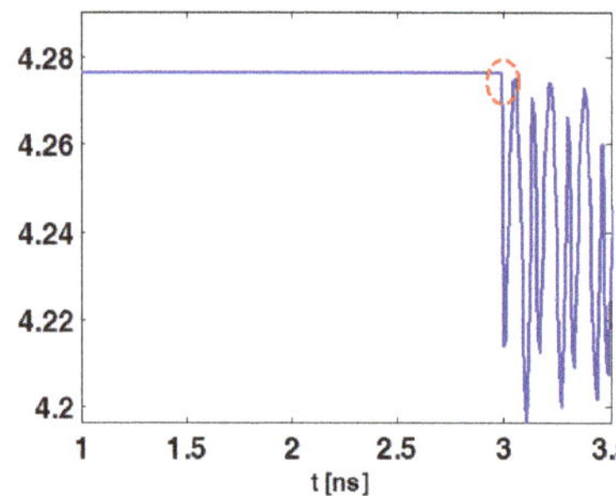


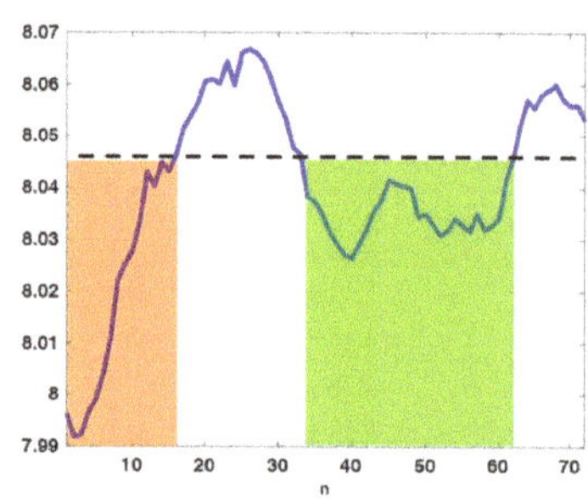


**Figure 5.** Illustrating entropy behaviour for the case of phantom B. The (**left**) panel shows the entropy $\epsilon_s(t_m)$ and the red dashed circle identifies the time-gating value (3 ns). The (**right**) panel shows $\hat{\epsilon}_s$, the orange and green shaded regions highlights the set of sensors' positions whose data can be retained.

Note that now each time trace underwent a different normalization with the normalizing factor being provided by the summation of the magnitude of its time samples. Once again, it follows that $\hat{s}_{Wn}(t_i) > 0$ and $\sum_{i=1}^{N_t}\hat{s}_{Wn}(t_i) = 1$, for each sensor's position. Hence, the $N_t$-dimensional vectors of the normalized windowed A-scans could be assimilated as above to a probability density function and the corresponding entropy can be computed as follows:

$$\hat{e}_s(\underline{r}_{on}) = -\sum_{l=1}^{N_t} \hat{s}_{Wn}(t_l)\log[\hat{s}_{Wn}(t_l)] \tag{28}$$

where $\underline{r}_{on}$ is the $n$-th sensor's position index. The rationale behind this further entropy step is the following. If data at a given position are mainly contributed by clutter and noise then a relatively high level of entropy is expected. This is because the signal magnitude along time should be nearly the same. Instead, when the target significantly contributes then the entropy should decrease. Accordingly, the subset of measurements that effectively "see" the target can be roughly identified by looking for where $\hat{e}_s(\underline{r}_{on})$ is below a threshold value. Say $n_i$ and $n_s$ indicate the positions in between $\hat{e}_s(\underline{r}_{on})$ is below the given threshold, then a windowing is applied to keep only the time traces collected over the positions indexed between $n_i$ and $n_s$. The entropy behaviour reported in Figure 5 (right panel) illustrates the previous discussion. In particular, only data whose positions belonged to the orange and green shaded regions should be retained during the image formation (the threshold was chosen heuristically). In particular, in the following reconstructions only sensors relative to the green zone were retained. Eventually, the second entropy step resulted in a selection of some of the rows of $\mathbf{s_W}$ reported in (26), so that the data to be used were:

$$\hat{\mathbf{s}}_\mathbf{W} = \begin{bmatrix} \mathbf{s}_{Wn_i} \\ \mathbf{s}_{Wn_{i+1}} \\ \vdots \\ \mathbf{s}_{Wn_{n_s}} \end{bmatrix} \tag{29}$$

These two-step entropy strategies allowed us to select the time-gating to apply to each time trace and to select the sub-set of data (across the different sensor positions) to employ in the reconstruction stage. However, this did not yet ensure that in the remained traces there was only the tumor signal. On the contrary, the latter could still be overshadowed by clutter due to the internal inhomogeneity of the breast. To mitigate this clutter residue, it was convenient to return back into the frequency domain (even because the detection algorithm worked in such a domain) by a DFT routine. Hence, (30) becomes:

$$\hat{\mathbf{S}}_\mathbf{W} = \begin{bmatrix} \mathbf{S}_{Wn_i} \\ \mathbf{S}_{Wn_{i+1}} \\ \vdots \\ \mathbf{S}_{Wn_{n_s}} \end{bmatrix} \tag{30}$$

which is a $\bar{N}_o \times N_f$ matrix, with $\bar{N}_o$ being the actual number of measurement positions. For the case at hand $\bar{N}_o = n_s - n_i$. Then, it was reasonable to assume that clutter magnitude was higher than tumor signals. Accordingly, a clutter-rejection subspace-based technique was adopted. In particular, the retained scattering matrix was first expressed in terms of its singular value decomposition (SVD):

$$\hat{\mathbf{S}}_\mathbf{W} = \mathbf{U}\mathbf{\Lambda}\mathbf{V}^\mathbf{H} \tag{31}$$

where $\mathbf{U}$ and $\mathbf{V}$ are unitary matrices containing the left and the right singular functions, respectively, and $\mathbf{\Lambda}$ is a diagonal matrix containing the singular values $\lambda_1, \lambda_2, \cdots, \lambda_P$, in decreasing order, with $P = \min[\bar{N}_o, N_f]$. Clutter could then be mitigated by disregarding the projection of the scattering matrix $\hat{\mathbf{S}}_\mathbf{W}$ onto the singular functions corresponding to the highest singular values. The number of projections to discard generally required a priori information on the clutter, which were in general not available. However, as shown in [35], a conservative choice is to discard the projections of the scattering matrix over the singular functions corresponding to the first or the first two highest singular values. Accordingly, the final de-cluttered data matrix was obtained as:

$$\mathbf{S_d} = \sum_{l=2 \text{ or } l=3}^{P} \lambda^l \mathbf{u}^l (\mathbf{v}^l)^H \quad (32)$$

with $\mathbf{u_l}$ and $\mathbf{v_l}$ being the $l$-th column vectors of $\mathbf{U}$ and $\mathbf{V}$, respectively. Eventually, $\mathbf{S_d}$ is the data matrix passed to the I-MUSIC stage.

### 4.4. Reconstruction Results

According to the sliced approach mentioned above, data collected at different heights were singularly processed to get the corresponding coronal slice reconstructions. In the sequel we just show only those ones obtained from data collected at the height corresponding to the centre of the tumor. In particular, although data were collected within the frequency band $[0.5, 5]$ GHz, in the following reconstructions only the band $B_w = [1, 3]$ GHz was exploited.

The rationale under the following examples is to appreciate the role played by the various steps the clutter rejection method consists of as well as the number of frequencies to be employed in the reconstructions.

The first example is shown in Figure 6 and refers to Phanton A whose MRI is reported in panel (f) to appreciate the breast internal morphology and for comparison purposes with respect to the microwave imaging. In that figure a blue dashed circle is also reported which identifies the spatial region used to perform the reconstructions. As can be seen, such a spatial region was larger than the the phantom coronal slice. In panels (a) to (e) the I-MUSIC indicator is reported. In particular, in panel (a) to (c) only 20 frequencies, uniformly taken within $B_w$, were exploited. More in detail, panel (a) shows the image obtained by pre-processing the data through only the time-gating procedure, by setting the time-gating at $t_{m_{min}}$ = 3 ns according to the first entropy step described above. As can be seen, the reconstruction just returned a hot spot roughly located at the centre of the imaging area. Since I-MUSIC tends to peak at the centre of targets, this means that only time-gating data were not enough to detect the tumor as data still appeared as if produced by a target whose equivalent centre was roughly in the centre of the scene. In panel (b), in order to improve clutter rejection, the subspace approach (achieved by discarding just the first projection of the scattering data matrix) was added to the time-gating. Once again, reconstruction peaks did not match with the expected tumor location, meaning that data were still dominated by a strong clutter contribution. Finally, in panel (c) we also enclosed the sensor selection procedure according to the second entropy computation described in the previous section. In particular, the 2D image was obtained by exploiting only the sensors whose index ranges from $n_i = 28$ to $n_s = 56$, which correspond to an angular coverage between 135° and 275°. For the case at hand, hence, the entropy procedure selected the part of the measurement circular line that is closer to the tumor. This is consistent with the adopted multimonostatic configuration since the scattered field data were collected only in reflection mode. As can be seen, now the tumor was clearly detected and this highlighted that the sensor selection (often overlooked in literature) was a crucial step in addressing imaging in a highly cluttered scenarios. Panel (c) also shows that the image was strongly populated by secondary lobe contributions. This might be due to the employed reduced number of sensors (arising from the second entropy step) which mainly impacted the side lobe structure of the point-spread function. According to the discussion reported above concerning the role of the number of frequencies in mitigating aliasing artefacts, and in general side lobe structure, the quality of reconstruction can be improved by using more frequencies. This was actually the case as can be appreciated looking at panels (d) and (e), where the number of frequencies was increased to $N_f = 40$ and $N_f = 100$, respectively. In particular, panel (e) shows an extremely clear tumor detection. Additionally, in that panel we marked through a yellow circle the actual tumor position.

A few comments are in order concerning the obtained reconstruction matching to what I-MUSIC is expected to return. As explained above, I-MUSIC is a radar approach and hence, as such, it is mainly asked to provide tumor detection and rough location. Therefore, it does not aim at reproducing the breast tissue profile as MRI does. At the microwave regime, this task can be attempted by exploiting more sophisticated reconstruction methods that perform the non-linear inversion. Ideally, $\Phi_{I-MUSIC}$ allows for very sharp tumor location (as compared to BF), as shown in the illustrative numerical examples reported above. However, because of uncertainties, clutter residues, and especially owing to the simplified model used while computing the steering vectors (we just used an equivalent homogeneous medium since the breast features are in general unknown), $\Phi_{I-MUSIC}$ results smeared and delocalized from the actual tumor position. Indeed, this is a drawback common to any radar approach that relies on an assumed scattering model. Nonetheless, this method is extremely quick and does not require a priori information about the used antenna, which was completely ignored in the imaging procedure [33,35]. Finally, we once again remark that the circular boundaries appearing in the reconstructions (i.e., panels (a) to (e)) just delimited the spatial region within which the reconstruction was carried out. The actual boundary of the breast was removed by the clutter rejection procedure. The relative (with respect to the image area) size of phantom coronal slice can be appreciated in panel (f) of Figure 6, where the boundary of the image spatial region has been overlapped to the phantom MRI. Eventually, the spots highlighted by the I-MUSIC does not in general indicate the actual tumor positions. However, they allows a clear tumor detection and to highlight in which quadrant (of the coronal slice) it appears.

The second reconstruction example refers to phantom B and is reported in Figure 7. In this case, the entropy procedure returned the same time-gating as above and almost the same observation angular coverage. Indeed, by considering four different slice heights, $n_i$ ranged from 28 to 34 while $n_s$ remained 56. The same discussion as above applies here, with the tumor being very well detected. However, we remark that since phantom B is more complex (from a morphological point of view) than phantom A (see panel (f)) and exhibits less "circular" symmetry, it is reasonable that clutter space dimension was increased. To this end, during the subspace projection clutter reduction stage the first two (instead of only the first one) projections of the scattering matrix were discarded in Equation (32). In general, it difficult to a priori set the number of projections to be discarded. Here, we used the conservative and heuristic approach to discard at most two projections. Rejecting more projections can help in further reducing clutter but the risk is that the tumor signal can be discarded as well.

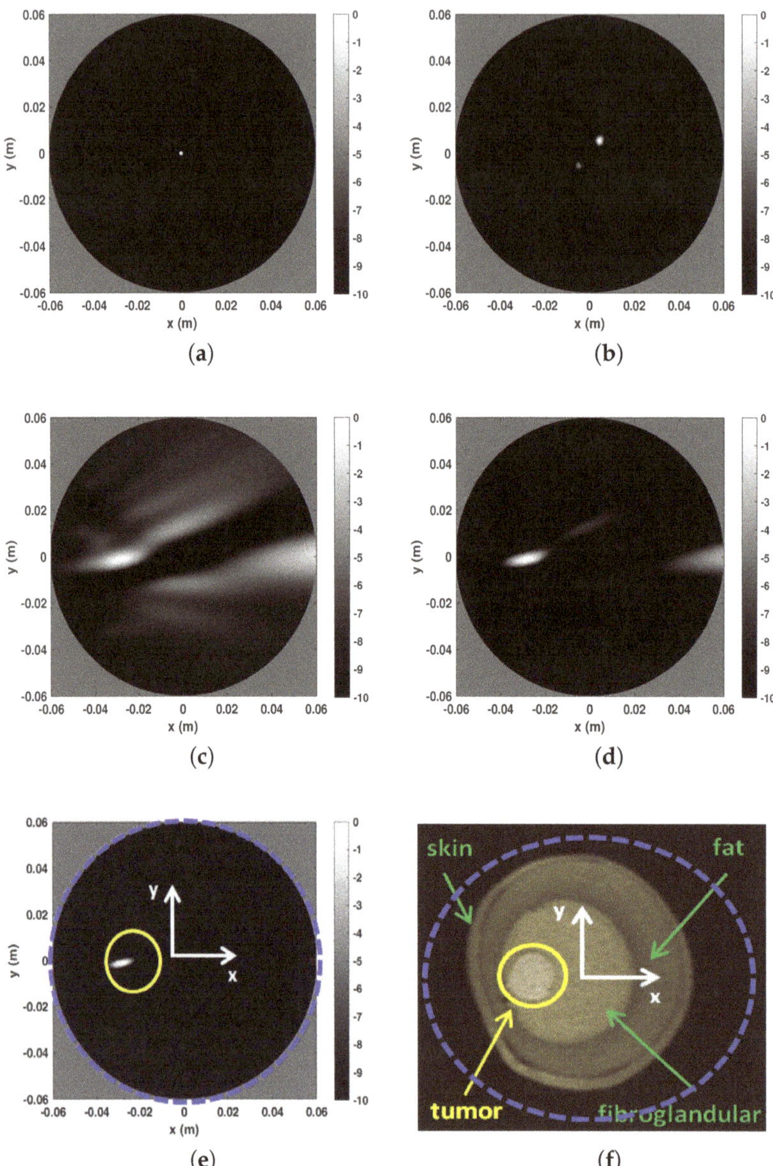


**Figure 6.** Reconstructions and MR image for phantom A. (**a**) Reconstruction with only time-gating, $N_f = 20$. (**b**) Reconstruction with time-gating + rejection of the first SVD projection of the scattering matrix, $N_f = 20$ (**c**) Reconstruction with time-gating + sensor selection + rejection of the first SVD projection of the scattering matrix, $N_f = 20$. (**d**) Reconstruction with time-gating + sensor selection + rejection of the first SVD projection of the scattering matrix, $N_f = 40$. (**e**) Reconstruction with time-gating + sensor selection + rejection of the first SVD projection of the scattering matrix, $N_f = 100$. (**f**) MR coronal slice image of Phantom A. In particular, in panel (**f**) the blue dashed circle indicates the circular boundary of the spatial region within which the reconstructions reported in the other panels have been achieved. This is highlighted even in panel (**e**). Moreover, in the latter, the yellow circle denotes the tumor location and size.

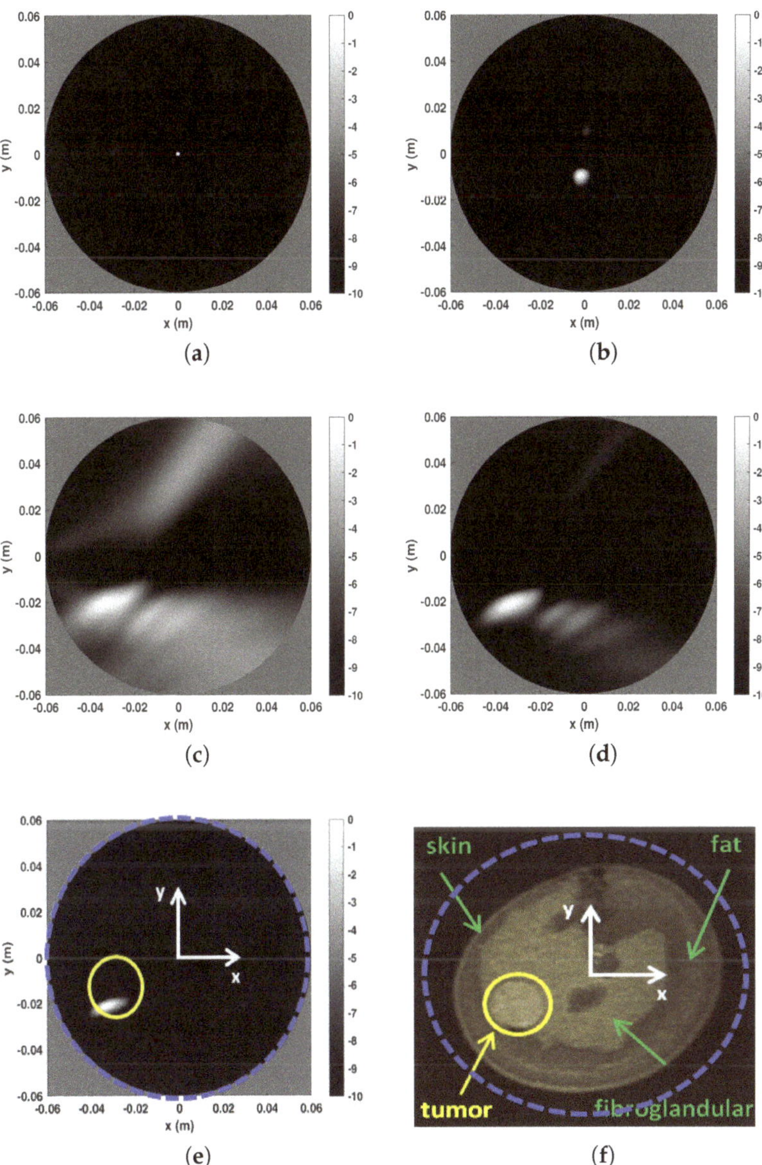


**Figure 7.** Reconstructions and MR image for phantom A. (**a**) Reconstruction with only time-gating, $N_f = 20$. (**b**) Reconstruction with time-gating + rejection of the first two SVD projections of the scattering matrix, $N_f = 20$ (**c**) Reconstruction with time-gating + sensor selection + rejection of the first two SVD projections of the scattering matrix, $N_f = 20$. (**d**) Reconstruction with time-gating + sensor selection + rejection of the first two SVD projections of the scattering matrix, $N_f = 40$. (**e**) Reconstruction with time-gating + sensor selection + rejection of the first two SVD projections of the scattering matrix, $N_f = 100$. (**f**) MR coronal slice image of Phantom B. In particular, in panel (**f**) the blue dashed circle indicates the circular boundary of the spatial region within which the reconstructions reported in the other panels have been achieved. This is highlighted even in panel (**e**). Moreover, in the latter, the yellow circle denotes the tumor location and size.

## 5. Conclusions

Microwave breast imaging requires to deal with a number of issues which go far beyond the need do devise suitable inversion algorithms. Indeed, under the simplified linear framework subtended by the so-called radar approaches, which aim at a mere detection and localization of tumors, data must be properly pre-processed (before the imaging stage) to make sure that the resulting signals are actually useful to pursue the objective.

In this regard, one of the problems to be faced is the need to estimate or compensate for the antenna frequency response, especially when coherent wide band radar imaging methods are employed to obtain the scene image. Indeed, on one hand, antenna frequency response shapes the actually received pulse signals and modifies the overall round-trip delay; both these effects must be taken into account while implementing the beam-forming image procedure. On the other hand, because of the close proximity set-up usually adopted in breast imaging, the antenna couples with the unknown breast and its response becomes different from the free-space case. To overcome this drawback, it is shown that incoherent methods, that do not simultaneously use the frequency data but rather process each frequency separately and then combine the outcomes, can be employed with a minor reduction of the performance, especially if incoherence is used in conjunction to a MUSIC like algorithm (I-MUSIC).

Another crucial aspect is the clutter that overwhelms the target signals and can impair imaging. In this paper we have introduced a new multi-step clutter rejection method that is based on two entropy computations for time-gating setting and the selection of the sensors whose signal are less corrupted by clutter, followed by a standard subspace rejection procedure based on the SVD computation of the scattering data matrix.

The effectiveness of the de-clutter plus I-MUSIC has demonstrated against experimental data collected by using a multimodal phantom we previously developed and characterized in [33]. The results show that, for the considered phantoms, the proposed method very well succeed in detecting and localizing the tumor, though the dielectric contrast with respect to the surrounding fibroglandural tissue was only 1.2:1. This contribution can be considered as completing [33], where we mainly focused on the phantom manufacturing and characterization and only barely described the microwave imaging procedure.

As a concluding remark, we would like to remark that microwave breast imaging is a very broad research field and by this paper we did not intend to give a comprehensive account of the huge available literature. We have just focused the spot on our specific perspective.

**Author Contributions:** Conceptualization, R.S. and G.R.; methodology, R.S., A.C. and A.D.; software, A.C. and A.D.; validation, A.C., G.R. and J.B.; formal analysis, A.C. and A.D.; data curation, G.R., A.C. and J.E.B.; writing—original draft preparation, R.S. and A.C.; supervision, R.S. All authors have read and agreed to the published version of the manuscript.

**Funding:** This research received no external funding.

**Institutional Review Board Statement:** Not applicable.

**Data Availability Statement:** No new data were created or analyzed in this study. Data sharing is not applicable to this article.

**Conflicts of Interest:** The authors declare no conflict of interest.

## References

1. Ferlay, J.; Soerjomataram, I.; Eser, R.D.S.; Mathers, C.; Rebelo, M.; Parkin, D.M.; Forman, D.; Bray, F. Cancer incidence and mortality worldwide: Sources, methods and major patterns in GLOBOCAN 2012. *Int. J. Cancer* **2015**, *136*, E359–E386. [CrossRef] [PubMed]
2. Levi, F.; Bosetti, C.; Lucchini, F.; Negri, E.; La Vecchia, C. Monitoring the decrease in breast cancer mortality in Europe. *Eur. J. Cancer Prev.* **2005**, *14*, 497–502. [CrossRef] [PubMed]
3. Myers, E.R.; Moorman, P.; Gierisch, J.M.; Havrilesky, L.J.; Grimm, L.J.; Ghate, S.; Davidson, B.; Mongtomery, B.R.C.; Crowley, M.J.; McCrory, D.C.; et al. Benefits and Harms of Breast Cancer Screening: A Systematic Review. *JAMA* **2015**, *314*, 1615–1634. [CrossRef] [PubMed]
4. Siegel, R.L.; Miller, K.D.; Jemal, A. Cancer statistics. *CA Cancer J. Clin.* **2016**, *66*, 7–30. [CrossRef]

5. Hellquist, B.N.; Czene, K.; Hjälm, A.; Nyström, L.; Jonsson, H. Effectiveness of population-based service screening with mammography for women ages 40 to 49 years with a high or low risk of breast cancer: Socioeconomic status, parity, and age at birth of first child. *Cancer* **2012**, *118*, 1170–1171. [CrossRef]
6. Preece, A.W.; Craddock, I.; Shere, M.; Jones, L.; Winton, H.L. Maria M4: Clinical evaluation of a prototype ultrawideband radar scanner for breast cancer detection. *J. Med. Imaging* **2016**, *3*, 033502-1–033502-7
7. O'Loughlin, D.; O'Halloran, M.; Moloney, B.M.; Glavin, M.; Jones, E.; Elahi, M.A. Microwave Breast Imaging: Clinical Advances and Remaining Challenges. *IEEE Trans. Biomed. Eng.* **2018**, *65*, 2580–2590. [CrossRef]
8. Kwon, S.; Lee, S. Recent Advances in Microwave Imaging for Breast Cancer Detection. *Int. J. Biomed. Imaging* **2016**, 5054–5912. [CrossRef]
9. Larsen, L.; Jacobi, J. Microwaves offer promise as imaging modality. *Diagn. Imaging* **1982**, *11*, 44–47.
10. Nikolova, N.K. Microwave imaging for breast cancer. *IEEE Microw. Mag.* **2011**, *12*, 78–94. [CrossRef]
11. Golnabi, A.H.; Meaney, P.M.; Epstein, N.R.; Paulsen, K.D. Microwave imaging for breast cancer detection: Advances in three-dimensional image reconstruction. In Proceedings of the 2011 Annual International Conference of the IEEE Engineering in Medicine and Biology Society, Boston, MA, USA, 30 August–3 September 2011; pp. 5730–5733.
12. Wang, L. Early Diagnosis of Breast Cancer. *Sensors* **2017**, *17*, 1572. [CrossRef] [PubMed]
13. Donelli, M.; Craddock, I.; Gibbins, D.; Sarafianou, M. A three-dimensional time domain microwave imaging method for breast cancer detection based on an evolutionary algorithm. *Prog. Electromagn. Res. M* **2011**, *18*, 179–195. [CrossRef]
14. Isernia, T.; Pascazio, V.; Pierri, R. On the local minima in a tomographic imaging technique. *IEEE Trans. Geosci. Rem. Sens.* **2001**, *39*, 1596–1607. [CrossRef]
15. Chew, W.C. *Waves and Fields in Inhomogeneous Media*; IEEE Press: New York, NY, USA, 1995.
16. Chen, J.; Yao, K.; Hudson, R. Source localization and beamforming. *IEEE Signal Process. Mag.* **2002**, *19*, 30–39. [CrossRef]
17. Hagness, S.C.; Taove, A.; Bridges, J.E. Two-dimensional FDTD analysis of a pulsed microwave confocal system for breast cancer detection: Fixed focus and antenna array sensors. *IEEE Trans. Biomed. Eng.* **1998**, *45*, 1470–1479. [CrossRef]
18. Lim, H.; Nhung, N.; Li, E.; Thang, N. Confocal microwave imaging for breast cancer detection: Delay-multiply-and-sum image reconstruction algorithm. *IEEE Trans. Biomed. Eng.* **2008**, *55*, 1697–1704. [PubMed]
19. Klemm, M.; Craddock, I.J.; Leendertz, J.A.; Preece, A.; Benjamin, R. Improved delay-and-sum beamforming algorithm for breast cancer detection. *Int. J. Ant. Propag.* **2008**, *2008*, 761402. [CrossRef]
20. Soldovieri, F.; Solimene, R. Ground Penetrating Radar Subsurface Imaging of Buried Objects. In *Radar Technology*; Kouemou, G., Ed.; IntechOpen: Rijeka, Croatia, 2010; Chapter 6.10.5772/7176. [CrossRef]
21. Lopez-Sanchez, J.M.; Fortuny-Guasch, J. 3-D Radar Imaging Using Range Migration Techniques. *IEEE Trans. Antennas Propag.* **2000**, *48*, 728–737. [CrossRef]
22. Stolt, R.H. Migration by Fourier Transform. *Geophysics* **1978**, *43*, 23–48. [CrossRef]
23. Gazdag, J.; Sguazzero, P. Migration of Seismic Data. *IEEE Proc.* **1984**, *72*, 1302–1315. [CrossRef]
24. Takeda, M.; Wamg, W.; Duan, Z.; Miyamoto, Y. Coherence holography. *Opt. Express* **2005**, *13*, 9629–9635. [CrossRef] [PubMed]
25. Goodman, J. *Introduction to Fourier Optics*; McGraw Hill: New York, NY, USA, 1968.
26. Soumekh, M. *Synthetic Aperture Radar Signal Processing with Matlab Algorithms*; Wiley-Interscience: New York, NY, USA, 1999.
27. Flores-Tapiaand, D.; Pistorius, S. Real time breast microwave radar image reconstruction using circular holography: A study of experimental feasibility. *Med. Phys.* **2011**, *38*, 5420–5431. [CrossRef] [PubMed]
28. Solimene, R.; Cuccaro, A.; Ruvio, G.; Tapia, D.F.; Halloran, M.O. Beamforming and Holography Image Formation Methods: An Analytic Study. *Opt. Express* **2016**, *24*, 9077–9093 [CrossRef] [PubMed]
29. Solimene, R.; Catapano, I.; Gennarelli, G.; Cuccaro, A.; Dell'Aversano, A.; Soldovieri, F. SAR imaging algorithms and some unconventional applications: A unified mathematical overview. *IEEE Sign. Process. Mag.* **2014**, *31*, 90–98. [CrossRef]
30. Ruvio, G.; Solimene, R.; D'Alterio, A.; Ammann, M.J.; Pierri, R. RF breast cancer detection employing a non-characterized vivaldi antenna and a MUSIC-like algorithm. *Int. J. RF Microw. Comput. Aided Eng.* **2013**, *23*, 598–609. [CrossRef]
31. Ruvio, G.; Solimene, R.; Cuccaro, A.; Ammann, M.J. Comparison of Non-Coherent Linear Breast Cancer Detection Algorithms Applied to a 2-D Numerical Breast Model. *IEEE Antennas Wirel. Propag. Lett.* **2013**, *41*, 853–856. [CrossRef]
32. Schmidt, R. Multiple Emitter Location and Signal Parameter Estimation. *IEEE Trans. Antennas Propag.* **1986**, *34*, 276–280. [CrossRef]
33. Ruvio, G.; Solimene, R.; Cuccaro, A.; Fiaschetti, G.; Fagan, A.J.; Cournane, S.; Cooke, J.; Ammann, M.J.; Tobon, J.; Browne, J.E. Multimodal Breast Phantoms for Microwave, Ultrasound, Mammography, Magnetic Resonance and Computed Tomography Imaging. *Sensors* **2020**, *20*, 2400. [CrossRef]
34. Solimene, R.; Ruvio, G.; Aversano, A.D.; Cuccaro, A.; Ammann, M.J.; Pierri, R. Detecting point-like sources of unknown frequency spectra. *Prog. Electromagn. Res. B* **2013**, *50*, 347–364. [CrossRef]
35. Ruvio, G.; Solimene, R.; Cuccaro, A.; Gaetano, D.; Browne, J.E.; Amman, M.J. Breast cancer detection using interferometric MUSIC: Experimental and numerical assessment. *Med. Phys.* **2014**, *41*, 102101–102111. [CrossRef]
36. Ruvio, G.; Cuccaro, A.; Solimene, R.; Brancaccio, A.; Basile, B.; Ammann, M.J. Microwave bone imaging: A preliminary scanning system for proof-of-concept. *IEEE Healthc. Technol. Lett.* **2016**, *3*, 218–221.
37. Solimene, R.; Cuccaro, A. Front wall clutter rejection methods in TWI. *IEEE Geosci. Remote Sens. Lett.* **2014**, *11*, 1158–1162. [CrossRef]

38. Tivive, F.H.C.; Amin, M.G.; Bouzerdoum, A. Wall clutter mitigation based on eigen-analysis in through-the-wall radar imaging In Proceedings of the 2011 17th International Conference on Digital Signal Processing (DSP), Corfu, Greece, 6–8 July 2011; pp. 6–8
39. Solimene, R.; D'Alterio, A. Entropy-Based Clutter Rejection for Intrawall Diagnostics. *Int. J. Geophys.* **2012**, *2012*, 418084 [CrossRef]

Journal of *Imaging*

MDPI

*Article*

# Evaluating Performance of Microwave Image Reconstruction Algorithms: Extracting Tissue Types with Segmentation Using Machine Learning

**Douglas Kurrant [1,*], Muhammad Omer [1], Nasim Abdollahi [2], Pedram Mojabi [2], Elise Fear [1] and Joe LoVetri [2]**

1 Department of Electrical and Computer Engineering, Schulich School of Engineering, University of Calgary, Calgary, AB T2N 1N4, Canada; muhammad.omer@circlecvi.com (M.O.); fear@ucalgary.ca (E.F.)
2 Department of Electrical and Computer Engineering, Faculty of Engineering, University of Manitoba, Winnipeg, MB R3T 5V6, Canada; abdollan@myumanitoba.ca (N.A.); pedram.mojabi@gmail.com (P.M.); joe.lovetri@umanitoba.ca (J.L.)
* Correspondence: djkurran@ucalgary.ca

**Abstract:** Evaluating the quality of reconstructed images requires consistent approaches to extracting information and applying metrics. Partitioning medical images into tissue types permits the quantitative assessment of regions that contain a specific tissue. The assessment facilitates the evaluation of an imaging algorithm in terms of its ability to reconstruct the properties of various tissue types and identify anomalies. Microwave tomography is an imaging modality that is model-based and reconstructs an approximation of the actual internal spatial distribution of the dielectric properties of a breast over a reconstruction model consisting of discrete elements. The breast tissue types are characterized by their dielectric properties, so the complex permittivity profile that is reconstructed may be used to distinguish different tissue types. This manuscript presents a robust and flexible medical image segmentation technique to partition microwave breast images into tissue types in order to facilitate the evaluation of image quality. The approach combines an unsupervised machine learning method with statistical techniques. The key advantage for using the algorithm over other approaches, such as a threshold-based segmentation method, is that it supports this quantitative analysis without prior assumptions such as knowledge of the expected dielectric property values that characterize each tissue type. Moreover, it can be used for scenarios where there is a scarcity of data available for supervised learning. Microwave images are formed by solving an inverse scattering problem that is severely ill-posed, which has a significant impact on image quality. A number of strategies have been developed to alleviate the ill-posedness of the inverse scattering problem. The degree of success of each strategy varies, leading to reconstructions that have a wide range of image quality. A requirement for the segmentation technique is the ability to partition tissue types over a range of image qualities, which is demonstrated in the first part of the paper. The segmentation of images into regions of interest corresponding to various tissue types leads to the decomposition of the breast interior into disjoint tissue masks. An array of region and distance-based metrics are applied to compare masks extracted from reconstructed images and ground truth models. The quantitative results reveal the accuracy with which the geometric and dielectric properties are reconstructed. The incorporation of the segmentation that results in a framework that effectively furnishes the quantitative assessment of regions that contain a specific tissue is also demonstrated. The algorithm is applied to reconstructed microwave images derived from breasts with various densities and tissue distributions to demonstrate the flexibility of the algorithm and that it is not data-specific. The potential for using the algorithm to assist in diagnosis is exhibited with a tumor tracking example. This example also establishes the usefulness of the approach in evaluating the performance of the reconstruction algorithm in terms of its sensitivity and specificity to malignant tissue and its ability to accurately reconstruct malignant tissue.

**Keywords:** breast imaging; microwave imaging; image reconstruction; segmentation; unsupervised machine learning; *k*-means clustering; Kolmogorov-Smirnov hypothesis test; statistical inference; performance metrics; contrast source inversion

**Citation:** Kurrant, D.; Omer, M.; Abdollahi, N.; Mojabi, P.; Fear, E.; LoVetri, J. Evaluating Performance of Microwave Image Reconstruction Algorithms: Extracting Tissue Types with Segmentation Using Machine Learning. *J. Imaging* **2021**, *7*, 5. https://doi.org/10.3390/jimaging7010005

Received: 20 November 2020
Accepted: 23 December 2020
Published: 7 January 2021

## 1. Introduction

Medical imaging with microwave tomography is investigated for breast health monitoring to complement X-ray mammography. For a typical imaging scenario, a multi-illumination approach is implemented by encircling the breast with antennas. The breast is successively illuminated by incident electromagnetic fields from different directions and the resulting scattered and transmitted fields are received by antennas positioned on the breast's periphery and recorded by the measurement system. Microwave tomography is a model-based imaging modality that extracts internal tissue information from these data to reconstruct an approximation of the actual spatial distribution of the dielectric properties over a reconstruction model consisting of discrete elements. With microwave tomography, bulk tissue characterization is the goal rather than more detailed depiction at the cellular level.

The dielectric properties of the breast tissues are represented by a complex permittivity where the real and imaginary components infer the ability of the tissue to store and absorb microwave energy, respectively [1]. The breast tissue types corresponding to skin, adipose (or fatty), transition, fibroglandular, and malignant tissues are characterized by their dielectric properties, which is supported by a number of large-scale studies [2–7]. Therefore, the complex permittivity profile that is reconstructed to form an image may be used to distinguish different tissue types. Estimating values of the dielectric properties of tissues over the model in order to reconstruct an image of the interior of the breast is achieved by solving an inverse scattering problem. The inverse problem is non-linear, so the model values are estimated iteratively using a process summarized in Figure 1.

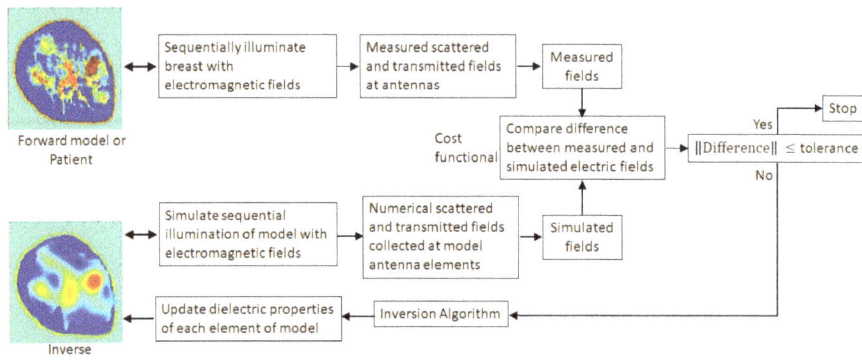


**Figure 1.** Microwave breast imaging procedure. A breast (represented by a forward model for a numerical study or measurements of a patient) is successively illuminated by incident fields from different directions. Microwave tomography is a model-based modality that extracts internal tissue information from the resulting scattered and transmitted fields to iteratively reconstruct an approximation of actual spatial distribution of dielectric properties of tissues in the breast interior. Different tissue types are distinguished from each other by their characteristic dielectric properties.

Evaluating approaches to medical image reconstruction requires application of effective metrics to compare different techniques and assess results. Microwave image reconstruction with tomography typically produces lower resolution images than clinical imaging methods such as X-ray. For simulations of known models or experiments with simple phantoms, direct comparisons between microwave images and known values (i.e., comparing the dielectric properties of the forward model with the inverse model shown in Figure 1) have been reported [8–10]. This includes examination of cross-sections through models, the average of the error at all points in the image, or the similarity between the spatial distribution of the known dielectric properties of the forward model and the dielectric properties estimated at each of the reconstruction model elements of the inverse model.

For more complex models or clinical cases, evaluation of images is often performed through visual comparison or interpretation based on the clinical history of the patient [11,12]. Quantitative assessment of microwave images is more consistent and precise than a qualitative approach. For evaluating variants of algorithms, assessing the accuracy of reconstructing different tissue types provides detailed insight into the algorithm's performance.

A more precise and consistent approach to image analysis may be carried out by automatically detecting regions of interest corresponding to various tissue types or anomalies. Accordingly, this necessitates methods capable of distinguishing between different tissue types and anomalies to assist with image interpretation and tumor localization. Moreover, segmenting reconstructed images into tissue types leads to the decomposition of the breast interior into disjoint tissue masks. Metrics are applied to compare masks extracted from reconstructed images and ground truth models. The quantitative results may be used to reveal the accuracy with which the geometric and dielectric properties are reconstructed in order to provide important insights into the performance of the reconstruction algorithm.

Segmenting images formed with microwave tomography can be challenging, as the images may have spurious artefacts and the interfaces that delineate tissue types may be blurred or incorrectly located. In addition, there may be a great deal of inhomogeneity amongst the same tissue type that is reconstructed, inconsistent mapping between estimated dielectric property values of the reconstructed model elements and the range of dielectric properties that characterize a tissue type, and differences in electrical properties reconstructed with variants of an algorithm [8,13–16].

The segmentation of images into different types of tissues is commonly accomplished using a simple thresholding technique (e.g., [16,17]), whereby reconstructed model elements are classified using ranges of values. However, this strategy assumes that there is a direct mapping between the dielectric property value of a model element estimated by the algorithm and the true dielectric property value of a corresponding tissue type. In practice, this is not necessarily the case, as the accuracy with which the dielectric profile is estimated is impacted by numerous factors, including the number of iterations, the distribution and density of the tissue properties, and measurement parameters (e.g., frequency, number of sensors). Another challenge related to the use of a threshold is that adjustment of the threshold value may significantly impact the specificity and sensitivity to various tissue types. Here, sensitivity and specificity do not refer to the performance of the microwave imaging algorithm in the context of a population of patients, but rather in terms of ability to accurately reconstruct malignant tissues. This problem is apparent when segmenting malignant from healthy tissues and is described in more detail in [17]. Collectively, these problems lead to inconsistent results that contribute to unreliable quantitative assessment of reconstructed images.

An unsupervised machine learning approach such as simulated annealing [18], or *k*-means clustering may be used for image segmentation. However, it is a challenge to determine the optimal number of clusters for the segmentation. Strategies for achieving this task include the elbow method [19], the average silhouette method [20], and the gap statistic method [21]. The elbow technique is a heuristic approach, and an "elbow" could not be unambiguously identified. For many of the images, a great deal of heterogeneity of the reconstructed dielectric properties was observed. This was particularly apparent for images formed from data generated from the heterogeneously dense, scattered density, and extremely dense breasts. The silhouette and gap methods lead to a large range of values that consistently implied a very large number of clusters to partition each image. Consequently, it was not possible to reliably implement any of these methods.

In order to address this problem, this paper presents an iterative approach that does not require the number of clusters to be pre-selected. This is accomplished with an unsupervised machine learning technique that is reinforced with hypothesis testing and statistical inference.

The proposed segmentation algorithm presented in Section 2 is comprised of an iterative clustering method that delineates the interior of the breast into regions dominated by fatty, transition, fibroglandular, and malignant tissues. This segmentation leads to the decomposition of the interior into disjoint tissue masks that are incorporated into a framework whereby both region and distance-based metrics assess image quality [22]. The metrics presented in Section 2 may be used for evaluating variants of reconstruction algorithms, as assessing the accuracy of reconstructing different tissue types provides detailed insight into the algorithm's performance. Specifically, the segmentation algorithm is applied to forward models and the corresponding microwave images reconstructed with the finite element method contrast source inversion (FEM-CSI) approach. Applying the metrics to the segmentation results allows for comparison between the reconstruction and the original model. Section 3 presents, analyzes, and discusses these results. Finally, conclusions and future explorations are presented in Section 4.

## 2. Methodology

### 2.1. Microwave Images

A high-level depiction of a typical microwave imaging algorithm is illustrated in Figure 1. Although not shown, the breast is encircled with antennas to permit the breast to be illuminated from a variety of locations and directions. Imaging is carried out in two steps. In the first step, the breast is illuminated successively with incident electromagnetic fields from each of the antennas. Hence, the breast is interrogated from multiple directions, and the resulting scattered and transmitted fields are received by antennas located on the breast's periphery and recorded by the measurement system (see [10,12,15,23–29], for examples). For a numerical experiment, an electromagnetic forward model comprised of tissues with dielectric properties reported from large-scale studies [2–7] is constructed with the techniques described in [30,31]. The model is sequentially illuminated with numerical incident fields, and the calculated scattered and transmitted fields received by the numerical antenna are stored.

Once the experimental data are collected, the reconstruction step using the inversion algorithm is carried out. This second step starts with a trial guess of the distribution. The electromagnetic model of the breast is initialized with this guess. An array of numerical antennas within a simulated measurement chamber that approximates the actual experimental system surrounds the breast and sequentially illuminates the breast with numerical incident fields. The resulting calculated scattered and transmitted fields received at the numerical antennas are recorded. A cost functional measures the discrepancy between the measured and calculated fields, and an inverse solver computes the optimal change in the parameter profile of the electromagnetic model necessary to reduce the discrepancy between these data. The trial solution is updated with these changes, and the forward solver recalculates the electric fields. The process continues in this iterative manner—updating and refining the reconstructed profile—until the calculated and measured fields match which, in turn, implies that the reconstructed profile matches the actual profile.

Various inverse solvers used have been proposed, including the finite element method contrast source inversion (FEM-CSI) [16,32,33], Gauss-Newton method, and conjugate gradient least squares (CGLS) algorithm [34], conjugate gradient method [13], a full-wave inversion method based on wavelet transform [35], wavelet expansion [36], the Distorted Born iterative method [8,37], and an inversion method based on an inexact Newton-type algorithm [38]. A significant challenge encountered when implementing these inverse solvers is that the inverse scattering problem, along with being non-linear, is severely ill-posed. This occurs due to the very large number of elements used by the reconstruction model to capture fine spatial features of the breast. Meanwhile, there are a very limited number of independent measurement data. Hence, the number of reconstruction elements (i.e., the dimension of the solution space) far exceeds the number of independent data resulting in non-unique solutions. An ill-posed inverse problem manifests as small perturbations of the measurement data leading to large errors in the reconstructions, and the convergence to false solutions that fit the data but differ significantly from the actual solution.

To alleviate the ill-posedness of the inverse problem, reconstruction techniques typically incorporate prior information into the objective function by using some form of regularization. The form of regularization used in this paper to improve image quality is to assimilate patient-specific information related to the electrical properties and anatomical structures of the breast into the inhomogeneous background [16,17,33]. The integration of the patient-specific information into the inhomogeneous background reduces the discrepancy between the background complex permittivity and the complex permittivity of the actual profile. In this manner, the patient-specific information serves to encourage convergence to the actual solution and generally reduces the degree of ill-posedness of the inverse scattering problem to improve the stability of the solution [16,39]. Moreover, the size of the solution space is reduced by constraining the size of the imaging domain (or reconstruction model) with knowledge of an estimation of the skin surface location.

Numerical experiments using realistic breast models based on MRI scans [30,40] are tested in this paper, which is depicted in Figure 1 as an electromagnetic forward model. The dielectric properties of the breast are reconstructed from scattered electromagnetic fields by solving an inverse scattering problem using a variant of the finite element method contrast source inversion (FEM-CSI) algorithm [16,33]. Structural information about the breast is introduced into the FEM-CSI algorithm as an inhomogeneous background $\epsilon_b(\mathbf{r})$. Results are formed by iteratively reconstructing the contrast profile given by,

$$\chi(\mathbf{r}) = \begin{cases} \frac{\epsilon(\mathbf{r})-\epsilon_b(\mathbf{r})}{\epsilon_b(\mathbf{r})}, & \mathbf{r} \in \mathcal{D} \\ 0, & \mathbf{r} \notin \mathcal{D} \end{cases}, \quad (1)$$

where $\chi(\mathbf{r})$ is the contrast profile, $\epsilon_b(\mathbf{r})$ is the inhomogeneous background profile, $\epsilon(\mathbf{r})$ is the complex permittivity profile, $\mathbf{r}$ is a position vector, and $\mathcal{D}$ is the imaging domain bound by boundary $\partial\mathcal{D}$.

The use of the background profile to incorporate prior structural information is illustrated in Figure 2. Figure 2a depicts the scenario where there is no structural prior information available, only knowledge of the dielectric properties of the immersion medium. This is equivalent to using the immersion background as the trial solution. This lack of prior information impacts the quality of the resulting microwave image, as the inversion algorithm converges to a solution having low image quality. On the other hand, Figure 2b portrays the case where prior structural information is available. The improvement in the quality of regularization leads to the convergence to a solution associated with a higher image quality relative to the case represented in Figure 2a.

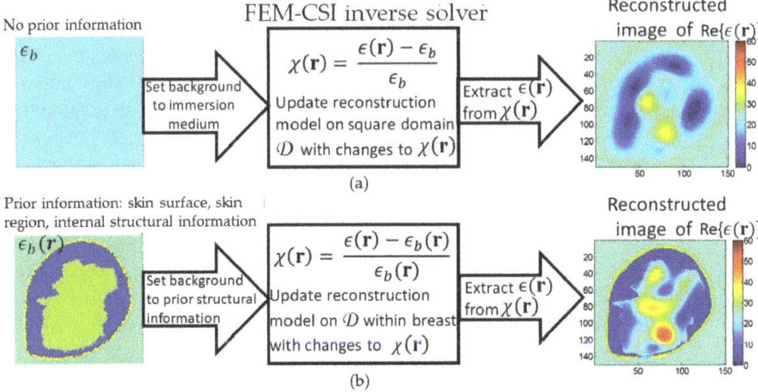


**Figure 2.** (**a**) With no prior information, background set to immersion medium dielectric properties, and contrast profile reconstructed over square imaging domain. (**b**) Prior information includes skin surface, skin region, and internal structural information. By identifying the breast surface, the imaging domain is constrained to the breast interior.

For this study, the FEM-CSI algorithm is terminated once the reconstructed image has stabilized. For example, this may be sensed using the methodology described in [16] or by adapting the technique presented in [41]. The complex permittivity profile is recovered from the contrast profile by using the background permittivity with the relation,

$$\epsilon(\mathbf{r}) = \epsilon_b(\mathbf{r})(\chi(\mathbf{r}) + 1). \tag{2}$$

Using Equations (1) and (2), a list of images of the reconstructed profile is created: the real component of the complex permittivity ($\text{Re}\{\epsilon(\mathbf{r})\}$), the imaginary component of the complex permittivity ($\text{Im}\{\epsilon(\mathbf{r})\}$), and the magnitude of the complex permittivity ($|\epsilon(\mathbf{r})|$), which is a non-linear mapping of the real and imaginary components. Each image is segmented separately using the algorithm described in the following sections.

### 2.2. Segmenting Interior into Healthy and Malignant Breast Tissue Types

The first aim of the segmentation algorithm is to recover the region containing model elements corresponding to malignant tissue (or tissues of interest). The current image of interest is denoted as $\mathcal{I}$. First, the region of interest (breast interior) is defined. The boundary $\partial\mathcal{D}$ of the imaging domain $\mathcal{D}$ given in Equation (1), where $\mathcal{D} \subset \mathcal{I}$, is identified. The boundary of a region of interest $\partial\mathcal{R}$ is constructed by uniformly contracting $\partial\mathcal{D}$ inward toward the center of $\mathcal{D}$ by some amount (e.g., 3.5 mm) using the morphological contraction method described in [42,43]. This allows artefacts on the periphery of the imaging domain to be excluded from analysis. The mask of the region $\mathcal{R}$ bound by $\partial\mathcal{R}$ is constructed such that,

$$mask_{\mathcal{R}} = \begin{cases} 1, & \mathbf{r} \in \mathcal{R} \\ 0, & otherwise \end{cases}. \tag{3}$$

Hence, the region of interest $\mathcal{R} \subset \mathcal{D}$ is extracted from $\mathcal{I}$, with

$$\mathcal{R} = mask_{\mathcal{R}} \odot \mathcal{I}. \tag{4}$$

All model elements outside $\mathcal{R}$ are assigned a value of $-100$. An example of $\mathcal{R}$ recovered from a reconstructed image that used this contraction method is shown in Figure 3a. Note that the immersion medium and skin are considered as background; only the region of the breast that is interior to the skin is partitioned into tissue types.

Next, the $k$-means clustering technique [44] is iteratively applied to $\mathcal{B}$, where $\mathcal{B} = \mathcal{R} \cup \mathcal{R}^c$. The number of clusters $k$ is initialized to three, and the $k$-means++ algorithm presented in [45] is used to initialize $k$ model elements as cluster centroids. This leads to the delineation of $\mathcal{R}$ into clusters $k$ = 2 and 3, while the background is outside of $\mathcal{R}$ and is assigned cluster $k$ = 1. This initial segmentation of $\mathcal{B}$ is shown in the left-most panel of Figure 3c. Note that the color bar for Figure 3c corresponds to the number of clusters used for the segmentation. An initial coarse estimate of the tumor region $\hat{\mathcal{T}}$ is identified with those model elements assigned the highest value, so $\hat{\mathcal{T}} = c_3$. Since cluster $c_2$ is within $\mathcal{R}$ but outside of $\hat{\mathcal{T}}$, $\hat{\mathcal{T}}^c = c_2$. Lastly, the background is outside of $\mathcal{R}$ and is always assigned to cluster $k$ = 1, which means that $\mathcal{R}^c = c_1$.

An iterative approach is used to refine $\hat{\mathcal{T}}$ and $\hat{\mathcal{T}}^c$, so that with each iteration, the number of clusters $k$ used in the $k$-means clustering algorithm is incremented by one. The iterative clustering technique is summarized by Figure 4. After each iteration, $\hat{\mathcal{T}}$ and $\hat{\mathcal{T}}^c$ are updated: $\hat{\mathcal{T}}$ corresponds to the cluster with the highest-valued integer (i.e., $\hat{\mathcal{T}} = c_{\max(k)}$), while the union of clusters $c_k$ with $k = \{2, 3, \ldots, max(k) - 1\}$ form $\hat{\mathcal{T}}^c$. At each iteration $k$, the mask $\hat{\mathcal{T}}^c$ is applied to the reconstructed image to extract model elements $v_{ck}$:

$$\begin{aligned} v_{ck} &= \left( \bigcup_{k=2}^{\max(k)-1} c_k \right) \odot \mathcal{I} \\ &= \hat{\mathcal{T}}^c \odot \mathcal{I}. \end{aligned} \tag{5}$$

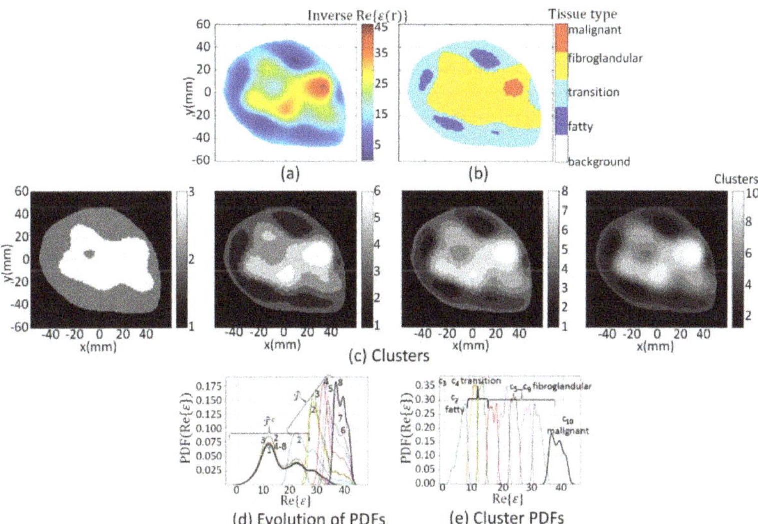


**Figure 3.** (**a**) Reconstructed component extracted from $\mathcal{I} = \mathrm{Re}\{\epsilon(\mathrm{r})\}$; (**c**) Evolution of clusters at $k$ = 3, 6, 8, and 10 when segmentation algorithm applied to $\mathcal{B}$; (**d**) Evolution of Probability Density Function (PDF) over data within $\hat{\mathcal{T}}^c$ and $\hat{\mathcal{T}}$ where numbers indicate iteration; (**e**) PDF over data within clusters $c_2$ (blue line) to $c_{10}$ (black line). Cluster $c_2$ corresponds to fatty tissue, $c_3 - c_4$ transition tissue, $c_5 - c_9$ fibroglandular tissues, and $c_{10}$ corresponds to malignant tissue, which are mapped to segmentation masks leading to tissue type image (**b**).

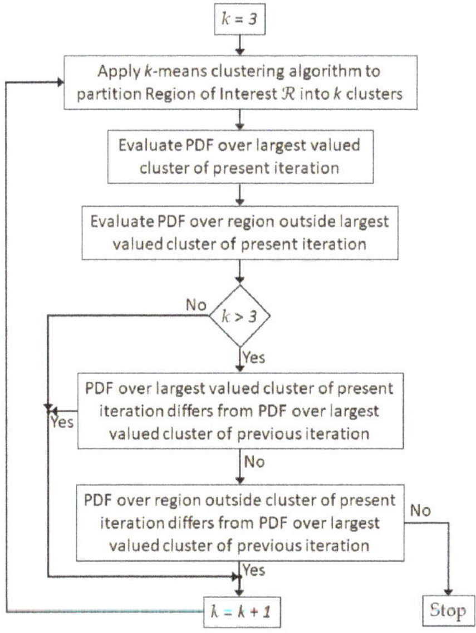


**Figure 4.** Flow diagram of segmentation algorithm used to refine partitioning of breast interior.

The iterative progression of the segmentation process is demonstrated in Figure 3c whereby clustering results are shown from left-to-right for $k$ = 3, 6, 8, and 10.

The empirical distribution function ($\mathrm{E}(\cdot)$) is applied to $v_{ck}$. When $k > 3$, a Kolmogorov-Smirnov (KS) two sample nonparametric hypothesis test evaluates the difference between the cumulative density functions (CDF) of the distributions of the two sample data [46,47]. The test is applied to $E(v_{ck})$ and $E(v_{ck-1})$ where $v_{ck-1}$ are model elements extracted over $\hat{\mathcal{T}}^c$ from the previous iteration. The test evaluates the null hypothesis ($H_{O1}$) that $v_{ck}$ and $v_{ck-1}$ come from the same distribution. Note that the test does not specify the form of the common distribution (e.g., normal distribution). Likewise, the mask $\hat{\mathcal{T}}$ is applied to the reconstructed image to extract model elements $v_{tk}$, where $v_{tk} = \hat{\mathcal{T}} \odot \mathcal{I}$. In this case, the KS two-sample test is performed on $E(v_{tk})$ and $E(v_{tk-1})$ to test the null hypothesis ($H_{O2}$) that $v_{tk}$ and $v_{tk-1}$ come from the same distribution. A significance level of 1% is used for both tests.

If either $H_{O1}$ or $H_{O2}$ is rejected, then the number of clusters is incremented by one, and the partitioning procedure is repeated until neither $H_{O1}$ nor $H_{O2}$ is rejected. When neither hypothesis is rejected, this step is terminated. The union of clusters $c_2 - c_{max\{k\}-1}$ form $\hat{\mathcal{T}}^c$, while $c_{max\{k\}}$ forms $\hat{\mathcal{T}}$. The probability density function (PDF) over data within $\hat{\mathcal{T}}^c$ and $\hat{\mathcal{T}}$ after each iteration is demonstrated in Figure 3d. Convergence of the PDFs is apparent after eight iterations (i.e., $k = 10$, since the segmentation process starts with $k = 3$), which leads to 10 disjoint clusters. Individual PDFs over data within each cluster $c_2 - c_8$ are shown in Figure 3e.

In terms of complexity, finding the global optimum of the *k*-means objective function is a Non-Deterministic Polynomial acceptable (or NP-hard) problem [48,49]. To avoid solving the NP-hard problem, as already indicated, the Lloyd's clustering algorithm [44] is used but offers a local search heuristic for *k*-means. Given enough time, the algorithm always converges after *i* iterations, but it may be a local minimum. Hence, the clustering algorithm is run multiple times *d* with different initializations of the centroids for each *k*. Then, the result that leads to the smallest objective function value is selected. The *k*-means++ initialization scheme is implemented to reduce the dependence of the initialization of the centroids on the convergence behavior [45].

The running time to implement the proposed segmentation technique is *O(IkidN)*; where *I* is the *n* by *m* image being processed, *k* is the number of clusters, *i* is the number of iterations of the *k*-means clustering algorithm needed until convergence, *d* is the number of times the clustering algorithm is repeated (i.e., find the result leading to the smallest valued objective function after running the algorithm *d* times), and *N* is the number of iterations of the segmentation algorithm required to partition the breast interior. This formulation is derived from [50] and [51], and it includes *N*, which is necessary to implement the segmentation algorithm. The process is repeated for the real component, imaginary component, and the magnitude of the complex permittivity.

For images with large dimensions (i.e., large *n* by *m*), parallel schemes may be implemented in python with the Scikit learn machine learning library (class sklearn.cluster.KMean) that use OpenMp to process small blocks of data in parallel, or Matlab in which the number of times *d* that the *k*-means algorithm is repeated is run in parallel. For the images presented, the data has an underlying clustering structure, and it was observed that the number of iterations *i* of the clustering algorithm until convergence was often small.

### 2.3. Mapping Clusters to Segmentation Masks and Tissue Types

So far, tissues corresponding to model elements with the highest values within the breast are identified by $\hat{\mathcal{T}} = c_{\max\{k\}}$. Cluster $c_1$ identifies the background $\mathcal{R}^c$. The remaining $k - 2$ clusters are mapped to segmentation masks as follows. Cluster $c_2$ bounds tissue having the lowest dielectric properties and corresponds to the lowest permittivity values within the breast interior. Consequently, it is reasonable to map $c_2$ to the segmentation mask corresponding to fatty tissue. Next, clusters $c_3$ and $c_4$ contain permittivity values that are higher than fatty tissue. The breast interior includes permittivity values that exceed the maximum value of adipose tissue but are lower than the minimum of the fibroglandular tissue range [3]. Therefore, $c_3$ and $c_4$ are mapped to a transition segmentation mask. When $max\{k\} > 4$, the union of $c_5$ to $c_{\max\{k\}-1}$ corresponds to segmentation mask $\hat{\mathcal{G}}$ associated with fibroglandular tissues. This is defined as:

$$\hat{\mathcal{G}} = \left(\cup_{k=5}^{\max(k)-1} c_k\right). \tag{6}$$

The final segmentation is comprised of masks formed by mapping clusters $k = 1, 2, \ldots max\{k\}$ to tissue types with the function

$$s(k) = \begin{cases} background,\ k = 1, \\ fatty,\ k = 2, \\ transition\ k = 3,\ 4, \\ fibroglandular,\ 4 < k < \max\{k\}, \\ malignant,\ k = \max\{k\}. \end{cases} \tag{7}$$

For the unusual case that there is only one iteration of the segmentation algorithm, clusters $c_k$, $k = 2, 3, 4$, are used to identify the fatty, fibroglandular, and malignant tissues, respectively.

The segmentation algorithm is applied to both the forward model and reconstructed images. The resulting segmentation masks are labeled as $\mathbf{ref_{mask}}$ and $\mathbf{rec_{mask}}$, respectively. To extract the corresponding property values, the reference mask is applied to the forward model. These segmented property values are referred to as the reference tissue, $\mathbf{ref_{tissue}}$. Likewise, the reconstructed masks are applied to the reconstructed images. These segmented property values are referred to as the reconstructed tissue, $\mathbf{rec_{tissue}}$, of the region. An example of the mapping of the clusters to tissue types is shown in Figure 3b. For this example, the ten clusters shown in the far-right panel of Figure 3c are mapped to segmentation masks and associated tissue types using Equation (7), resulting in the segmented image shown in Figure 3b. Videos demonstrating the iterative refinement of the clusters and segmentation process are provided in the supplemental materials [52].

### 2.4. Quality Assessment

To measure the image reconstruction performance quantitatively, five region-based metrics are applied to assess the overlap between $\mathbf{ref_{mask}}$ and $\mathbf{rec_{mask}}$. A distance-based metric is also used to evaluate shape fidelity.

First, the accuracy of the geometry of a tissue group is evaluated with [16]

$$\text{Fidelity}(\mathbf{ref_{mask}}, \mathbf{rec_{mask}}) = \frac{\mathbf{ref_{mask}}^{\mathrm{T}}\mathbf{rec_{mask}}}{||\mathbf{ref_{mask}}||_2||\mathbf{rec_{mask}}||_2}, \tag{8}$$

where the two 2D masks to be compared are first vectorized. The Fidelity value varies from 0 (no similarity) to 1 (perfect similarity). Distortion of the structure and the presence of artefacts decrease the value of this metric. This metric is useful for evaluating the reconstruction of the fibroglandular region.

The next metric evaluates the accuracy with which both the geometric and dielectric properties of the underlying structures are reconstructed. This is measured using the normalized cross-correlation function (xCorrDiel) given by Equation (8), except that $\mathbf{ref_{mask}}$ and $\mathbf{rec_{mask}}$ are replaced with $\mathbf{ref_{tissue}}$ and $\mathbf{rec_{tissue}}$. In addition to sensing distortion and artefacts, this metric measures how accurately the electric properties are reconstructed within the structure.

The Dice similarity coefficient describes spatial overlap, and is given by [53]

$$\text{Dice}(\mathbf{ref_{mask}}, \mathbf{rec_{mask}}) = \frac{(\mathbf{ref_{mask}} \cap \mathbf{rec_{mask}})}{\frac{1}{2}(|\mathbf{ref_{mask}}| + |\mathbf{rec_{mask}}|)} = \frac{2|\mathbf{ref_{mask}} \cap \mathbf{rec_{mask}}|}{|\mathbf{ref_{mask}}| + |\mathbf{rec_{mask}}|} \tag{9}$$

where $|\cdot|$ is the cardinality of non-zero model elements within a mask.

The fourth metric assesses the proportion of malignant tissue correctly reconstructed within the tumor region (or ratio of tumor detected—RD). This is measured with [16]

$$\mathrm{RD}(\mathbf{ref_{mask}}, \mathbf{rec_{mask}}) = \frac{(\mathbf{ref_{mask}} \cap \mathbf{rec_{mask}})}{|\mathbf{ref_{mask}}|} \tag{10}$$

where $|\mathbf{ref_{mask}} \cap \mathbf{rec_{mask}}|$ denotes taking the cardinality of non-zero model elements that are in both the reference and reconstructed masks. Values close to zero imply that the algorithm is insensitive to malignant tissue, as a very small proportion of the lesion is reconstructed within the tumor region. Conversely, values close to 1 imply that the reconstruction algorithm is sensitive to malignant tissue, as most of the malignant tissue is reconstructed within the tumor region.

The final metric is artefact rejection (AR), which measures the proportion of tissue incorrectly reconstructed as malignant tissue outside the tumor region. AR is given by [16],

$$\mathrm{AR}(\mathbf{ref_{mask}}, \mathbf{rec_{mask}}) = 1 - \frac{|\mathbf{rec_{mask}}| - (\mathbf{ref_{mask}} \cap \mathbf{rec_{mask}})}{|\mathbf{ref_{mask}}|}. \tag{11}$$

A small value of AR indicates that a large proportion of tissue has been incorrectly reconstructed as malignant tissue outside the tumor region. Conversely, values close to 1 imply that only a small proportion of the malignant tissue is reconstructed outside the tumor region. The metrics given by Equations (8), (10) and (11) are described in more detail in [16,17].

The evaluation metrics given by Equations (9)–(11) are based on the region overlap between the reference and reconstructed segmentation masks. Theses metrics are relatively insensitive to under or over estimation of the tumor region [54], so they may not be appropriate for evaluating shape fidelity. Hence, a distance-based evaluation metric referred to as the Hausdorff distance ($\mathrm{H_A}$) described and analyzed in [54] provides an alternative perspective. With this measure, points extracted from the interfaces (or edges) of the reconstructed and reference masks are denoted as $\mathbf{rec} = \{a_1, a_2, \ldots, a_{Na}\}$ and $\mathbf{ref} = \{b_1, b_2, \ldots, b_{Nb}\}$, respectively. Accordingly, the Hausdorff distance evaluates how closely the shape of the reconstructed mask matches the shape of the reference mask. A variant of the Hausdorff distance between **rec** to **ref**, referred to as the average Hausdorff distance, is used for this study and is given by [55]

$$\mathrm{H_A}(\mathbf{ref}, \mathbf{rec}) = \max\{\mathrm{h}(\mathbf{rec}, \mathbf{ref}), \mathrm{h}(\mathbf{ref}, \mathbf{rec})\}. \tag{12}$$

where

$$\mathrm{h}(\mathbf{ref}, \mathbf{rec}) = \frac{1}{\mathrm{N_a}} \sum\nolimits_{\mathrm{a} \in \mathbf{rec}} \left\{ \min_{\mathrm{b} \in \mathbf{ref}} \|\mathrm{a} - \mathrm{b}\| \right\}. \tag{13}$$

As a pre-processing step suggested by [56], prior to computing Equation (12), the points are translated such that the center of the region enclosed by the corresponding closed contour is at the origin.

To complement the quantitative measures, qualitative assessment of images is enhanced by constructing contours from the edge points used to evaluate the average Hausdorff distances. Then, the contours are superimposed onto the forward model and reconstructed masks.

## 3. Results and Discussion

Three general case studies are used to demonstrate the utility of the proposed image analysis framework. For the first set of cases presented in Section 3.1, the forward model used to generate the numerical electromagnetic data for the study remains the same. Therefore, the shape, size, density, and tissue distribution of the breast is constant, but the degree of structural detail of the prior information (i.e., the regularization) used by the FEM-CSI algorithm varies. This leads to reconstructed images having a wide variety of image quality. The segmentation and application of metrics is shown to provide quantitative evaluation of the impact that the degree of structural detail of prior information has on image quality.

For the second set of cases that is presented in Section 3.2, the forward model used to generate the numerical data varies, but the degree of prior information used by the FEM-CSI algorithm is kept constant. Image quality is impacted primarily due to the differences in the shape, size, density, and tissue distribution of the breast being imaged, not the prior information. This demonstrates that the segmentation technique and the quantitative assessment leads to consistent results across breasts with a variety of shapes and tissue distributions.

Finally, in Section 3.3, tumor tracking cases demonstrate the potential for using the segmentation algorithm to extract clinically useful information.

### 3.1. Varying Structural Detail in Prior Information

The electromagnetic model (model 1) that is used for the first set of cases is a heterogeneously scattered breast constructed from an MRI slice [40]. The segmentation algorithm is applied to the real component of the complex permittivity of the forward model. The boundary, $\partial\mathcal{D}$, is set to the interface between the immersion medium and the skin surface. The boundary of the region of interest $\partial\mathcal{R}$ is formed by uniformly contracting $\partial\mathcal{D}$ inward towards the center of the model by 3.5 mm. Mask, $mask_{\mathcal{R}}$, is formed from the region bound by $\partial\mathcal{R}$ using Equation (3), and is applied to the forward model to recover data $\mathcal{R}$ with Equation (4). Figure 5a shows $\mathcal{R}$ extracted from the forward model of model 1. The same procedure is used to recover $\mathcal{R}$ over $mask_{\mathcal{R}}$ for the remainder of cases in this study.

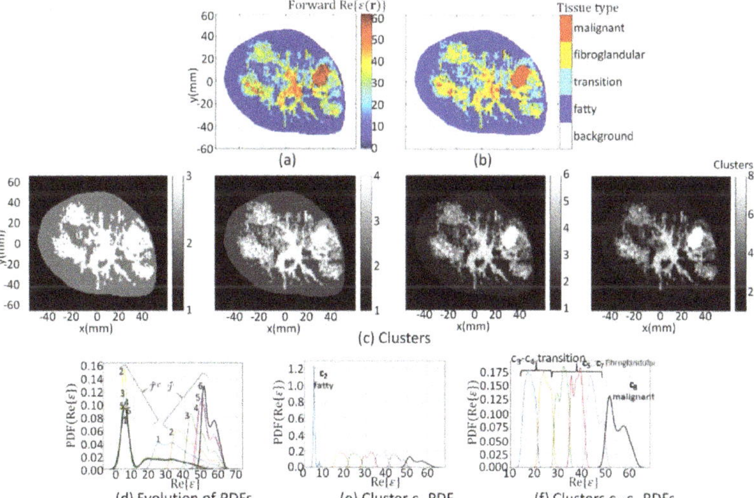


**Figure 5.** Model 1 forward model segmentation results. (**a**) $\mathcal{R}$ extracted from forward model; (**c**) Evolution of clusters at $k$ = 3, 4, 6, and 8; (**d**) Evolution of PDF over data within $\hat{\mathcal{T}}^c$ and $\hat{\mathcal{T}}$ where numbers indicate iteration; (**e**) PDF over data within cluster $c_2$, and (**f**) clusters $c_3$ (blue line) to $c_8$ (black line). Cluster $c_2$ corresponds to fatty tissue, $c_3 - c_4$ corresponds to transition tissue, $c_5 - c_7$ fibroglandular tissues, and $c_8$ corresponds to malignant tissue, which are mapped to segmentation masks leading to tissue type image (**b**).

The segmentation algorithm is applied to $\mathcal{B}$ (where $\mathcal{B} = \mathcal{R} \cup \mathcal{R}^c$) and converges after six iterations, leading to $\mathcal{B}$ being partitioned into eight disjoint clusters. The union of clusters $c_2 - c_{max\{k\}-1}$ form $\hat{\mathcal{T}}^c$, while $c_{max\{k\}}$ forms $\hat{\mathcal{T}}$. The PDF over data within $\hat{\mathcal{T}}$ and $\hat{\mathcal{T}}^c$ after each iteration is shown in Figure 5d, demonstrating the convergence that terminates the segmentation process. Individual PDFs over data within each cluster $c_2 - c_8$ are shown in Figure 5e,f. Finally, clusters are mapped to segmentation masks and associated tissue types using Equation (7), resulting in the segmented image shown in Figure 5b The forward model segmentation results are used as a reference and are compared with the segmentation results of the corresponding reconstructed images.

Numerical electromagnetic data are generated with the model 1 forward model. For the first case (3.1a), detailed patient-specific prior information is provided. Accordingly, the inhomogeneous background $\epsilon_b(\mathbf{r})$ in (1) emulates the structural information that would be recovered from an MRI image. This process is described in more detail in [16].

The FEM-CSI algorithm reconstructs the contrast profile $\chi(\mathbf{r})$; then, Equations (1) and (2) are employed to recover a list of images from (**r**), given by Re{$\epsilon$(**r**)}, Im{$\epsilon$(**r**)}, and $|\epsilon(\mathbf{r})|$. These images are shown Figure 6a. The tissue type and cluster images formed when the segmentation algorithm is applied are shown in Figure 6b,c, respectively. More detailed results in a format similar to Figure 5 showing the evolution of the PDF over data within $\hat{\mathcal{T}}$ and $\hat{\mathcal{T}}^c$ and the clusters after each iteration are furnished by Supplementary Materials Figures S1–S10. Moreover, the detailed results for all of the cases examined in Section 3.1 and video demonstrations are also available from the repository described in [52].

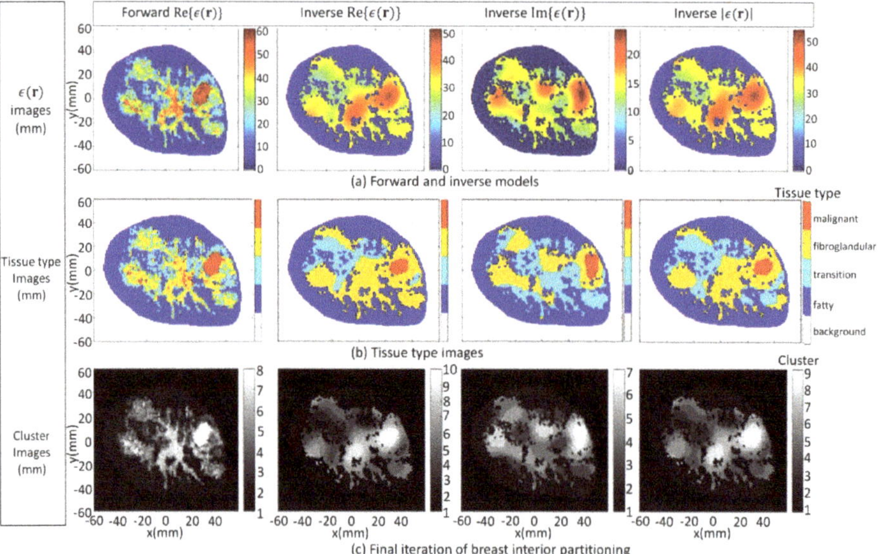


**Figure 6.** Case 3.1a forward model and reconstruction results when algorithm applied to model 1 data and $\epsilon_b(\mathbf{r})$ is set to detailed internal structure (**a**); Tissue type images (**b**); Final iteration of segmentation algorithm (**c**).

For the second case (3.1b), the inhomogeneous background $\epsilon_b(\mathbf{r})$ in Equation (1) is set to information extracted from radar-based techniques described in [16,57–59] and has less detail relative to the first case. Specifically, structural information related to the skin, fat, and glandular regions is provided along with estimates of the mean dielectric properties over these regions. The corresponding images reconstructed by the FEM-CSI algorithm are shown in Figure 7a and exhibit a lower degree of quality relative to the first case. The tissue type and cluster images are shown in Figure 7b,c, respectively.

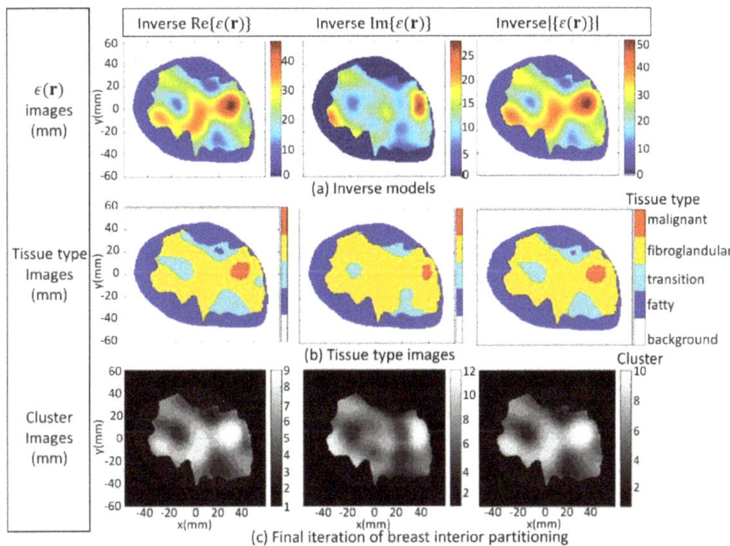


**Figure 7.** Case 3.1b reconstruction results when algorithm applied to model 1 data and $\epsilon_b(\mathbf{r})$ is set to structural information related to skin, fat, and glandular regions extracted by radar-based technique (**a**); Tissue type images (**b**); Final iteration of segmentation algorithm (**c**).

For the third and final case (3.1c), the inhomogeneous background $\epsilon_b(\mathbf{r})$ in Equation (1) incorporates structural information related to the skin region along with a homogenous breast interior with complex dielectric properties estimated with [16,57–59]. The reconstructed results shown in Figure 8a exhibit the lowest degree of quality of the three cases studied in this section, and they are the most challenging to segment. The tissue type mapping and cluster images are shown in Figure 8b,c, respectively.

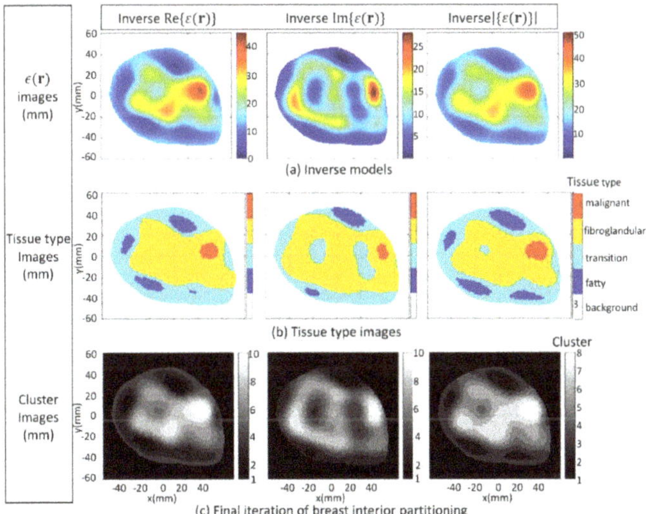


**Figure 8.** Case 3.1c reconstruction results when algorithm applied to model 1 data and $\epsilon_b(\mathbf{r})$ is set to structural information related to skin region extracted by radar-based technique (**a**); Tissue type images (**b**); Final iteration of segmentation algorithm (**c**).

The consistency of the proposed approach becomes particularly useful when segmenting images for which interfaces that delineate tissue types are blurred or are incorrectly located. This is evident for all three cases when segmenting the malignant from fibroglandular tissue and when segmenting the fibroglandular tissues from the breast interior for the third case. In addition to blurred interfaces, differences in electrical properties reconstructed that depends on the degree of structural detail of the prior information used by the FEM-CSI algorithm is also observed for the three cases. Regardless of these challenges, the proposed segmentation methodology gives reasonable estimates of glandular and tumor regions in all reconstructions. The qualitative image analysis is shown for all three cases in Figure 9. The regional and distance-based metrics are applied to the glandular and tumor regions, leading to the quantitative results shown in Tables 1 and 2, respectively.

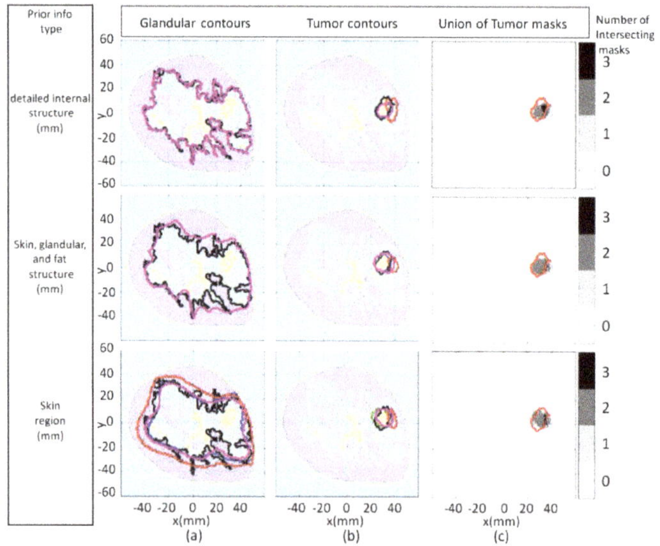


**Figure 9.** Model 1 qualitative image analysis of reconstructed images formed using various prior information detail. Glandular mask contours (**a**), and tumor mask contours (**b**) with contours extracted from forward model (black-line), reconstructed $\text{Re}\{\epsilon(\mathbf{r})\}$ (blue-line), $\text{Im}\{\epsilon(\mathbf{r})\}$ (red-line), and $|\epsilon(\mathbf{r})|$ (pink-line). Forward model contour (red-line) superimposed onto union of reconstructed tumor masks (**c**).

**Table 1.** Model 1: Glandular region metrics—varying degree of prior information.

| Case | Metric | Real | Imaginary | Magnitude |
|---|---|---|---|---|
| 3.1a (detailed internal structure) | Fidelity | 0.95 | 0.95 | 0.95 |
| | Dice | 0.95 | 0.95 | 0.95 |
| | xcorrDiel | 0.91 | 0.89 | 0.91 |
| | $H_A$ | 0.66 | 0.66 | 0.66 |
| 3.1b (regional internal structure) | Fidelity | 0.85 | 0.85 | 0.85 |
| | Dice | 0.85 | 0.85 | 0.85 |
| | xcorrDiel | 0.85 | 0.82 | 0.85 |
| | $H_A$ | 5.68 | 5.68 | 5.68 |
| 3.1c (skin region) | Fidelity | 0.85 | 0.81 | 0.86 |
| | Dice | 0.85 | 0.79 | 0.86 |
| | xcorrDiel | 0.83 | 0.75 | 0.83 |
| | $H_A$ | 4.39 | 7.09 | 4.06 |

**Table 2.** Model 1: Tumor region metrics—varying degree of prior information.

| Case | Metric | Real | Imaginary | Magnitude |
|---|---|---|---|---|
| 3.1a (detailed internal structure) | RD | 0.63 | 0.20 | 0.60 |
| | AR | 0.95 | 0.37 | 0.87 |
| | Dice | 0.75 | 0.22 | 0.69 |
| | $H_A$ | 1.56 | 1.63 | 1.44 |
| 3.1b (regional internal structure) | RD | 0.77 | 0.01 | 0.69 |
| | AR | 0.88 | 0.65 | 0.78 |
| | Dice | 0.82 | 0.01 | 0.73 |
| | $H_A$ | 1.20 | 2.73 | 1.32 |
| 3.1c (skin region) | RD | 0.71 | 0.05 | 0.86 |
| | AR | 0.87 | 0.60 | 0.61 |
| | Dice | 0.77 | 0.06 | 0.76 |
| | $H_A$ | 1.44 | 2.42 | 1.49 |

The effectiveness of the metrics incorporating segmentation results is evident from the results shown in Tables 1 and 2. As expected, the values of the metrics demonstrate that reducing the structural detail in the prior information leads to a degradation of reconstruction of the glandular structure. However, reducing this structural detail also impacts the quality of the reconstruction of the tumor region in a more complicated manner. For this set of examples, the specificity (implied by value of AR) degrades and the sensitivity improves (implied by value of RD) with decreasing amounts of structural prior information. Furthermore, each component of the reconstruction is impacted differently. Namely, the quality of the imaginary component in terms of sensitivity (RD) and tumor shape ($H_A$) benefits from a greater detail of prior structural information relative to the real component. These examples demonstrate the utility of having a framework that effectively provides a quantitative assessment of regions that contain a specific tissue. In particular, the regional and distance metrics provide valuable insight into a complex issue such as the evaluation of the impact that the degree of structural detail of prior information has on image quality.

A key motivation for developing the proposed segmentation methodology is to resolve the challenges that arise when using thresholding techniques. The challenges are demonstrated by applying the thresholding technique implemented by the studies described in [16,17] to the reconstructed images in this section. Specifically, threshold values are set to 95%, 90%, 85% and 80% of the maximum reconstructed value within the breast interior. In Figure 10, the black contour extracted from the forward model serves as a ground truth for comparison with the thresholded tumor contours. Likewise, metrics are applied to the reference and reconstructed tumor masks resulting from thresholding and are presented in Table 3.

The results shown in Figure 10 and Table 3 demonstrate the challenge of determining an appropriate threshold value to use with the threshold-based segmentation technique. Namely, adjustments of the threshold values demonstrate the trade-off between sensitivity and specificity that classification problems experience when using a methodology that depends on a fixed threshold value. For example, setting the segmentation threshold value for malignant tissue too low (e.g., 80%) leads to an improvement in sensitivity (i.e., high RD value) at the expense of the deterioration of the specificity (i.e., decrease in AR). This occurs because model elements that are within the fibroglandular structure are incorrectly attributed to malignant tissue. Likewise, setting the threshold value too high (e.g., 95%) impacts sensitivity by incorrectly assigning reconstructed tissue to fibroglandular tissue when it is, in fact, malignant tissue. Accordingly, the choice of what value of threshold to use is not obvious and, to complicate matters, it has been observed that the maximum value of the reconstructed tissue using FEM-CSI depends on the number of iterations.

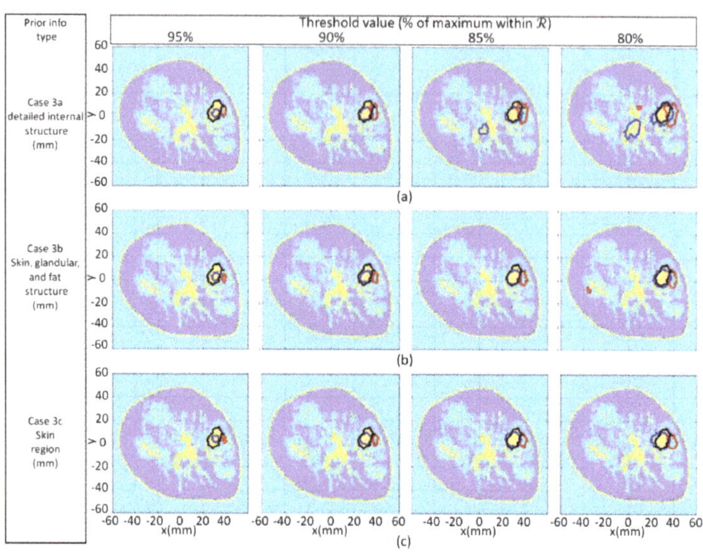


**Figure 10.** Model 1 qualitative image analysis of reconstruction images using various threshold values applied to cases 3.1a (**a**), 3.1b (**b**), and 3.1c (**c**). For each case, contours associated with tumor masks from forward model, reconstructed $\mathrm{Re}\{\epsilon(\mathbf{r})\}$, and $\mathrm{Im}\{\epsilon(\mathbf{r})\}$ shown with black, blue, and red lines, respectively, superimposed onto forward model.

**Table 3.** Model 1 tumor region metrics: tumor region extracted with threshold technique using various values of threshold.

| Case | Metric | 95% | 90% | 85% | 80% |
|---|---|---|---|---|---|
| 3.1a (detailed internal structure) Real component | RD | 0.26 | 0.64 | 0.76 | 0.90 |
| | AR | 0.98 | 0.94 | 0.39 | −0.62 |
| | Dice | 0.41 | 0.75 | 0.64 | 0.51 |
| | $H_A$ | 3.46 | 1.47 | 1.28 | 1.78 |
| 3.1a (detailed internal structure) Imaginary component | RD | 0.05 | 0.11 | 0.23 | 0.33 |
| | AR | 0.79 | 0.65 | 0.44 | 0.16 |
| | Dice | 0.07 | 0.15 | 0.26 | 0.31 |
| | $H_A$ | 3.35 | 2.28 | 1.44 | 1.48 |
| 3.1b (regional internal structure) Real component | RD | 0.26 | 0.53 | 0.72 | 0.83 |
| | AR | 1.00 | 0.98 | 0.90 | 0.66 |
| | Dice | 0.40 | 0.68 | 0.79 | 0.76 |
| | $H_A$ | 3.53 | 1.95 | 1.34 | 1.16 |
| 3.1b (regional internal structure) Imaginary component | RD | 0.00 | 0.00 | 0.02 | 0.05 |
| | AR | 0.86 | 0.74 | 0.61 | 0.41 |
| | Dice | 0.00 | 0.00 | 0.03 | 0.61 |
| | $H_A$ | 4.35 | 3.33 | 2.39 | 1.83 |
| 3.1c (skin region) Real component | RD | 0.21 | 0.52 | 0.73 | 0.82 |
| | AR | 1.00 | 0.99 | 0.84 | 0.61 |
| | Dice | 0.35 | 0.68 | 0.78 | 0.74 |
| | $H_A$ | 4.01 | 2.13 | 1.32 | 1.50 |
| 3.1c (skin region) Imaginary component | RD | 0.00 | 0.00 | 0.03 | 0.06 |
| | AR | 0.88 | 0.76 | 0.65 | 0.54 |
| | Dice | 0.00 | 0.01 | 0.04 | 0.08 |
| | $H_A$ | 4.47 | 3.41 | 2,63 | 2.11 |

In contrast, the proposed technique does not rely on assumed dielectric property values of the reconstructed tissues. Moreover, the proposed iterative approach does not require the number of clusters to be pre-selected, as the unsupervised machine learning technique is reinforced with hypothesis testing and statistical inference to automatically determine the number of clusters.

The convenience of using this strategy is evident when observing the variation in the final number of clusters, as shown in the bottom row of Figures 6–8. The examples demonstrate that pre-selecting the number of clusters beforehand is not practical. Furthermore, using the proposed strategy leads to a more precise and consistent approach to image analysis compared to alternative methods by automatically detecting regions of interest in the image corresponding to various tissue types or anomalies. This advantage is particularly evident when comparing the metric values in Table 2 with those in Table 3. In Table 3, there is a significant variation in the values of all metrics across all reconstruction components and test cases, depending on the threshold value used. The variation in the metric values leads to inconsistent results that contribute to unreliable quantitative assessment of reconstructed images.

It is also observed that the threshold technique requires different threshold values in order to achieve the same results as the proposed automatic segmentation method. For example, for case 3.1a, the thresholding technique requires values of approximately 90% and less than 85% to segment the real and imaginary components, respectively. Different threshold values are also needed depending on the image component and the case examined. This observation demonstrates that using the proposed technique leads to a simplification of the segmentation process that may result in improved consistency and reliability of results. Moreover, it is not necessary for the user to make a decision on a threshold value to use or to iteratively fine tune threshold values depending on the image component or reconstructed image. This observation also demonstrates the flexibility of the proposed technique and its ability to automatically adapt to a scenario (e.g., image quality).

### 3.2. *Varying Breast Shape and Tissue Distribution*

The second part of the study is comprised of three cases, namely breast models with different shapes and tissue distributions. The degree of prior information used by the FEM-CSI algorithm is kept constant, so image quality is impacted primarily due to the shape, size, and tissue distribution of the breast being imaged. The inhomogeneous background $\epsilon_b(\mathbf{r})$ in Equation (1) is extracted from ultrasound data described in [60]. An electromagnetic model (model 3.2a) described in [40] of a heterogeneously dense breast that is constructed from an MRI slice is used for the first case.

When applied to the forward model, the segmentation algorithm converges after five iterations, leading to $\mathcal{B}$ being partitioned into seven disjoint clusters. These clusters are mapped to masks and associated tissue types using Equation (7). The forward model segmentation results are used as a reference and are compared with the segmentation results of the corresponding reconstructed images. Numerical electromagnetic data are generated with forward model 3.2a. The FEM-CSI algorithm iteratively reconstructs the contrast profile [17] and the corresponding images, given by $\mathrm{Re}\{\epsilon(\mathbf{r})\}$, $\mathrm{Im}\{\epsilon(\mathbf{r})\}$, and $|\epsilon(\mathbf{r})|$, are shown in Figure 11a. The tissue type and cluster images are shown in Figure 11b,c, respectively. The qualitative image analysis is shown in Figure 12. The regional and distance-based metrics lead to the quantitative results shown in Table 4.

Model 3.2b is an electromagnetic model of a fatty breast that is constructed from a sequence of MRI slices described in [30]. The segmentation algorithm is applied to the forward model and converges after four iterations. The FEM-CSI algorithm iteratively reconstructs the contrast profile [17]. Results obtained when the segmentation algorithm is applied to the forward model and the reconstructed images are shown in Figure 13. The qualitative image analysis is shown in Figure 14, while regional and distance-based metrics are summarized in Table 5.

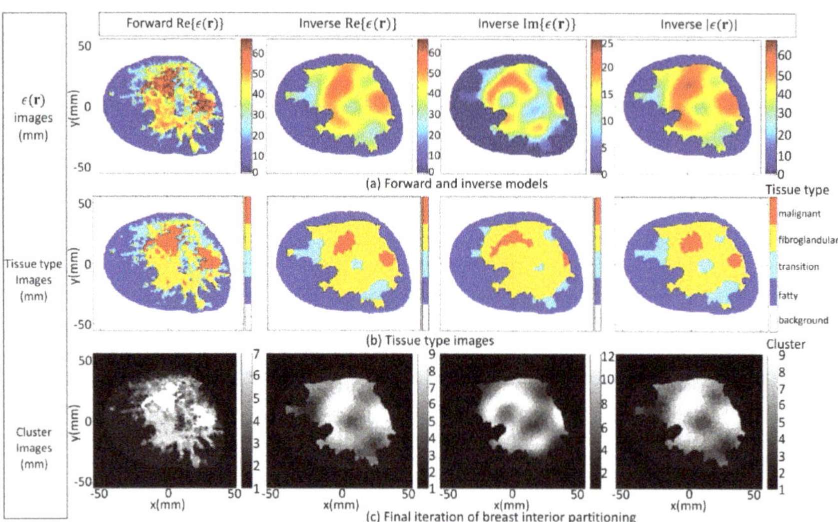


**Figure 11.** Model 3.2a forward model and reconstruction results (**a**); Tissue type images (**b**); Final iteration of segmentation algorithm (**c**).

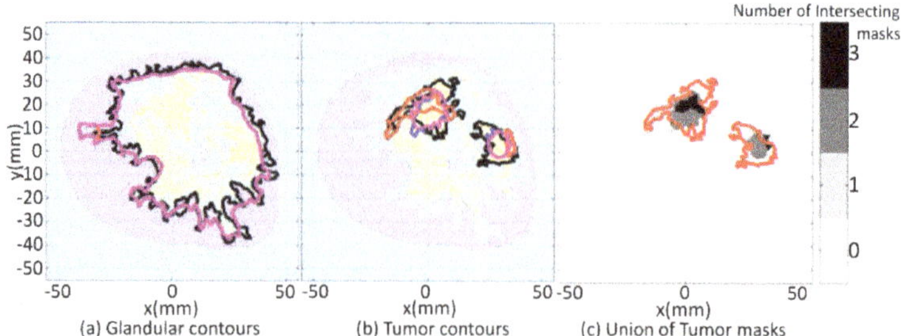


**Figure 12.** Model 3.2a qualitative image analysis. Glandular mask contours (**a**), and tumor mask contours (**b**) with contours extracted from forward model (black-line), reconstructed $\mathrm{Re}\{\epsilon(\mathbf{r})\}$ (blue-line), $\mathrm{Im}\{\epsilon(\mathbf{r})\}$ (red-line), and $|\epsilon(\mathbf{r})|$ (pink-line). Forward model contour (red line) superimposed onto union of reconstructed tumor masks (**c**).

**Table 4.** Model 3.2a quantitative results.

| Region | Metric | Real | Imaginary | Magnitude |
|---|---|---|---|---|
| Glandular | Fidelity | 0.90 | 0.90 | 0.90 |
| | Dice | 0.90 | 0.90 | 0.90 |
| | xcorrDiel | 0.91 | 0.88 | 0.91 |
| | $H_A$ | 1.66 | 1.64 | 1.66 |
| Tumor 1 | RD | 0.44 | 0.35 | 0.50 |
| | AR | 0.92 | 0.78 | 0.96 |
| | Dice | 0.58 | 0.45 | 0.65 |
| | $H_A$ | 3.80 | 3.66 | 3.71 |
| Tumor 2 | RD | 0.40 | 0.09 | 0.36 |
| | AR | 0.94 | 0.93 | 0.94 |
| | Dice | 0.55 | 0.15 | 0.51 |
| | $H_A$ | 2.85 | 4.52 | 3.39 |

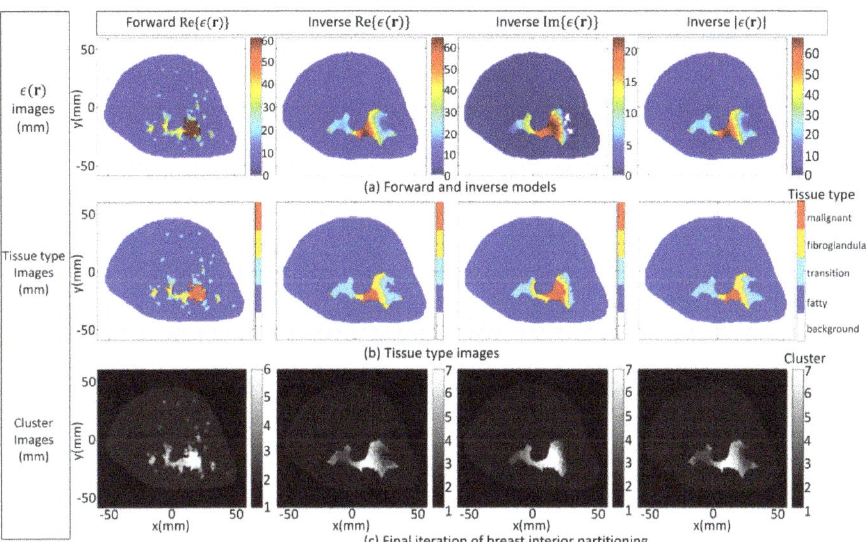


**Figure 13.** Model 3.2b forward model and reconstruction results (**a**); Tissue type images (**b**); Final iteration of segmentation algorithm (**c**).

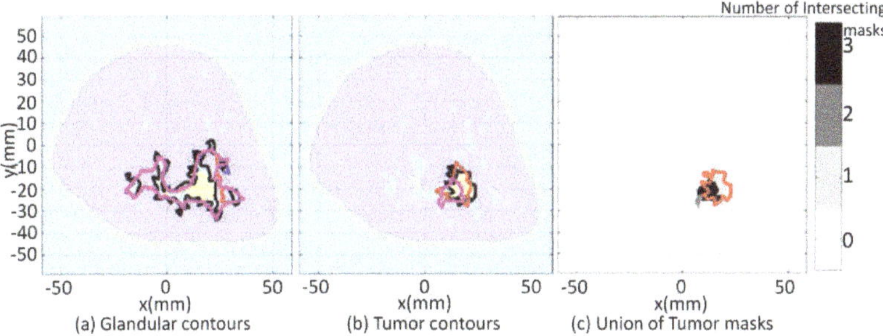


**Figure 14.** Model 3.2b qualitative image analysis. Glandular mask contours (**a**), and tumor mask contours (**b**) with contours extracted from forward model (black-line), reconstructed Re$\{\epsilon(\mathbf{r})\}$ (blue-line), Im$\{\epsilon(\mathbf{r})\}$ (red-line), and $|\epsilon(\mathbf{r})|$ (pink-line). Forward model contour (red line) superimposed onto union of reconstructed tumor masks (**c**).

**Table 5.** Model 3.2b quantitative results.

| Region | Metric | Real | Imaginary | Magnitude |
|---|---|---|---|---|
| Glandular | Fidelity | 0.61 | 0.65 | 0.62 |
| | Dice | 0.59 | 0.64 | 0.60 |
| | xcorrDiel | 0.72 | 0.79 | 0.72 |
| | $H_A$ | 3.34 | 2.54 | 3.29 |
| Tumor | RD | 0.34 | 0.76 | 0.34 |
| | AR | 0.71 | 0.61 | 0.71 |
| | Dice | 0.41 | 0.71 | 0.41 |
| | $H_A$ | 1.62 | 1.10 | 1.62 |

Model 3.2c is used as the final case studied for this part of the study, and it is an electromagnetic model of a dense breast that is constructed from a sequence of MRI slices [30]. The segmentation algorithm is applied to the forward model and converges after four iterations. The FEM-CSI algorithm iteratively reconstructs the contrast profile [17].

The results obtained when the segmentation algorithm is applied to the forward model and the reconstructed images are shown in Figure 15. The qualitative image analysis is shown in Figure 16, and a summary of the regional and distance-based metrics is provided in Table 6.

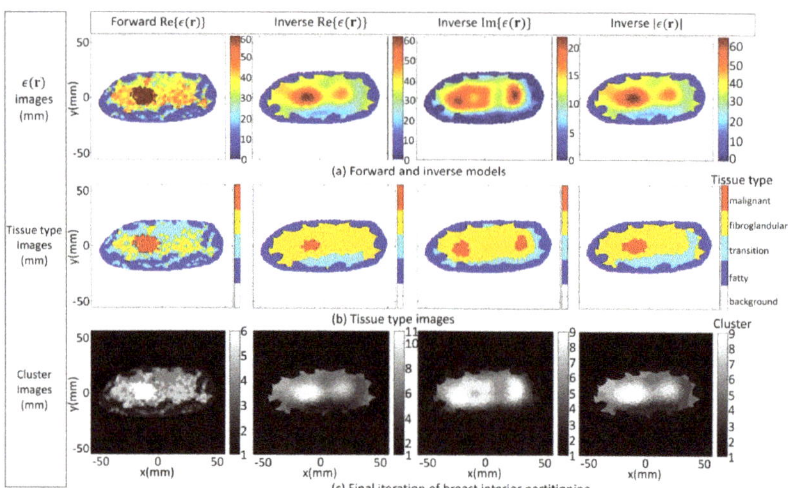


**Figure 15.** Model 3.2c forward model and reconstruction results (**a**); Tissue type images (**b**); Final iteration of segmentation algorithm (**c**).

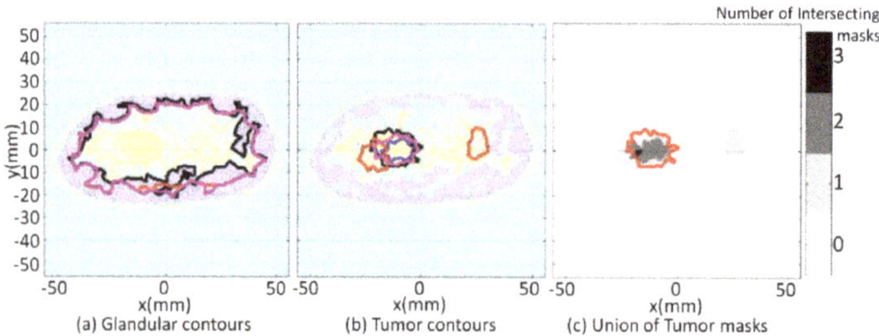


**Figure 16.** Model 3.2c qualitative image analysis. Glandular mask contours (**a**), and tumor mask contours (**b**) with contours extracted from forward model (black-line), reconstructed $\text{Re}\{\epsilon(\mathbf{r})\}$ (blue-line), $\text{Im}\{\epsilon(\mathbf{r})\}$ (red-line), and $|\epsilon(\mathbf{r})|$ (pink-line). Forward model contour (red line) superimposed onto union of reconstructed tumor masks (**c**).

**Table 6.** Model 3.2c quantitative results.

| Region | Metric | Real | Imaginary | Magnitude |
|---|---|---|---|---|
| Glandular | Fidelity | 0.87 | 0.88 | 0.87 |
| | Dice | 0.87 | 0.87 | 0.87 |
| | xcorrDiel | 0.92 | 0.92 | 0.92 |
| | $H_A$ | 2.13 | 2.16 | 2.13 |
| Tumor | RD | 0.37 | 0.16 | 0.69 |
| | AR | 1.00 | 0.65 | 0.95 |
| | Dice | 0.54 | 0.21 | 0.79 |
| | $H_A$ | 2.96 | 2.56 | 1.36 |

For this section, the tissue distribution of each model varied, but the prior knowledge of internal structural information was kept the same. Even with considerable variation in breast density and tissue distribution between models, it was demonstrated that the segmentation algorithm is robust to these variations. As observed with the cases in Section 3.1, the final number of clusters that the algorithm converges to varies, depending on the tissue distribution of the breast and image component being segmented. Unlike thresholding segmentation techniques that require pre-selected thresholds, or an unsupervised machine learning approach such as $k$-means clustering that requires a pre-selected number of clusters, the proposed image segmentation does not require prior information. Consequently, it is not data-specific, unlike these other techniques, and it was able to reliably and consistently segment the reconstructed images into tissue types to permit the quantitative assessment of regions that contain a specific tissue.

These results also provide insight into the impact that the breast density and tissue distribution has on the performance of the FEM-CSI algorithm. Specifically, reconstruction of the real and imaginary components of the malignant tissue was effectively assessed. For the imaginary component, the metrics suggest that the reconstruction algorithm is more sensitive to malignant tissue (i.e., higher RD value) and reconstructed the tumor region more accurately (lower $H_A$ value) for the fatty breast compared to the other two cases. On the other hand, for the real component, the metrics suggest that the reconstruction algorithm is equally sensitive to the malignant tissue for all three tissue distributions. However, similar to the imaginary component, the tumor region of the real component was reconstructed more accurately for the fatty breast scenario. For the dense breast, the advantages of analyzing the magnitude of the reconstructed image is evident, as there is both an improvement in sensitivity and accuracy of the tumor region that is reconstructed compared to the quality of the real and imaginary components.

Similar to the test cases studied in Section 3.1, the examples investigated in this section demonstrate the utility of having a framework that effectively provides a quantitative assessment of regions that contain a specific tissue to provide valuable insight into a complex issue. Namely, the evaluation of the impact that the tissue distribution and breast density have on image quality and the performance of the reconstruction algorithm can be effectively assessed. These insights are not necessarily revealed or as obvious with a qualitative assessment such a visual examination and image comparisons.

The test cases also demonstrate the practical utility of mapping clusters to distinct tissue types. The tissue mapped images may be used to assist with image interpretation and to more readily identify anomalies.

### 3.3. Tumor Tracking

The contrast in dielectric properties between healthy and malignant tissues reported in the large-scale studies [2–7] may be exploited with microwave imaging in order to image malignant tissue. This is supported with clinical studies described in [10,12,24,25] that demonstrate the utility of microwave tomography for breast screening and therapy monitoring. Consequently, the final part of the study is comprised of two tumor tracking examples to demonstrate that the segmentation technique may assist with extracting clinically useful information. Similar to the second part of the study described in Section 3.2, the degree of structural detail of the prior information used by the FEM-CSI algorithm is the same for each case. For both cases, the inhomogeneous background $\epsilon_b(\mathbf{r})$ in (1) is set to information extracted from the radar-based technique described in [16,57–59]. Model 1, which is also used in Section 3.1, is the forward model used to generate the numerical electromagnetic data.

For the first case (3.3a), a large tumor region is present in the forward model, as shown in Figure 17. The segmentation algorithm is applied to the forward model and converges after five iterations, so $\mathcal{B}$ is partitioned into seven disjoint clusters. These clusters are mapped to segmentation masks and associated tissue types using Equation (7). The forward model segmentation results are used as a reference and are compared with the segmentation results of the corresponding reconstructed images.

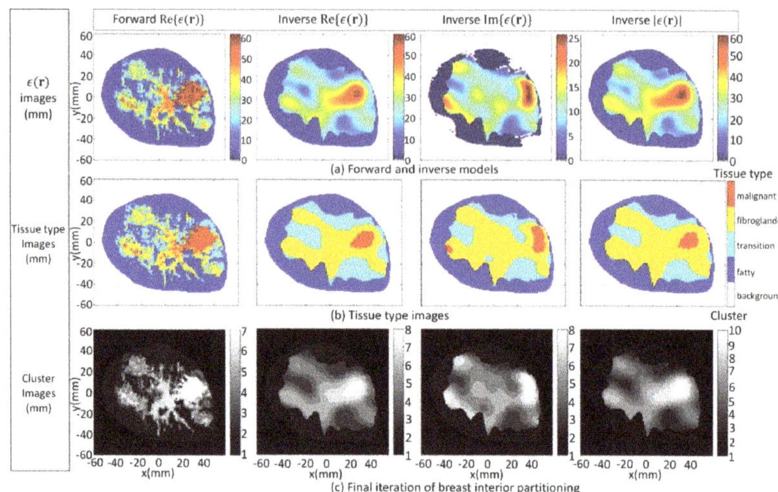


**Figure 17.** Model 1 forward model with large tumor embedded in fibroglandular tissues and reconstruction results (**a**); Tissue type images (**b**); Final iteration of segmentation algorithm (**c**).

The FEM-CSI algorithm iteratively reconstructs the contrast profile [17]. The corresponding images are shown in Figure 17a. The tissue type and cluster images are shown in Figure 17b,c, respectively.

For the second case (3.3b), the size of the tumor region is reduced, but its location within the forward model is approximately the same as the first case. The results when the segmentation algorithm is applied to the forward model and the reconstructed images are shown in Figure 7 (Section 3.1).

The qualitative image analysis is shown for each case in Figure 18. The region and distance-based metrics are applied to the reference and reconstructed masks of the tumor regions, leading to the quantitative results shown in Table 7.

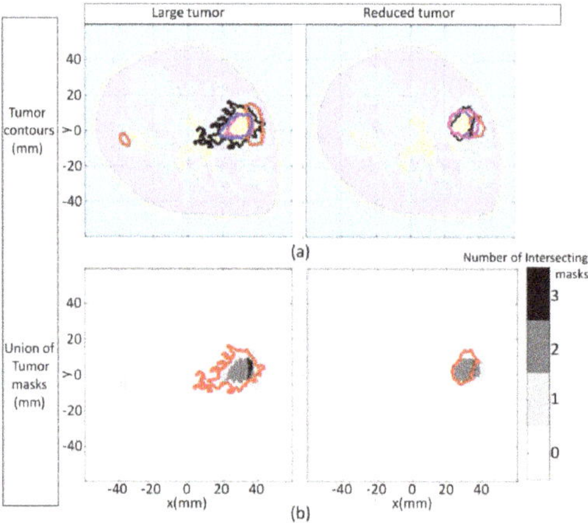


**Figure 18.** Model 1 tumor tracking qualitative image analysis. Contours for large tumor and reduced tumor cases (**a**) with contours extracted from forward model (black line), reconstructed $\mathrm{Re}\{\epsilon(\mathbf{r})\}$ (blue line), $\mathrm{Im}\{\epsilon(\mathbf{r})\}$ (red line), and $|\epsilon(\mathbf{r})|$ (pink line). Forward model contour (red line) superimposed onto union of masks formed with malignant tissue reconstructed from FEM-CSI $\mathrm{Re}\{\epsilon(\mathbf{r})\}$, $\mathrm{Im}\{\epsilon(\mathbf{r})\}$, $|\epsilon(\mathbf{r})|$ (**b**).

**Table 7.** Model 1 tumor tracking quantitative results.

| Case | Metric | Real | Imaginary | Magnitude |
|---|---|---|---|---|
| 3.3a—Large tumor | RD | 0.46 | 0.17 | 0.30 |
| | AR | 1.00 | 0.76 | 1.00 |
| | Dice | 0.63 | 0.24 | 0.46 |
| | $H_A$ | 4.90 | 6.67 | 5.71 |
| 3.3b—Reduced tumor | RD | 0.77 | 0.01 | 0.69 |
| | AR | 0.88 | 0.65 | 0.78 |
| | Dice | 0.82 | 0.01 | 0.73 |
| | $H_A$ | 1.20 | 2.73 | 1.32 |

The potential for using the algorithm to provide clinically useful information is demonstrated with this set of tumor tracking examples. Microwave tomography typically produces lower resolution images than clinical imaging methods such as X-ray. Hence, segmenting medical images formed with microwave tomography for tumor tracking examples can be challenging as the interfaces that delineate tissue types may be blurred. This is particularly challenging when malignant tissue is embedded in glandular tissue. Contributing to the challenge is the possibility that there may be a great deal of inhomogeneity amongst the glandular tissue. Regardless of these challenges, the proposed segmentation procedure demonstrated the ability to delineate the reconstructed tissue from the glandular tissue.

Once the tissue regions are extracted, metrics are applied for quantitative analysis in order to assess the results. The metrics shown in Table 7 infer that for the large tumor reconstruction scenario, the algorithm is less sensitive but has a higher specificity to the malignant tissue relative to the reduced tumor scenario. The values of the average Hausdorff distance shown in Table 7 indicate that the reconstruction algorithm did not reconstruct the shape of the malignant region as accurately compared to the reduced tumor scenario. The metrics collectively suggest that there is inadequate information furnished from the images to make a judgement with respect to whether a significant reduction in the size of the malignant region has occurred (in response to some treatment, for example).

Similar to the test cases examined in the previous sections, this set of cases demonstrate the practical convenience of mapping clusters to distinct tissue types. The tissue mapped images may be used to assist with image interpretation and to more readily make inferences on the location of the malignant tissue within the glandular structure. This example also demonstrates the utility of providing a framework for assessing the performance of the reconstruction algorithm. For example, the metrics may be used to inform researchers with regard to adjustments to the reconstruction algorithm or measurement system parameters such as an increase in the number of sensors to improve the sensitivity and overall performance of the reconstruction algorithm.

## 4. Conclusions

A medical image segmentation technique has been presented that partitions microwave breast images into regions of interest corresponding to distinct tissue types in order to facilitate the evaluation of image quality. A key advantage for using the algorithm over other approaches is that it supports a quantitative analysis of microwave images without prior assumptions such as knowledge of the expected dielectric property values that characterize each tissue type. Unlike supervised machine learning approaches that require copious amounts of data to effectively train a model, it can be used for scenarios where there is a scarcity of data. It also addresses a significant difficulty encountered by many unsupervised machine learning approaches in that it does not require a predetermined number of clusters to partition the image. The proposed technique is not data-specific, as it was able to segment a variety of images with different image quality. Moreover, it was able to reliably and consistently segment images derived from breasts with various tissue distributions and densities into tissue types to permit quantitative assessment of regions that contain a specific tissue.

The segmentation into tissue types leads to the decomposition of the breast interior into disjoint tissue masks. An array of region and distance-based metrics were applied to compare masks extracted from reconstructed images and ground truth models. The quantitative results revealed the accuracy with which the geometric and dielectric properties are reconstructed. The incorporation of the segmentation results into an evaluation framework with metrics was demonstrated and effectively furnished quantitative assessment of tissue-specific regions. The examples demonstrated the utility of having this framework to provide valuable insight into a complex issue. Namely, the impact that changes in tissue distribution and breast density have on image quality and the performance of the reconstruction algorithm can be effectively assessed. These insights are not necessarily revealed or as obvious with a qualitative assessment such a visual examination and image comparisons.

It is anticipated that this framework may also be applied to the analysis of the data acquisition environment to quantify changes in image quality to inform researchers on the number and location of sensors, the incident field frequency, measurement chamber design, and the orientation of the receivers relative to the data acquisition surface. For this study, the numerical breast models were used for the forward model and furnished the reference regions to compare with the tissues segmented from the image. However, when using clinical data, the reference model may be the patient at a previous point in time to quantify how a region changed over time in response to a treatment. The reference model for clinical or experimental data may also be an inverse model obtained with variations on the same algorithm or a different reconstruction algorithm (comparing the FEM-CSI inverse solver with the Distorted Born iterative method, for example).

In addition to facilitating a quantitative analysis of images, the tissue masks facilitate supplying qualitative information to assist in the interpretation of the microwave images. This qualitative information is augmented with images showing the location of estimated tissue interfaces that provide a visual means to quickly interpret an image or the performance of an inversion algorithm.

More broadly, the presented technique provides a general framework that may be applied to an extensive range of medical imaging modalities. This may be particularly useful for developing modalities for which users do not have much experience with the reconstructed images, as well as when there is scarcity of data available for supervised learning. Initial investigations into the application of the technique to ultrasound images has assisted with studies reported in [17,60]. The diverse range of potential applications that may implement the presented image analysis technique also includes liquid biopsy analysis [61–63].

Future work includes integrating this segmentation approach with performance metrics (e.g., [16,17,39,60]), and composite tissue-type and probability images [64].

**Supplementary Materials:** The following are available online at https://www.mdpi.com/2313-433X/7/1/5/s1, Detailed results for all cases presented in Section 3 including the clusters after each iteration and the evolution of the PDF of the data over $\hat{\mathcal{T}}$ and $\hat{\mathcal{T}}^c$ are available online at https://github.com/djkurran/Segmentation-unsupervised-machine-learning [52]. A list of figures available in the repository is as follows: Figure S1: Model 1 forward model segmentation results, Figure S2: Case 3.1a Segmentation results of reconstruction derived from detailed internal structure prior—Real component, Figure S3 Case 3.1a Segmentation results of reconstruction derived from detailed internal structure prior—Imaginary component, Figure S4 Case 3.1a Segmentation results of reconstruction derived from detailed internal structure prior—Magnitude, Figure S5 Case 3.1b Segmentation results of reconstruction derived from regional internal structure prior—Real component, Figure S6 Case 3.1b Segmentation results of reconstruction derived from regional internal structure prior—Imaginary component, Figure S7 Case 3.1b Segmentation results of reconstruction derived from regional internal structure prior—Magnitude, Figure S8 Case 3.1c Segmentation results of reconstruction derived from skin region prior—Real component, Figure S9 Case 3.1c Segmentation results of reconstruction derived from skin region prior—Imaginary component, Figure S10 Case 3.1c Segmentation results of reconstruction derived from skin region prior—Magnitude.

**Author Contributions:** Conceptualization, D.K. and M.O.; Methodology, D.K.; Software and Visualization, D.K.; Validation, D.K., M.O., N.A., P.M., E.F., J.L.; Formal Analysis, D.K.; Resources, D.K., M.O, N.A., P.M.; Data Curation, D.K., M.O., N.A., P.M.; Supervision, E.F., J.L.; Writing—original draft, D.K.; Writing—review and editing, D.K., M.O., N.A., P.M., E.F., J.L.; Funding Acquisition, E.F., J.L. All authors have read and agreed to the published version of the manuscript.

**Funding:** This work was supported by the Canadian Breast Cancer Foundation Under Grant R1612.

**Institutional Review Board Statement:** Not applicable.

**Informed Consent Statement:** Not applicable.

**Data Availability Statement:** Publically available datasets related to the scattered density, and heterogeneously dense categorized breasts were analyzed in this study. These data can be found here: https://github.com/djkurran/MWSegEval/testData. Use [40] when citing these data. Moreover, publically available datasets related to the fatty and extremely dense categorized breasts analyzed in this study for Case 3.2 are available here: Omer, M., Fear, E. Anthropomorphic breast model repository for research and development of microwave breast imaging technologies. *Sci Data* **5**, 180257 (2018). https://doi.org/10.1038/sdata.2018.257. Use [65] when citing these model data.The novel computer code and software developed by the authors that integrates the unsupervised machine learning and thresholding segmentation techniques into an image processing toolbox are available in the publically available repository: https://github.com/djkurran/MWSegEval. A wiki page associated with this repository hosts a detailed on-line manual for the toolbox. Use [22] when citing the toolbox.

**Acknowledgments:** D.K. would like to thank Cameron Kaye (M.D.) of the University of Manitoba for sharing his insights related to the practical application of the Kolmogorov-Smirnov hypothesis test, and Jèrèmie Bourqui of the University of Calgary for the development of tools used to automatically generate numerical electromagnetic reflection data used for this study.

**Conflicts of Interest:** The authors declare no conflict of interest.

## References

1. Pastorino, M. *Microwave Imaging*; John Wiley & Sons: Hoboken, NJ, USA, 2010.
2. Lazebnik, M.; Popovic, D.; McCartney, L.; Watkins, C.B.; Lindstrom, M.J.; Harter, J.; Sewall, S.; Ogilvie, T.; Magliocco, A.; Breslin, T.M.; et al. A large-scale study of the ultrawideband microwave dielectric properties of normal, benign and malignant breast tissues obtained from cancer surgeries. *Phys. Med. Biol.* **2007**, *52*, 6093. [CrossRef] [PubMed]
3. Lazebnik, M.; McCartney, L.; Popovic, D.; Watkins, C.B.; Lindstrom, M.J.; Harter, J.; Sewall, S.; Magliocco, A.; Booske, J.H.; Okoniewski, M.; et al. A large-scale study of the ultrawideband microwave dielectric properties of normal breast tissue obtained from reduction surgeries. *Phys. Med. Biol.* **2007**, *52*, 2637. [CrossRef] [PubMed]
4. Martellosio, A.; Pasian, M.; Bozzi, M.; Perregrini, L.; Mazzanti, A.; Svelto, F.; Summers, P.E.; Renne, G.; Preda, L.; Bellomi, M. Dielectric properties characterization from 0.5 to 50 GHz of breast cancer tissues. *IEEE Trans. Microw. Theory Tech.* **2017**, *65*, 998–1011. [CrossRef]
5. Cheng, Y.; Fu, M. Dielectric properties for non-invasive detection of normal, benign, and malignant breast tissues using microwave theories. *Thorac. Cancer* **2018**, *9*, 459–465. [CrossRef] [PubMed]
6. Hussein, M.; Awwad, F.; Jithin, D.; El Hasasna, H.; Athamneh, K.; Iratni, R. Breast cancer cells exhibits specific dielectric signature in vitro using the open-ended coaxial probe technique from 200 MHz to 13.6 GHz. *Sci. Rep.* **2019**, *9*, 4681. [CrossRef] [PubMed]
7. Sugitani, T.; Kubota, S.I.; Kuroki, S.I.; Sogo, K.; Arihiro, K.; Okada, M.; Kadoya, T.; Hide, M.; Oda, M.; Kikkawa, T. Complex permittivities of breast tumour tissues obtained from cancer surgeries. *Appl. Phys. Lett.* **2014**, *104*, 253702. [CrossRef]
8. Shea, J.D.; Kosmas, P.; Hagness, S.C.; van Veen, B.D. Three-dimensional microwave imaging of realistic numerical breast phantoms via a multiple-frequency inverse scattering technique. *Med. Phys.* **2010**, *37*, 4210–4226. [CrossRef]
9. Golnabi, A.H.; Meaney, P.M.; Paulsen, K.D. Tomographic microwave imaging with incorporated prior spatial information. *IEEE Trans. Microw.* **2013**, *61*, 104–116. [CrossRef]
10. Meaney, P.M.; Fanning, M.W., Raynolds, T.; Fox, C.J.; Fang, Q.; Kogel, C.A.; Poplack, S.P.; Paulsen, K.D. Initial clinical experience with microwave breast imaging in women with normal mammography. *Academ. Radiol.* **2007**, *14*, 207–218. [CrossRef]
11. Colgan, T.J.; Hagness, S.C.; van Veen, B.D. A 3-D level set method for microwave breast imaging. *IEEE Trans. Biomed. Eng.* **2015**, *62*, 2526–2534. [CrossRef] [PubMed]
12. Meaney, P.M.; Kaufman, P.A.; Muffly, L.S.; Click, M.; Poplack, S.P.; Wells, W.A.; Schwartz, G.N.; di Florio-Alexander, R.M.; Tosteson, T.D.; Li, Z.; et al. Microwave imaging for neoadjuvant chemotherapy monitoring: Initial clinical experience. *Breast Cancer Res.* **2015**, *15*, R15. [CrossRef] [PubMed]
13. Fhager, A.; Hashemzadeh, P.; Persson, M. Reconstruction quality and spectral content of an electromagnetic time-domain inversion algorithm. *IEEE Trans. Biomed. Eng.* **2006**, *53*, 1594–1604. [CrossRef] [PubMed]

14. Meaney, P.M.; Yagnamurthy, N.K.; Paulsen, K.D. Pre-scaled two-parameter Gauss-Newton image reconstruction to reduce property recovery imbalance. *Phys. Med. Biol.* **2002**, *47*, 1101–1119. [CrossRef]
15. Gilmore, C.; Zakaria, A.; Pistorius, S.; LoVetri, J. Microwave imaging of human forearms: Pilot study and image enhancement. *J. Biomed. Imaging* **2013**, *2013*. [CrossRef] [PubMed]
16. Kurrant, D.; Baran, A.; LoVetri, J.; Fear, E. Integrating prior information into microwave tomography Part 1: Impact of detail on image quality. *Med. Phys.* **2017**, *44*, 6461–6481. [CrossRef]
17. Abdollahi, N.; Kurrant, D.; Mojabi, P.; Omer, O.; Fear, E.; LoVetri, J. Incorporation of ultrasonic prior information for improving quantitative microwave imaging of breast. *IEEE J. Multiscale Multiphys. Comput. Tech.* **2019**, *4*, 98–110. [CrossRef]
18. Zakaria, A.; Anastasia, A.; LoVetri, J. Estimation and use of prior information in FEM-CSI for biomedical microwave tomography. *IEEE Antennas Wireless Propag. Lett.* **2012**, *11*, 1606–1609. [CrossRef]
19. Thorndike, R.L. Who belongs in the family? *Psychometrika* **1953**, *18*, 267–276. [CrossRef]
20. Kaufman, L.; Rousseeuw, P.J. *Finding Groups in Data: An introduction to Cluster Analysis*; John Wiley & Sons, Inc.: New York, NY, USA, 1990.
21. Tibshirani, R.; Walther, G.; Hastie, T. Estimating the number of clusters in a data set via the gap statistic. *J. Statist. Soc. B.* **2001**, *63*, 411–423. [CrossRef]
22. Kurrant, D.; Abdollahi, N.; Omer, M.; Fear, E.; LoVetri, J. MWSegEval—An image analysis toolbox for microwave breast images. *SoftwareX* **2020**, submitted for review.
23. Jeon, S.I.; Kim, B.R.; Son, S.H. Clinical trial of microwave tomography imaging. In Proceedings of the IEEE URSI Asia-Pacific Radio Science Conference (URSI AP-RASC), Seoul, Korea, 21–25 August 2016; pp. 1–2.
24. Meaney, P.M.; Fanning, M.W.; Li, D.; Poplack, S.P.; Paulsen, K.D. A clinical prototype for active microwave imaging of the breast. *IEEE Trans. Microw. Theory Tech.* **2000**, *48*, 1841–1853.
25. Grzegorczyk, T.M.; Meaney, P.M.; Kaufman, P.A.; Paulsen, K.D. Fast 3-D tomographic microwave imaging for breast cancer detection. *IEEE Trans. Med. Imaging* **2012**, *31*, 1584–1592.
26. Son, S.H.; Simonov, N.; Kim, H.J.; Lee, J.M.; Jeon, S.I. Preclinical prototype development of a microwave tomography system for breast cancer detection. *ETRI J.* **2010**, *32*, 901–910.
27. Pagliari, D.J.; Pulimeno, A.; Vacca, M.; Tobon, J.A.; Vipiana, F.; Casu, M.R.; Solimene, R.; Carloni, L.P. A low-cost, fast, and accurate microwave imaging system for breast cancer detection. In Proceedings of the 2015 IEEE Biomedical Circuits and Systems Conference (BioCAS), Atlanta, GA, USA, 22–24 October 2015; pp. 1–4.
28. Nemez, K.; Baran, A.; Asefi, M.; LoVetri, J. Modeling error and calibration techniques for a faceted metallic chamber for magnetic field microwave imaging. *IEEE Trans. Microw. Theory Techn.* **2017**, *65*, 4347–4356.
29. Asefi, M.; Baran, A.; LoVetri, J. An experimental phantom study for air-based quasi-resonant microwave breast imaging. *IEEE Trans. Microw. Theory Tech.* **2019**, *67*, 3946–3954.
30. Omer, M.; Fear, E.C. Automated 3D method for the construction of flexible and reconfigurable numerical breast models from MRI scans. *Med. Biol. Eng. Comput.* **2018**, *56*, 1027–1040. [CrossRef] [PubMed]
31. Zastrow, E.; Davis, S.K.; Lazebnik, M.; Kelcz, F.; van Veen, B.; Hagness, S.C. Development of anatomically realistic numerical breast phantoms with accurate dielectric properties for modeling microwave interactions with the human breast. *IEEE Trans. Biom. Eng.* **2008**, *55*, 2792–2800. [CrossRef] [PubMed]
32. Zakaria, A.; Gilmore, C.; LoVetri, J. Finite-element contrast source inversion method for microwave imaging. *Inverse Probl.* **2010**, *26*, 115010.
33. Baran, A.; Kurrant, D.J.; Zakaria, A.; Fear, E.C.; LoVetri, J. Breast imaging using microwave tomography with radar-based tissue-regions estimation. *Prog. Electromagn. Res.* **2014**, *149*, 161–171.
34. Rubaek, T.; Meaney, P.M.; Meincke, P.; Paulsen, K.D. Nonlinear microwave imaging for breast cancer screening using Gauss-Newton's method and the CGLS inversion algorithm. *IEEE Trans. Antennas Propag.* **2007**, *55*, 2320–2331.
35. Scapaticci, R.; Catapano, I.; Crocco, L. Wavelet-based adaptive multiresolution inversion for quantitative microwave imaging of breast tissues. *IEEE Trans. Antennas Propag.* **2012**, *60*, 3717–3726. [CrossRef]
36. Palmeri, R.; Bevacqua, M.; Scapaticci, R.; Morabito, A.; Crocco, L.; Isernia, T. Biomedical imaging via wavelet-based regularization and distorted iterated virtual experiments. In Proceedings of the 2017 International Conference on Electromagnetics in Advanced Applications (ICEAA), Verona, Italy, 11–15 September 2017; pp. 1381–1384.
37. Miao, Z.; Kosmas, P. Multiple-frequency DBIM-TwIST algorithm for microwave breast imaging. *IEEE Trans. Antennas Propag.* **2017**, *65*, 2507–2516. [CrossRef]
38. Maffongelli, M.; Poretti, S.; Salvadè, A.; Monleone, R.D.; Meani, F.; Fedeli, A.; Pastorino, M.; Randazzo, A. Preliminary test of a prototype of microwave axial tomograph for medical applications. In Proceedings of the 2015 IEEE International Symposium on Medical Measurements and Applications (MeMeA), Torino, Italy, 7–9 May 2015; pp. 46–51.
39. Kurrant, D.; Baran, A.; LoVetri, J.; Fear, E. Integrating prior information into microwave tomography Part 2: Impact of errors in prior information on microwave tomography image quality. *Med. Phys.* **2017**, *44*, 6482–6503. [CrossRef] [PubMed]
40. Kurrant, D.; Fear, E. Regional estimation of the dielectric properties of inhomogeneous objects using near-field reflection data. *Inverse Probl.* **2012**, *28*, 075001. [CrossRef]
41. Kaye, C.; Jeffrey, I.; LoVetri, J. Novel stopping criteria for optimization-based microwave breast imaging algorithms. *J. Imaging.* **2019**, *55*, 22. [CrossRef]

42. Haralick, R.; Shapiro, L.G. *Computer and Robot Vision*; Addison-Wesley: Boston, MA, USA, 1992; Volume I, pp. 158–205.
43. Van den Boomgard, R.; van Balen, R. Methods for fast morphological image transforms using bitmapped images. *Comput. Vis. Graph. Image Process. Graph. Models Image Process.* **1992**, *54*, 254–258. [CrossRef]
44. Lloyd, S. Least squares quantization in PCM. *IEEE Trans. Inf. Theory* **1982**, *28*, 129–137. [CrossRef]
45. Arthur, D.; Vassilvitskii, S. *K-Means++: The Advantages of Careful Seeding*; Stanford University: Stanford, CA, USA, 2007; pp. 1027–1035.
46. Massey, F.J. The Kolmogorov-Smirnov test for goodness of fit. *J. Am. Stat. Assoc.* **1951**, *46*, 68–78. [CrossRef]
47. Miller, L.H. Table of percentage points of Kolmogorov statistics. *J. Am. Stat. Assoc.* **1956**, *51*, 111–121. [CrossRef]
48. Garey, M.; Johnson, D.; Witsenhausen, H. The complexity of the generalized Lloyd-Max problem (Corresp.). *IEEE Trans. Inf. Theory* **1982**, *28*, 255–256. [CrossRef]
49. Kleinberg, J.; Papadimitriou, C.; Raghavan, P. A microeconomic view of data mining. *Data Min. Knowl. Discov.* **1998**, *2*, 311–324. [CrossRef]
50. Hartigan, J.A.; Wong, M.A. Algorithm AS 136: A k-means clustering algorithm. *J. R. Stat. Soc. Ser. C.* **1979**, *28*, 100–108.
51. Manning, C.D.; Raghavan, P.; Schütze, H. *Introduction to Information Retrieval*; Cambridge University Press: New York, NY, USA, 2008; ISBN 978-0521865715OCLC19078612.
52. Kurrant, D. Supplemental Materials for Evaluating MicrowaveImage Reconstruction Algorithm Performance: Extracting Tissue Types with Segmentation Using Machine Learning. GitHub Repository. 2020. Available online: https://github.com/djkurran/Segmentation-unsupervised-machine-learning (accessed on 1 January 2021).
53. Thada, V.; Jaglan, V. Comparison of Jaccard, Dice, cosine similarity coefficient to find best fitness value for web retrieved documents using genetic algorithm. *IJIET* **2013**, *2*, 202–205.
54. Huttenlocher, D.P.; Klanderman, G.A.; Rucklidge, W.J. Comparing images using the Hausdorff distance. *IEEE Trans. Pattern Anal. Mach. Intell.* **1993**, *15*, 850–863.
55. Dubuisson, M.P.; Jain, A.K. A modified Hausdorff distance for object matching. In Proceedings of the 12th International Conference on Pattern Recognition, Jerusalem, Israel, 9–13 October 1994; pp. 566–568.
56. Shapiro, M.D.; Blaschko, M.B. *On Hausdorff Distance Measures*; UM-CS-2004-071; University of Massachusetts: Amherst, MA, USA, 2004.
57. Kurrant, D.J.; Fear, E.C. Technique to decompose near-field reflection data generated from an object consisting of thin dielectric layers. *IEEE Trans. Antennas Propag.* **2012**, *60*, 3684–3692. [CrossRef]
58. Kurrant, D.; Fear, E. Defining regions of interest for microwave imaging using near-field reflection data. *IEEE Trans. Microw. Theory Techn.* **2013**, *61*, 2137–2145. [CrossRef]
59. Kurrant, D.; Bourqui, J.; Fear, E. Surface estimation for microwave imaging. *Sensors* **2017**, *17*, 1658. [CrossRef]
60. Omer, M.; Mojabi, P.; Kurrant, D.; LoVetri, J.; Fear, E. Proof of-concept of the incorporation of ultrasound-derived structural information into microwave radar imaging. *IEEE J. Multiscale Multiphys. Comput. Tech.* **2018**, *3*, 129–139.
61. Loeian, M.S.; Mehdi Aghaei, S.; Farhadi, F.; Rai, V.; Yang, H.W.; Johnson, M.D.; Aqil, F.; Mandadi, M.; Rai, S.N.; Panchapakesan, B. Liquid biopsy using the nanotube-CTC-chip: Capture of invasive CTCs with high purity using preferential adherence in breast cancer patients. *Lab Chip.* **2019**, *19*, 1899–1915. [CrossRef]
62. Nagrath, S.; Sequist, L.V.; Maheswaran, S.; Bell, D.W.; Irimia, D.; Ulkus, L.; Smith, M.R.; Kwak, E.L.; Digumarthy, S.; Muzikansky, A.; et al. Isolation of rare circulating tumour cells in cancer patients by microchip technology. *Nature* **2007**, *450*, 1235–1239.
63. Khosravi, F.; Trainor, P.J.; Christopher, L.; Kloecker, G.; Wickstrom, E.; Rai, S.N.; Panchapakesan, B. Static micro-array isolation, dynamic time series classification, capture and enumeration of spiked breast cancer cells in blood: The nanotube–CTC chip. *Nanotechnology* **2016**, *27*, 44LT03. [CrossRef] [PubMed]
64. Mojabi, P.; Abdollahi, N.; Omer, M.; Kurrant, D.; Jeffrey, I.; Fear, E.; LoVetri, J. Tissue-type imaging for ultrasound-prior microwave inversion. In Proceedings of the 2018 18th International Symposium on Antenna Technology and Applied Electromagnetics (IEEE ANTEM), Waterloo, ON, Canada, 19–22 August 2018; pp. 1–3.
65. Omer, M.; Fear, E. Anthropomorphic breast model repository for research and development of microwave breast imaging technologies. *Sci. Data* **2018**, *5*, 180257. [CrossRef] [PubMed]

Journal of
*Imaging*

MDPI

*Article*

# An Iterative Algorithm for Semisupervised Classification of Hotspots on Bone Scintigraphies of Patients with Prostate Cancer

**Laura Providência [1,2], Inês Domingues [2,*] and João Santos [2,3]**

[1] Faculdade de Ciências, Universidade do Porto, 4169-007 Porto, Portugal; lauraprovid@hotmail.com
[2] Medical Physics, Radiobiology and Radiation Protection Group, IPO Porto Research Centre (CI-IPOP), 4200-072 Porto, Portugal; joao.santos@ipoporto.min-saude.pt
[3] Instituto de Ciência Biomédicas Abel Salazar, Rua de Jorge Viterbo Ferreira n° 228, 4050-313 Porto, Portugal
* Correspondence: inesdomingues@gmail.com

**Abstract:** Prostate cancer (PCa) is the second most diagnosed cancer in men. Patients with PCa often develop metastases, with more than 80% of this metastases occurring in bone. The most common imaging technique used for screening, diagnosis and follow-up of disease evolution is bone scintigraphy, due to its high sensitivity and widespread availability at nuclear medicine facilities. To date, the assessment of bone scans relies solely on the interpretation of an expert physician who visually assesses the scan. Besides this being a time consuming task, it is also subjective, as there is no absolute criteria neither to identify bone metastases neither to quantify them by a straightforward and universally accepted procedure. In this paper, a new algorithm for the false positives reduction of automatically detected hotspots in bone scintigraphy images is proposed. The motivation relies in the difficulty of building a fully annotated database. In this way, our algorithm is a semisupervised method that works in an iterative way. The ultimate goal is to provide the physician with a fast, precise and reliable tool to quantify bone scans and evaluate disease progression and response to treatment. The algorithm is tested in a set of bone scans manually labeled according to the patient's medical record. The achieved classification sensitivity, specificity and false negative rate were 63%, 58% and 37%, respectively. Comparison with other state-of-the-art classification algorithms shows superiority of the proposed method.

**Keywords:** bone scintigraphy; prostate cancer; machine learning; semisupervised classification; false positives reduction



**Citation:** Providência, L.; Domingues, I.; Santos, J. An Iterative Algorithm for Semisupervised Classification of Hotspots on Bone Scintigraphies of Patients with Prostate Cancer. *J. Imaging* **2021**, *7*, 148. https://doi.org/10.3390/jimaging7080148

Academic Editors: Leonardo Rundo, Carmelo Militello, Vincenzo Conti, Fulvio Zaccagna and Changhee Han

Received: 31 July 2021
Accepted: 13 August 2021
Published: 17 August 2021



## 1. Introduction

According to the World Health Organization, prostate cancer (PCa) is the second most commonly diagnosed cancer in men, accounting for more than 1.4 million new cases and more than 375,000 deaths worldwide in 2020. Patients with advanced prostate cancer often develop metastases, which are caused by primary tumor cells that escape from the prostate gland and spread through the lymphatic system or the bloodstream to other areas of the body. The most frequent site for metastatic growth of prostate cancer is the bone, and almost all patients with advanced prostate cancer show histological skeletal involvement, being estimated that 84% to 90% of patients with metastatic disease had bone metastases [1–3]. Even though the bone metastases are seldom the cause of death, they are the leading cause of morbidity and a major challenge in the management of patients, leading to a diminished quality of life. The presence of bone metastases, specially in higher extents, is an indicator of progression of the disease and typically correlates with a poor prognosis [4,5]. Currently there is no cure for metastatic prostate cancer, but it can often still be treated to slow down its growth. A precise detection and up-take quantification of bone metastases is essential to provide the physicians the accurate staging they require to choose the appropriate treatment for an individual patient, to monitor the evolution of the disease and to evaluate the treatment efficiency.

The most common diagnostic procedure used for screening, assessment of treatment and follow-up of patients with bone metastases is whole-body bone scintigraphy (BS) [6], due to its relatively high sensivity, ranging from 70% to 78% [7–9], and widespread availability at relatively low cost. Bone scintigraphy, also known as bone scan, is a nuclear medicine imaging technique used in screening for several skeleton related pathological conditions, including bone metastases. In a bone scintigraphy, a bone-seek radioisotope, that is, a substance that collects in the bones following the normal physiological processes, is injected intravenously into the patient. The radioactive isotope will flow through the body and will have a tendency to accumulate in areas of high bone metabolic activity. Following the radiopharmaceutical administration, a time period of 2 to 4 h [10] is observed to allow biodistribution and up-take and then a simultaneous image of the anterior (AP) and posterior (PA) views is acquired in a gamma-camera. Because the radioisotope has accumulated in the regions of bone, the scans will reveal brighter areas, which indicate an increased rate of bone metabolic activity such as abnormal growth caused by metastases. These areas are referred to as hotspots, and may indicate not only the presence of bone metastases, but also other conditions such as trauma, microarthritis, benign degeneration, or bone infections [11]. The biggest disadvantage in the use of bone scintigraphy to detect bone metastases is, therefore, its low specificity. Because it evaluates the distribution of active bone formation in the skeleton and identifies the sites where metabolic reactions are occurring, it detects several suspicious uptakes of nonmetastatic origin, which lead to high a false positive rate of BS to detect bone metastases. To date, the assessment of bone scans relies solely on the interpretation of an expert physician who visually assesses the scan. Besides this being a time consuming task, it is also extremely subjective, as there is no absolute and clear criteria neither to differentiate bone metastases from benign bone lesions, neither to quantity them. This means that, up to this date, the disease stage as well as the response to treatment is subjected to a certain degree of uncertainty, implying that the process of determining whether or not the patient condition is regressing is sometimes subjective. Given the high occurrence of metastatic PCa, there should be by now a more practical and, most importantly, more objective criteria to evaluate quantitatively a bone scintigraphy.

This work aims to create an algorithm capable of classifying hotspots from bone scintigraphy images, and is manly motivated by the call for a method whose development does not require a fully labeled database. A labeled data set of hotspots is rare and most likely unavailable for most researchers, and therefore one propose a semisupervised method that only requires knowledge about the type of bone scan the hotspot is extracted from. Comparison with other state-of-the-art classification algorithms shows superiority of the proposed method, achieving a sensitivity of 0.63, a specificity of 0.58 and a false negative rate of 0.37. This algorithm was able to decrease the false positive rate from 0.73 after detection to almost half, 0.42 after the false positive attenuation.

The main contributions of the present work include:

- The proposal of a new, iterative, semisupervised algorithm for attenuation of false positive metastases;
- Extensive experiments on a real dataset of scintigraphies from 102 patients with prostate cancer;
- A suggestion for a hotspots detection technique;
- Comparison with nonsupervised and one-class classifiers.

The remaining of this paper is organised as follows: Section 2 reviews the state of the art; Section 3 gives a detailed description of the here proposed semisupervised iterative algorithm; Section 4 gives the materials and methods, including the database, the hotspots detection technique, the extracted features, the competing classification algorithms for false positives reduction and the evaluation methodology; Section 5 presents the results, and Section 6 presents a discussion of these results. The document finishes with some conclusions and directions for future work in Section 7.

## 2. Related Work

The literature found on this topic shows there has been some effort to develop a computer-aided diagnosis system capable of automatically detecting and quantifying bone metastases in bone scintigraphies.

Brown et al. [12] developed a computer-aided system to automatically segment and quantify bone scan lesions. The bone lesion segmentation was accomplished by doing an atlas-based anatomic segmentation to divide the body into 6 different regions, followed by the application of region specific threshold to detect the hotspots. The method achieved a median sensitivity of 94.1%, specificity of 89.2% and accuracy of 89.4%. After the detection of the hotspots, the resulting images were reviewed by a nuclear medicine physician who removed false positive lesions; the hotspots classified as malign could then be used to assess the severity of the disease and disease response to treatment. Despite the good results, this algorithm is not fully automatic, as it requires the intervention of a physician to remove false positives (nonmetastases related bone uptakes) from the scans. This is a huge downside as the automatic differentiation between malignant and nonmalignant bone uptakes is an essential requirement in a bone metastases evaluation algorithm, as it is a task that is not trivial even for the most experienced physician and thus brings a lot of subjectivity to the final assessment. A classification algorithm capable of automatically distinguish metastases from benign lesion is thus needed.

Sadik et al. [13,14] developed a fully automated classification system for the detection of metastases that used artificial neural networks. Both works intended to classify the whole-body bone scan as a whole, regarding the presence or absence of bone metastases, and not the hotspots individually. The final classifier would return a value between 0 and 1, that reflected the probability of the patients having metastases. The algorithm proposed in [13] achieved sensitivity of 90% and a specificity of 74%, while the one proposed in [14] achieved a higher specificity of 89%, keeping the same sensitivity of 90%.

Papandrianos and his team [15–17] have published three papers describing the work they have made on this field, devoted to the development of Convolutional Neural Networks (CNN) models for automatic classification of whole-body scans from patients with bone metastases. Just like Sadik et al., the authors intended to classify the body scans as a whole, and not the hotspots individually. In [15,16] they were dealing with a two-class classification problem regarding the presence (malignant scan) or absence (healthy scan) of bone metastases in patients with breast and prostate cancer, respectively. The best CNN architectures in [15,16] achieved an accuracy of 92.0/97.4%, a sensitivity of 94.0/96.5% and a specificity of 92.0/96.8%. The major problem with these models is that in the clinical practice the division of the bone scans into healthy or malign is oversimplified, as it ignores the fact that some patients suffer from benign conditions which will reveal several suspicious uptakes of nonmetastatic origin in the final images. As they aimed to cope with a two-class classification problem, all scans from patients containing degenerative lesions and other nonmalignant bone uptakes were removed in a manual preselection process. This is a major drawback, as a fully automatic algorithm to assess whole body scintigraphy should also be able to classify false positive bone uptakes as benign lesions. In [17] the same authors investigated a way to partial solve this problem, by developing a similar CNN based algorithm to classify bone scintigraphy images as healthy, malignant or degenerative, leading to a three-class classification problem. The best CNN architecture achieved a sensitivity of 92.7% and a specificity of 96.0%. Although the automatic distinction between malignant and nonmalignant images is an improvement over the previous models, it does not offer a solution for the cases in which one patient has bone uptakes with both malignant and nonmalignant origins, which is one of the major problems in visual bone scintigraphy assessment. In fact, neither of the papers proposed by Papandrianos et al. or Sadik et al present an algorithm that is capable of quantifying the bone lesions individually, which is essential when an objective assessment of the disease staging is needed. It is not enough to build an algorithm that is able to distinguish images that present solely malignant lesions from those that present solely benign lesions. A suitable algorithm must be able to quantify and classify each lesion individually.

The only algorithms developed to classify bone lesions individually are the ones found in the Master theses of Dang [18] and Belcher [19]. In both works, a CNN was developed to classify hotspots in bone scintigraphy images for prostate cancer, by determining whether they had a high or low risk of being bone metastases from PCa metastatic cancer. The final CNN from [18] had an accuracy, true positive rate and AUC (Area Under the ROC Curve) of 89.0%, 98.0% and 0.96, respectively. To measure the CNN performance, [19] only used the area of the ROC curve, for which was obtained a score of 0.974. Despite appearing to be a promising approach to the classification of hotspots in bone scintigraphy images, the previously described works use supervised techniques, which rely on an extensive number of labeled data. The access to such a large data set was only possible due to EXINI Diagnostic AB, which is a Sweden based company that uses artificial intelligence to develop automated analysis platforms for medical images like cardiac, brain and bone scans [20]. It has shown to be quite popular among researchers working in the quantification of bone metastases. EXINI has developed the aBSI (automated Bone Scan Index), a software only medical device that provides a fully quantitative assessment of a patient's skeletal disease on a bone scan, as the fraction of the total skeleton weight [21]. As it is a closed-source software, little is known about its operating principles, except that it was trained to classify hotspots as lesions using a collection of more than 40,000 hotspots derived from bone scans of patients with a variety of metastatic cancers. It is able to segment the skeleton, identify hotspots, quantify their intensity and classify them as lesions [22].), which provided them with a database composed by more than ten thousand labeled hotspots from bone scans. Such large scale annotated data sets are, however, rare in the medical context. Training a CNN from scratch to perform bone lesion classification would require thousands of labeled images, a task that would not only be extremely complex and time consuming, but also dependent on the availability of experienced physicians. Furthermore, the labelling would be subject to the subjectivity inherent in the classification of lesions detected in bone scintigraphy.

The algorithms developed so far for the assessment of whole-body bone scans either use fully supervised learning algorithms, which require access to a (big) labeled data set, or rely on some sort of manual removal of false positives. Here, we propose a semisupervised method for the classification of automatically detected hotspots in bone scintigraphy images.

## 3. hotBSI: Semisupervised Iterative Algorithm for Hotspots Classification

The core and main contribution of the present paper is the hotBSI (hotspots on Bone Scintigraphy Images) algorithm. This algorithm was derived from the need of hotspots false positive reduction scintigraphy images, in the presence of not completely labeled database. Section 3.1 explains the workings of hotBSI, wihch can be used with any classifier of choice. The classifiers used in the present work are listed in Section 3.2.

### 3.1. hotBSI Description

An initial classifier $C_0$ was first trained, in the presence of noise, to distinguish between malign from nonmalign hotspots. It should be pointed out that this classifier is trained under a lot of noise, as it was assumed that every detection in a bone scan belonging the *malign* category belonged to malign class, which is not true, as the majority of the detections in these scans are actually nonmalign. The next stage involves an iterative process through the following steps:

1. The last trained classifier, $C_{i-1}$, is used to classify the detections on the scans belonging to the malign class. For each detected region, the classifier returns the likelihood that the region comes from the *malign* or *nonmalign* class;
2. For each patient in the malign category:
    (a) The detection with the highest likelihood of being malignant is selected;
    (b) All other detections with likelihood of being malignant *higher than a predetermined threshold* (if any) are also selected.
3. A new training data set is created, so that detections made on nonmalign scans are considered as false positives (and labeled as 0) and the above selected regions are considered as true-positives (or malign hotspots, labeled as 1);
4. Train a new classifier $C_i$ with the new training data set.

The algorithm runs during a predetermined number of iterations (set as 100 in the current experiments). Other stopping criteria will be pursued in the future. A schematic description of hotBSI is given in Algorithm 1 and Figure 1. The value of the threshold was set to 0.8.

**Algorithm 1** hotBSI algorithm

Inputs:
NM - feature set from all the hotspots extracted from the nonmalign images
M - feature set from all the hotspots extracted from the malign images
T - threshold (default as 0.8)
NrIt - number of iterations (default as 100)

Output:
C - a classifier to classify new hotspots as nonmalign or malign

Train an initial classifier, $C_0$, with the input features (NM $\cup$ M)
**for** i = 1:NrIt **do**
    Empty M
    **for** each patient in the *malign* set **do**
        Use $C_{i-1}$ to predict the probabilities of the detections to be a metastases ($P_{\text{met}}$)
        Identify the hotspot with the highest likelihood of being a metastasis ($P_{\text{max}}$)
        **for** d = 1 : number of detected hotspots for the current patient **do**
            **if** $P_{\text{met}}(d) == P_{\text{max}} \mid\mid P_{\text{met}}(d) > T$ **then**
                Add the hotspot to M
    Create a new training set, NM $\cup$ M
    Train a new classifier $C_i$ with the new training data set
**return** $C_{NrIt}$

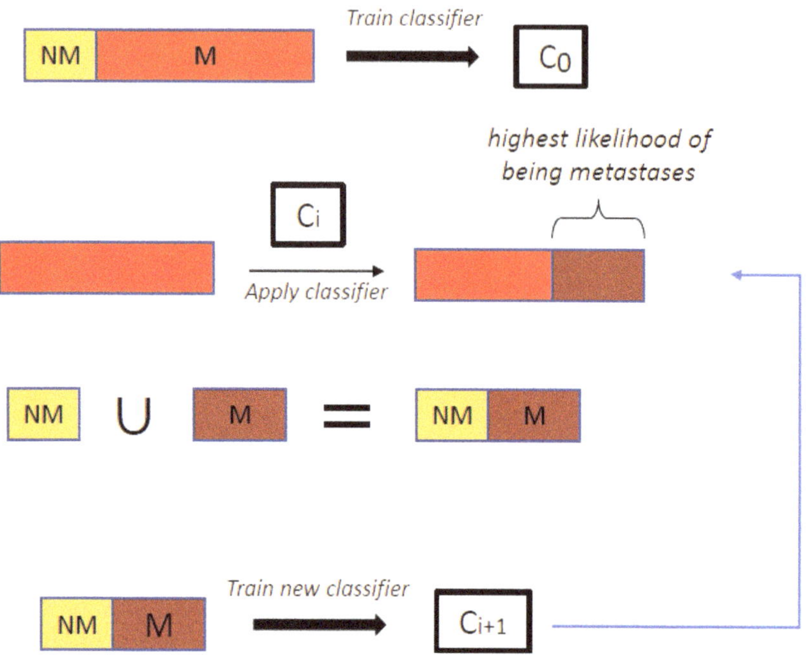


**Figure 1.** hotBSI algorithm. NM stands for detections labeled as nonmalignant, while M stands for detections labeled for training in a given iteration as malignant.

### 3.2. Learning Algorithms

The classifiers were trained using four different supervised learning algorithms: (i) a support vector machine (SVM) trained with a linear kernel with scale 1, where the values obtained with the linear SVM score function (bias = 1.08) were transformed into posterior probabilities using the sigmoid function with slope −1.40 and intercept 0.06; (ii) *k*-nearest neighbors (KNN), trained with five nearest neighbors with uniform weighting and the Euclidean distance function as the distance metric; (iii) decision trees (DTs), trained with a minimum of 10 samples per branch node, a maximum number of splits equal to the number of samples minus one and the Gini's diversity index as the split criterion and (iv) linear discriminant analysis (LDA) with 'Delta' (linear coefficient threshold) and 'Gamma' (amount of regularization) both equal to 0.

## 4. Materials and Methods

This section encompasses several details related with the implementation and evaluation. The database is described in Section 4.1. The methodology is given in Section 4.2, including the method for the detection of hotspots, the list of extracted features, and the state of the art classification techniques used for comparison with the proposed iterative method. Lastly, the evaluation methodology is presented in Section 4.3.

### 4.1. Database

The database consists of 195 bone scintigraphy images from 102 patients with prostate cancer with suspected bone metastatic disease. The equipment used for scanning patients was either a *Millennium MG* (GE Medical Systems), which digitally record anterior and posterior scans with a resolution of 1024 × 256 pixels, or a *BrightView* (Philips Healthcare), which digitally records anterior and posterior scans with a resolution of 1024 × 512 pixels. The pixel depth (maximum number of counts which could be stored in a pixel) is 16-bits for every image. For each bone scan, a medical report describing the condition of the patient in question written by a nuclear medicine physician is available. All data was provided by Instituto Português de Oncologia do Porto Francisco Gentil (IPO Porto). The data was collected and held anonymously and the developed algorithms did not contain information concerning the patients, but rather information extracted from the data during the algorithm development. This project was authorized by IPO-Porto Healthcare Ethics Committee.

The scans were organized into three categories: (i) *healthy*, if no suspicious bone uptake was detected, (ii) *benign*, if bone hotspots with no metastatic origin are present or (iii) *malign*, if bone metastases exist. Table 1 summarizes the available database, including the number of bone scans per category. It is important to point out that images from the malign category can also present benign hotspots.

**Table 1.** Database summary. The database consists of a total of 195 bone scans divided into one of three categories: healthy, if no suspicious bone uptakes were detected, benign if bone hotspots with benign origins are present, or malign, if the images have bone metastases.

| Bone Scan Type | No of Bone Scans |
|---|---|
| Healthy | 37 |
| Benign | 72 |
| Malign | 86 |
| **Total** | **195** |

### 4.2. Methodology

The methodology proposed in this paper for the automatic false positives reduction of hotspots in bone scintigraphy images involves a three step process (Figure 2): detection of the hotspots, extraction of features from the detected hotspots and training an algorithm for the classification of the detected regions.

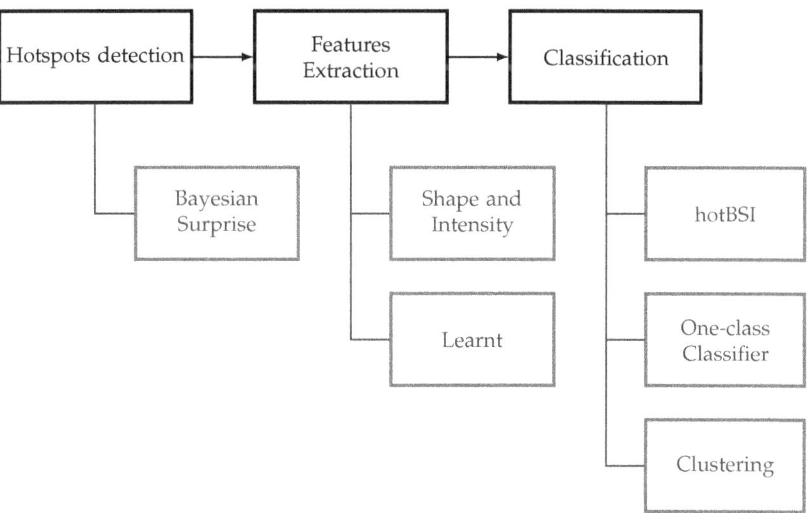


**Figure 2.** Methodology overview.

#### 4.2.1. Hotspots Detection

Although a customised hotspots detection algorithm was developed, we note that the here proposed algorithm, hotBSI, is independent of the detection algorithm and can be used with any detector of choice.

The present detector is based on the approach proposed in [23], where a technique based on Bayesian surprise is used to detect calcifications in mammogram images. The algorithm takes advantage of the fact that the hotspots are bright regions (that is, regions with higher grey levels) surrounded by pixels with lower grey values. The first step of the algorithm consists in applying a mask to the original image to exclude the background and keep solely the body of the patient. The mask was obtained by binarizing the original grayscale image by thresholding using the Otsu's method [24] Then, the hotspots were detected through the following steps (Figure 3):

1. Consider a square patch of the masked image with half-radius $r_{in}$;
2. Consider the region surrounding the patch described in 1, defined by a radius $r_{out} = \sqrt{2} \cdot r_{in}$ and with centre coinciding with that of the inner patch;
3. Calculate the mean grey level of both the inner patch and the surrounding region;
4. Compare the mean grey levels: if the absolute difference of the two values is higher than a certain threshold $\delta$, the inner patch is considered a hotspot.

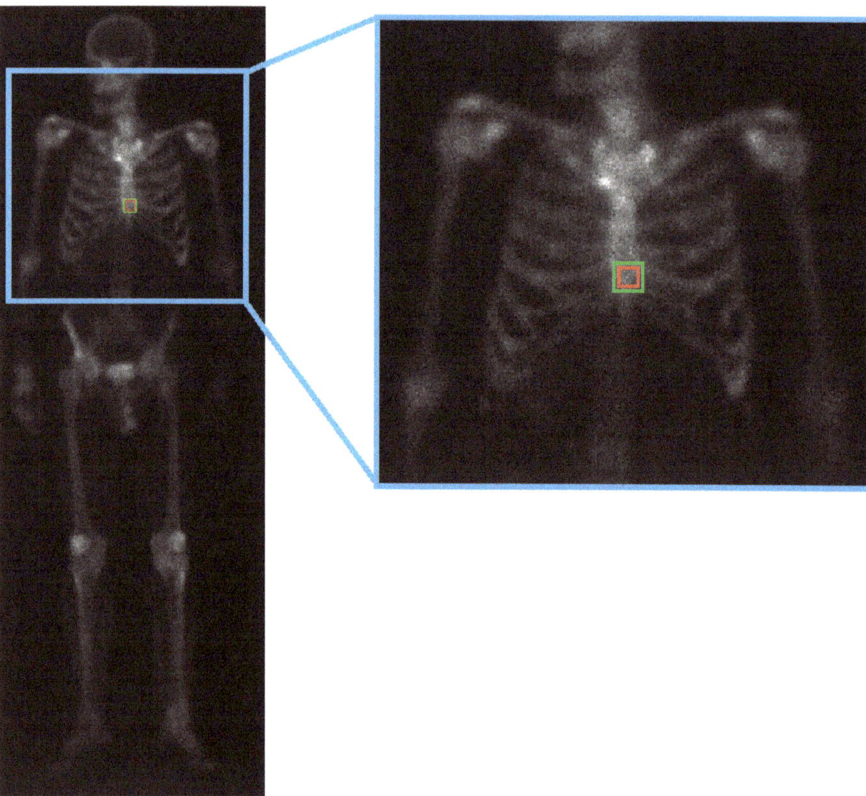

**Figure 3.** Detection illustration. Note that for illustration purposes, only one region is being tested in the current image. In the full detection algorithm, all of the regions within the mask are evaluated.

The steps were repeated for every patch in the masked image with the following empirically obtained values: $r_{in} = 5$ cm and $\delta = 20$. The final threshold $\delta$ was chosen to obtain as few false positives as possible, while at the same time not losing any malignant hotspot. In this way, a considerable amount of hotspots not related to bone metastases are detected with this algorithm. These hotspots can be due to some kind of benign bone condition or can be due to normal and healthy physiological processes. Since the patient condition is determined through the assessment of the malign bone lesions, the number of false positive detections should be reduced. This was achieved through the development of classification algorithms, which require the extraction of features from the detected regions.

#### 4.2.2. Feature Extraction

The detection algorithm is followed by a feature extraction stage which obtains the features from the hotspots that will serve as input to a classification algorithm. Two types of features were extracted: handcrafted low-level features and learnt high-level features.

Sixteen (16) shape and four (4) intensity handcrafted features were first extracted from each automatically detected region. The list of the handcrafted features can be found in Table A1 in the Appendix A.

High-level features were extracted using the convolutional base of a pretrained CNN. Since the used CNN requires input images of size $n \times n \times 3$, each automatically detected patch was converted into RGB by replicating the grey image in each channel. The detections were also resized so that their size matched the one required by the input layer of the network in question, $224 \times 224 \times 3$. Next, a pretrained ResNet18 network was used to extract features from the regions (we refer to [25] for a review of deep learning). The "pool5" layer was used as the output layer to extract a 512-dimensional vector for each possible hotspot (see Figure A1 of Appendix B).

#### 4.2.3. Methods Used for Comparison

Two state-of-the-art methods were used as a comparison with the hotBSI algorithm here proposed. Given the lack of a fully annotated database, which precluded the use of supervised learning methods, an unsupervised and a semisupervised learning algorithms were used.

For the unsupervised method, a clustering technique with the *k*-means clustering algorithm was used. A *k*-means clustering algorithm with two clusters was initially applied to the training set, and a model for the classification of new data was built by assuming that each final cluster represented a class and by assigning each hotspot from the test set to the nearest cluster centroid. By choosing two clusters, it was expected that the data could be partitioned into a cluster of nonmalign data and a cluster of malign data (metastases). The distance metric used for defining the initial clusters, as well as to assign new data to these clusters, was the square Euclidean distance.

The semisupervised method was a one-class classification (OCC) algorithm. The hotspots extracted from the nonmalign set (false positives) were used to train an one-class support vector machine algorithm (OC-SVM), and a model which classified new hotspots as nonmalign or as outliers (here considered to be metastases—true positives) was obtained. The OC-SVM algorithm used was the one proposed by [26] and was trained with an outlier fraction of 5%, a Gaussian kernel function with a Kernel scale parameter of 1.81 and a Sequential Minimal Optimization (SMO) as an optimization routine.

### *4.3. Evaluation Methodology*

In the present work, detections automatically made in scans from the bone scan category *Healthy* and *Benign* were considered as false positives, whereas detections extracted made in scans with the bone scan category *Malign* were considered as true positives.

A test set was created with the detections extracted from scans of 30 patients randomly chosen from the *Healthy*, *Benign* and *Malign* bone scan categories (10 patients per category). This test data set (and only this test data set) was manually labeled, identifying the true detections (malign) and the false positive ones (nonmalign). The number of patients and detections per class for the training and the test set are presented in Table 2. We do acknowledge the imbalanced nature of the data set and intend to experiment on ways to deal with this issue in the future [27,28].

**Table 2.** Split of the dataset.

| Bone Scan Category | No. of Patients | | No. of Detections | |
|---|---|---|---|---|
| | Training | Test | Training | Test |
| Nonmalign | 89 | 20 | 1941 | 393 |
| Malign | 76 | 10 | 5620 | 918 |
| **Total** | **65** | **30** | **7561** | **1311** |

The algorithms were evaluated using common performance metrics such as sensitivity, specificity, accuracy, precision, false positive rate (FPR), F1-score and AUC (area under the ROC curve). In addiction, the false negative rate (FNR) is also calculated, as it was considered that a low FNR was of special importance for this particular classifier.

Since the goal of this algorithm is to be used in the clinical practice to aid physicians in the diagnose and follow-up of patients with metastatic cancer, it is important that the final algorithm has a FNR as low as possible. A high FNR would mean that the algorithm was classifying a lot of malign hotspots as nonmalign, which could be dangerous to the patient, as it was failing to diagnose them with the disease and preventing them from having access to an early treatment.

## 5. Results

In this section, the results are reported. The performance of the detection algorithm is firstly shown (Section 5.1), followed by the analysis of the efficiency of the different classification algorithms to remove false positive detections (Section 5.2). For each classification model, the results obtained when using both handcrafted and high-level features are presented.

### *5.1. Detection Results*

The algorithm described in Section 4.2.1 successfully detected all the hotspots corresponding to metastases (see Table 3). Figure 4 illustrates the detection algorithm in bone scintigraphy images from the nonmalign set, while Figure 5 illustrates the detection algorithm in bone scintigraphy images from the malign set. Comparing the results with the respective patient's medical reports, it can be concluded that the algorithm successfully detected all the hotspots corresponding to metastases. On the other hand, this algorithm presents a high rate of false positive detections: approximately 73% of the detected hotspots were not metastases. Observing the figures, it can be seen that most of the detected hotspots are healthy or benign (that is, nonmalign), while only a small percentage of the detected hotspots are actually metastases.

**Table 3.** Results of the detection phase.

| | |
|---|---|
| TP | 1.00 |
| FN | 0.00 |
| FP | 0.73 |
| Sensitivity | 0.00 |
| FNR | 0.00 |
| Precision | 0.58 |
| F1 | 0.73 |
| FPPI | 32 |

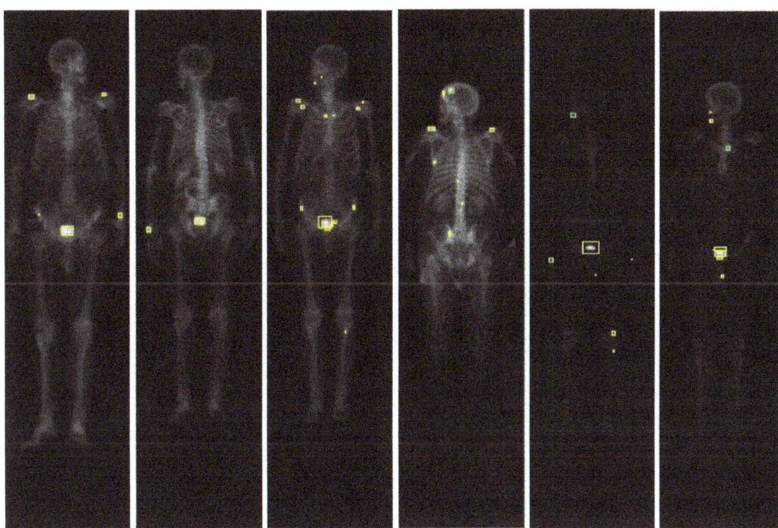

**Figure 4.** Results of the detection algorithm in bone scintigraphy images from the nonmalign set. The colours of the bounding boxes were manually chosen for the purposes of illustration, according to the respective medical report of the patient: red represents metastases, green represents benign bone lesions and yellow represents false positives (hotspots that are neither malign nor benign lesions).

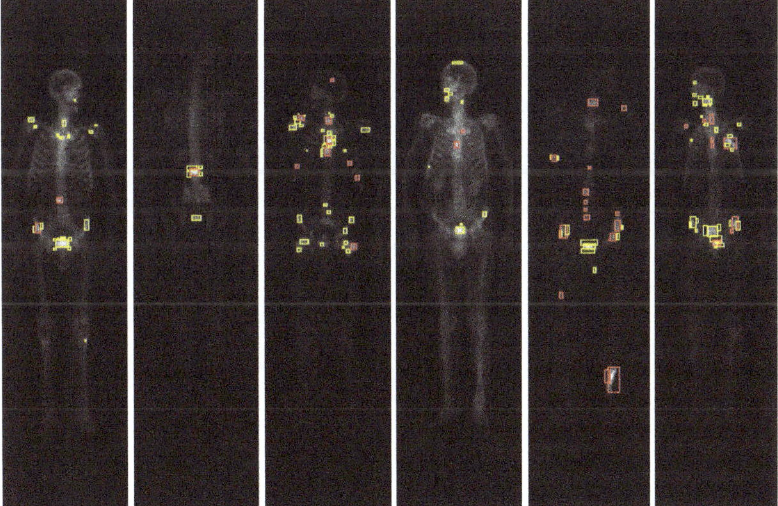

**Figure 5.** Results of the detection algorithm in bone scintigraphy images from the malign set. The colours of the bounding boxes were manually chosen for the purposes of illustration, according to the respective medical report of the patient: red represents metastases, green represents benign bone lesions and yellow represents false positives (hotspots that are neither malign nor benign lesions).

### *5.2. False Positive Attenuation Results*

The proposed algorithm, hotBSI, was used to classify the hotspots from the test set. Tables 4 and 5 gather the performance results for the hotBSI trained with SVM/KNN and DTs/DLA, respectively. Results obtained with the *k*-means and one-class classification algorithms are shown in Table 6. In all tables, results for both handcrafted (HC) and ResNet18 (RN18) features are presented.

The confusion matrices obtained with all the algorithms can be found in Figures A2–A7 of Appendix C.

**Table 4.** Results with the hotBSI trained with support vector machine and $k$-nearest neighbors.

| Classifier | SVM | | KNN | |
|---|---|---|---|---|
| | HC | RN18 | HC | RN18 |
| Sensitivity | 0.13 | 0.63 | 0.85 | 0.67 |
| Specificity | 0.83 | 0.58 | 0.17 | 0.51 |
| Accuracy | 0.65 | 0.59 | 0.35 | 0.55 |
| FNR | 0.86 | 0.37 | 0.15 | 0.32 |
| FPR | 0.18 | 0.42 | 0.83 | 0.49 |
| Precision | 0.23 | 0.35 | 0.27 | 0.34 |
| F1 | 0.17 | 0.46 | 0.41 | 0.45 |
| AUC | 0.50 | 0.66 | 0.52 | 0.62 |

**Table 5.** Results with the hotBSI trained with decision trees and linear discriminant analysis.

| Classifier | DTs | | LDA | |
|---|---|---|---|---|
| | HC | RN18 | HC | RN18 |
| Sensitivity | 0.92 | 0.80 | 0.83 | 0.70 |
| Specificity | 0.14 | 0.33 | 0.19 | 0.43 |
| Accuracy | 0.35 | 0.46 | 0.36 | 0.51 |
| FNR | 0.08 | 0.20 | 0.17 | 0.30 |
| FPR | 0.86 | 0.66 | 0.81 | 0.56 |
| Precision | 0.28 | 0.31 | 0.28 | 0.31 |
| F1 | 0.43 | 0.44 | 0.41 | 0.43 |
| AUC | 0.46 | 0.57 | 0.44 | 0.59 |

**Table 6.** Results with OCC and Kmeans.

| Classifier | $k$-Means | | OCC | |
|---|---|---|---|---|
| | HC | RN18 | HC | RN18 |
| Sensitivity | 0.17 | 0.08 | 0.08 | 0.26 |
| Specificity | 0.86 | 0.92 | 0.90 | 0.72 |
| Accuracy | 0.67 | 0.70 | 0.68 | 0.60 |
| FNR | 0.83 | 0.92 | 0.92 | 0.74 |
| FPR | 0.14 | 0.92 | 0.10 | 0.28 |
| Precision | 0.30 | 0.28 | 0.23 | 0.26 |
| F1 | 0.22 | 0.13 | 0.12 | 0.14 |
| AUC | – | – | 0.51 | 0.50 |

## 6. Discussion

This work had as main goal the development of an algorithm capable of automatically identifying metastases in bone scintigraphy images from patients with prostate cancer. If successful, this algorithm could be used in the clinical practice to quantify bone scans and work as an aiding tool for the diagnosis and follow-up of patients with bone metastases. Despite consensus on the need for such an algorithm, and despite efforts of the scientific community to develop one, such a diagnosis tool is currently unavailable in the medical community.

The current work differs from the ones developed so far in the same topic in the sense that it does not resort to a fully supervised data set to train the classifier. An algorithm that proves to be successful even without the access to a labeled data set can be extremely useful in the clinical context, where access to a labeled database is often difficult to achieve.

Here, an algorithm for the automatic detection of hotspots in bone scans was suggested, followed by the development of an algorithm capable of classifying the detected hotspots as malign or nonmalign. The detection algorithm proved to be successful on a database of patients with prostate cancer, as all malign hotspots were correctly identified. This was guaranteed by choosing a threshold value that would ensure that no metastases candidates were left undetected. This came at the cost of a high false positive detection rate, meaning that most of the hotspots detected by the algorithm were nonmalign. As the patient condition is determined through the assessment of the malign bone lesions, an algorithm for the attenuation of the false positive was developed. The evaluation metrics considered the most relevant for the current classifier and the respective values obtained for the proposed algorithm are now discussed.

### 6.1. Area under the ROC Curve (AUC)

The AUC values, usually close or equal to 0.50, translate the low to none capacity of most classifiers to distinguish between nonmalign and malign hotspots. The highest AUC score was obtained with the hotBSI trained with SVM and ResNet18 features (AUC = 0.66).

### 6.2. Sensitivity and Specificity

High values of sensitivity and specificity were only obtained when the classifier was biased toward one class: high sensitivity scores (>0.85) were always accompanied by a low specificity score, which meant that it was considering almost every hotspot to belong to the positive (malign) class; on the other hand, high specificity scores (>0.85) were always accompanied by a low sensitivity score, meaning that it was assigning the majority of hotspots to the negative (nonmalign) class. Neither situation is desirable for the final algorithm. The classifiers with more balanced scores in terms of sensitivity and specificity were (i) the hotBSI trained with SVM and ResNet18 features (sensitivity = 0.63, specificity = 0.58) and (ii) the hotBSI trained with KNN and ResNet18 features (sensitivity = 0.67, specificity = 0.51).

### 6.3. False Negative Rate (FNR)

An important evaluation metric for an algorithm whose goal is to classify hotspots in patients who might have bone metastases is the false negative rate. It is desirable that this value is as low as possible, as a low FNR would mean that the classifier was incorrectly labelling a lot of malign hotspots (metastases) as nonmalign; this would result in an algorithm that would label patients with metastatic cancer as healthy, which would be dangerous is the clinical context. Very low FNR only happened with classifiers that were assigning almost every hotspot to the malign class: taking a look at the hotBSI trained with decision trees it can be observed that a FNR rate of 0.08 was obtained. Although at first glance this may seem like an almost perfect result, further analysis on the remaining metrics lead us to conclude that this FNR only happens because the classifier is assigning almost every hotspot to the malign class and, therefore, it had a low probability of missing metastases (sensitivity = 0.92, specificity = 0.14). Such a classifier is obviously not acceptable, as it has no discriminatory power. Classifiers that obtained lower FNR while keeping more acceptable values for the other metrics include (i) the hotBSI trained with discriminant analysis and ResNet18 features (FNR = 0.30), (ii) the hotBSI trained with KNN and ResNet18 features (FNR = 0.33) and (iii) the hotBSI trained with SVM and ResNet18 features (FNR = 0.37).

### 6.4. False Positive Rate Reduction

As mentioned in Section 5.1, the detection algorithm presented a false positive rate of 73.07%. After applying the classifiers to these detections, the lowest FPR scores were obtained with (i) the hotBSI trained with SVM and handcrafted features (FPR = 0.18), (ii) the OCC trained with handcrafted and ResNet18 features (FPR = 0.10 and FPR = 0.28, respectively) and (iii) *k*-means with handcrafted features (FPR = 0.14). This low values are, however, only due the fact that these algorithms were classifying most of the metastases as nonmalign, which is not desirable, as it will lead to a high FNR. The classifier that presented the lowest FPR while keeping an acceptable value for the FNR was the hotBSI trained with SVM and ResNet18 features (FPR = 0.42). This represents a decrease of 30.59% compared to the FPR score obtained with initial detection algorithm, when no classifiers had been yet applied.

### 6.5. Comparison with the State-of-the-Art Algorithms

Table 7 gathers the best results obtained with the hotBSI algorithm, as well as the best results obtained with the *k*-means and one-class classifier. The best hotBSI algorithm was considered to be the one trained with SVM and ResNet18 features; the best *k*-means and one-class algorithms were considered to be the ones trained with handcrafted and ResNet18 features, respectively. The proposed algorithm shows superiority in almost every metric, in particular in the AUC (0.66 compared to 0.50 from the OCC classifier), sensitivity (0.63 compared with 0.17 and 0.26 from the *k*-means and OCC classifiers, respectively) and the false negative rate (0.37 compared with 0.83 and 0.74 from the *k*-means and OCC classifiers, respectively). It should be noted that the only two metrics in which the state-of-the-art algorithms performed better were accuracy and specificity. This is clearly explained by noting that this happens since these algorithms are classifying most of the hotspots as nonmalign (note the low sensitivity from the same classifiers); as a consequence, they will present a high specificity, as if most of the hotspots are being classified as nonmalign there is a better chance that the algorithm will correctly classify nonmalign hotspots as nonmalign. Besides the low specificity, this comes with a cost of a high false negative rate, as a lot of malign hotspots are being incorrectly classified as nonmalign. The better scores in accuracy are also easily explained by looking at the percentage of nonmalign and malign hotspots present in the test set: 73% of these hotspots were from the nonmalign category, while only 27% were from the malign category. Because the *k*-means and OCC classifiers are manly assigning hotspots to the negative (nonmalign) class, and because most of the test set is composed by hotspots from this class, they will get a high accuracy score, even if most of the data is wrongly classified. Having all of this into account, it can be concluded that the proposed algorithm performs better than the state-of-the-art algorithms at the task of hotspots classification and, therefore, at the task of false positive attenuation.

**Table 7.** Comparison of the best hotBSI with the best state-of-the-art algorithms.

| | **hotBSI (RN18)** | ***k*-Means (HC)** | **OCC (RN18)** |
|---|---|---|---|
| Sensitivity | 0.63 | 0.17 | 0.26 |
| Specificity | 0.58 | 0.86 | 0.72 |
| Accuracy | 0.59 | 0.67 | 0.60 |
| FNR | 0.37 | 0.83 | 0.74 |
| FPR | 0.42 | 0.14 | 0.28 |
| Precision | 0.35 | 0.30 | 0.26 |
| F1 | 0.46 | 0.22 | 0.14 |
| AUC | 0.66 | – | 0.50 |

## 7. Conclusions

An algorithm for the classification of automatically detected hotspots in bone scintigraphy images of patients with prostate cancer was proposed. Such an algorithm can be used in combination with computer-assisted PCa detection approaches such as the one described in [29], making it extremely useful in the medical community, as it provides the physicians with an aiding tool to quantify whole-body bone scans from patients with bone metastases.

The biggest challenge when building such an algorithm is the lack of a labeled data set. Here, we tried to overcome that problem by developing an algorithm that only requires knowledge about the type of bone scan from which the hotspot is extracted from. Comparison with state-of-the-art algorithms shows superiority of the proposed method. However, analysis of the performance metrics obtained for the hotBSI shows that this algorithm is still not ready to be used in the clinical practice: the not so high scores for sensitivity, specificity and AUC are still a concern; the false negative rate, despite clearly inferior to the state-of-the-art algorithms, is also still high. Improvements on the algorithm are therefore need. These include:

- Finding features that are more discriminative, for instance, by using a different pre-trained network, by extracting features from different layers or by extracting features from autoencoders;
- Using other classifiers to train the hotBSI;
- Apply variations in the hotBSI, for example, by choosing a stopping criteria in the iteration that is not the number of iterations;
- Retrain the algorithm with a more balanced data set.

Once an algorithm with a performance that is considered good enough to be used in the clinical practice is obtained, a quantitative image biomarker can be used to automatically quantify a bone scintigraphy of new patients with prostate cancer. Literature shows that the most adequate image biomarker for quantifying a bone scan is the Bone Scan Index (BSI) [22,30–33].

The final goal is to build a software that can be used in the clinical context, that is capable of not only quantifying a given bone scintigraphy of a patient with prostate cancer, but also give information about disease progression, response to treatment and disease prognosis. Such a software will make the process of assessing a bone scan more objective, simpler and faster, and will for sure be an asset in the medical community.

**Author Contributions:** Conceptualization, J.S.; methodology, I.D.; validation, I.D.; formal analysis, L.P.; investigation, L.P. and I.D.; resources, I.D.; data curation, L.P. and I.D.; writing—original draft preparation, L.P.; writing—review and editing, I.D. and J.S.; supervision, I.D. and J.S.; project administration, J.S.; funding acquisition, J.S. All authors have read and agreed to the published version of the manuscript.

**Funding:** This work is partially financed by National Funds through the Portuguese funding agency, FCT—Fundação para a Ciência Tecnologia within project UIDP/00776/2020.

**Institutional Review Board Statement:** The study was conducted according to the guidelines of the Declaration of Helsinki. The study was approved by IPO Porto Healthcare Ethics Committee. Protocol code CES.274/020, date of approval 1 October 2020.

**Informed Consent Statement:** Informed consent was obtained from all subjects involved in the study.

**Conflicts of Interest:** The authors declare no conflict of interest. The funders had no role in the design of the study; in the collection, analyses, or interpretation of data; in the writing of the manuscript, or in the decision to publish the results.

## Appendix A. Handcrafted Features

The full list of handcrafted features is given in Table A1.

**Table A1.** Name and description of the handcrafted features. Top part of the table corresponds to shape measurements, while the bottom half are pixel value measurements.

| Property | Description |
|---|---|
| Area | No. of pixels in the region |
| AxisLengthRatio | Ratio between *MajoraxisLength* and *MinoraxisLength* |
| BoundingBox | Position and size of the smallest box containing the region |
| Centroid | Center of mass of the region |
| ConvexArea | Number of pixels in *ConvexImage* [1] |
| Eccentricity | Eccentricity of the ellipse $\epsilon$ [2] |
| EquivDiameter | Diameter of a circle with the same area as the region |
| EulerNumber | No. of objects in the region minus the no. of holes in those objects |
| Extent | Ratio of pixels in the region to pixels in the total bounding box |
| FilledArea | Number of on pixels in *FilledImage* [3] |
| InvCircularity | Inverse of the roundness [4] of the object |
| MajoraxisLength | Length (in pixels) of the major axis of $\hat{\epsilon}$ [5] |
| MinoraxisLength | Length (in pixels) of the minor axis of $\hat{\epsilon}$ |
| Orientation | Angle between the $x$-axis and the major axis of $\hat{\epsilon}$ |
| Perimeter | Distance around the boundary of the region |
| Solidity | Proportion of the pixels in the convex hull that are also in the region |
| MaxIntensity | Value of the pixel with the greatest intensity in the region |
| MeanIntensity | Mean of all the intensity values in the region |
| MinIntensity | Value of the pixel with the lowest intensity in the region |
| WeightedCentroid | Center of the region based on location and intensity value |

[1] *ConvexImage*: Image that specifies the *ConvexHull* [6], with all pixels within the hull filled in (binary image). [2] $\epsilon$: ellipse that has the same second-moments as the region. [3] *FilledImage* Image the same size as the bounding box of the region, returned as a binary. [4] The roundness of an object is defined as $\frac{4 \cdot \text{Area} \cdot \pi}{\text{Perimeter}^2}$. [5] $\hat{\epsilon}$: ellipse that has the same normalized second central moments as the region. [6] *ConvexHull*: Smallest convex polygon that can contain the region.

## Appendix B. Learnt Features

A schematic representation of the architecture of the ResNet18 network is given in Figure A1.

**Figure A1.** Diagram of ResNet18 with highlighted "pool5" layer.

## Appendix C. Confusion Matrixes

The confusion matrices obtained during the false positive attenuation phase (Section 5.2) are given in Figures A2–A7.

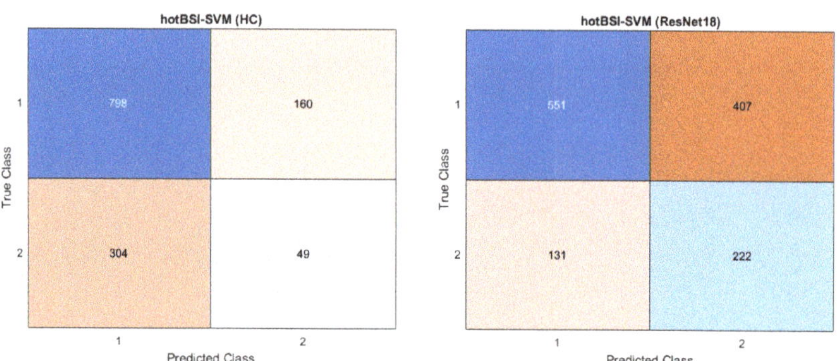


**Figure A2.** Confusion matrices for the hotBSI-SVM trained with handcrafted (**left**) and ResNet18 (**right**) features.

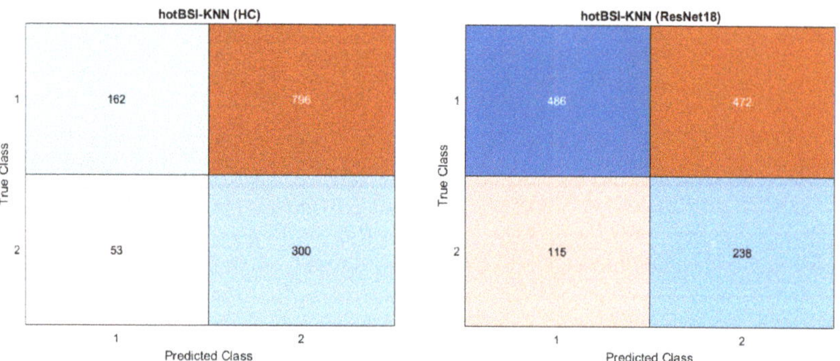


**Figure A3.** Confusion matrices for the hotBSI-KNN trained with handcrafted (**left**) and ResNet18 (**right**) features.

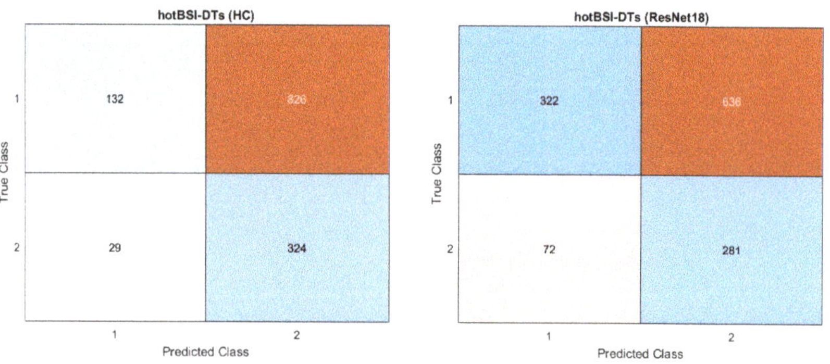


**Figure A4.** Confusion matrices for the hotBSI-DTs trained with handcrafted (**left**) and ResNet18 (**right**) features.

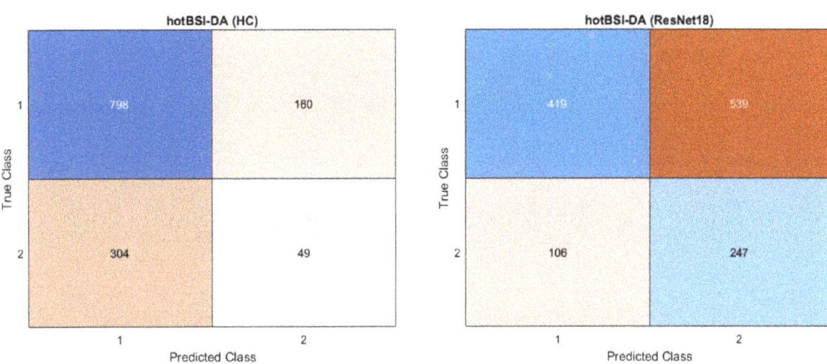


**Figure A5.** Confusion matrices for the hotBSI-DA trained with handcrafted (**left**) and ResNet18 (**right**) features.

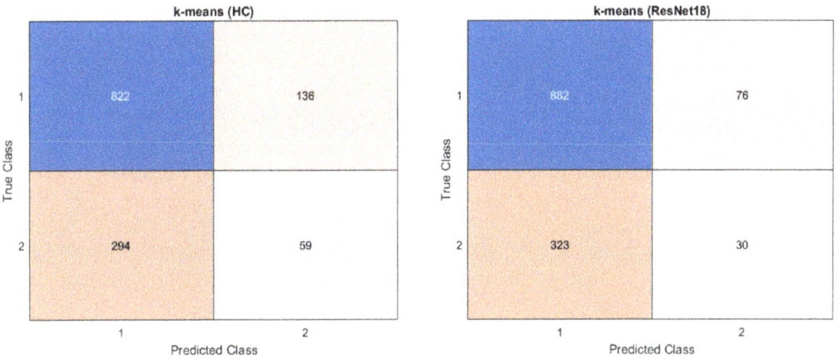


**Figure A6.** Confusion matrices for the k-means algorithm trained with handcrafted (**left**) and ResNet18 (**right**) features.

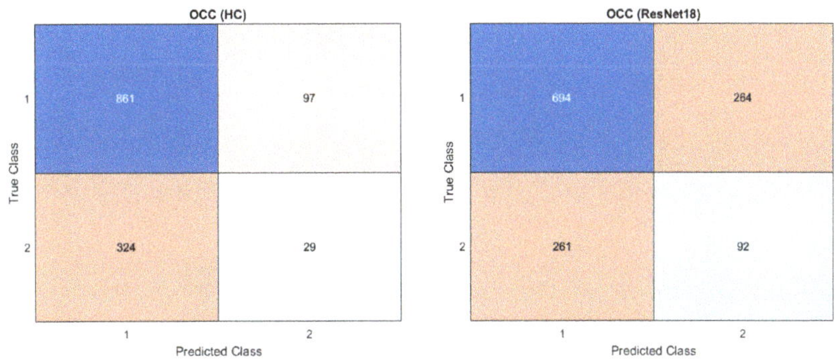


**Figure A7.** Confusion matrices for the OCC algorithm trained with handcrafted (**left**) and ResNet18 (**right**) features.

## References


1. Msaouel, P.; Pissimissis, N.; Halapas, A.; Koutsilieris, M. Mechanisms of bone metastasis in prostate cancer: Clinical implications *Best Pract. Res. Clin. Endocrinol. Metabol.* **2008**, *22*, 341–355. [CrossRef]
2. Bubendorf, L.; Schöpfer, A.; Wagner, U.; Sauter, G.; Moch, H.; Willi, N.; Gasser, T.C.; Mihatsch, M.J. Metastatic patterns of prostate cancer: An autopsy study of 1589 patients. *Hum. Pathol.* **2000**, *31*, 578–583. [CrossRef] [PubMed]
3. Gandaglia, G.; Abdollah, F.; Schiffmann, J.; Trudeau, V.; Shariat, S.; Kim, S.; Perrotte, P.; Montorsi, F.; Briganti, A.; Trinh, Q.D.; et al. Distribution of Metastatic Sites in Patients With Prostate Cancer: A Population-Based Analysis. *Prostate* **2014**, *74*. [CrossRef]
4. Norgaard, M.; Jensen, A.O.; Jacobsen, J.B.; Cetin, K.; Fryzek, J.P.; Sørensen, H.T. Skeletal Related Events, Bone Metastasis and Survival of Prostate Cancer: A Population Based Cohort Study in Denmark (1999 to 2007). *J. Urol.* **2010**, *184*, 162–167. [CrossRef] [PubMed]
5. Soloway, M.S.; Hardeman, S.W.; Hickey, D.; Todd, B.; Soloway, S.; Raymond, J.; Moinuddin, M. Stratification of patients with metastatic prostate cancer based on extent of disease on initial bone scan. *Cancer* **1988**, *61*, 195–202. [CrossRef]
6. Brenner, A.I.; Koshy, J.; Morey, J.; Lin, C.; DiPoce, J. The Bone Scan. *Semin. Nuclear Med.* **2012**, *42*, 11–26. [CrossRef] [PubMed]
7. Ohta, M.; Tokuda, Y.; Suzuki, Y.; Kubota, M.; Makuuchi, H.; Tajima, T.; Nasu, S.; Yasuda, S.; Shohtsu, A. Whole body PET for the evaluation of bony metastases in patients with breast cancer: Comparison with 99Tcm-MDP bone scintigraphy. *Nuclear Med. Commun.* **2001**, *22*, 875–879. [CrossRef]
8. Even-Sapir, E.; Metser, U.; Mishani, E.; Lievshitz, G.; Lerman, H.; Leibovitch, I. The detection of bone metastases in patients with high-risk prostate cancer: 99mTc-MDP Planar bone scintigraphy, single- and multi-field-of-view SPECT, 18F-fluoride PET, and 18F-fluoride PET/CT. *J. Nuclear Med.* **2006**, *47*, 287–297.
9. O'Sullivan, G.J. Imaging of bone metastasis: An update. *World J. Radiol.* **2015**, *7*, 202. [CrossRef]
10. Mettler, F.A.; Guiberteau, M.J. *Essentials of Nuclear Medicine and Molecular Imaging*; Elsevier: Amsterdam, The Netherlands, 2019.
11. Purden, J. Nuclear medicine 2: Principles and technique of bone scintigraphy. *Nursing Times* **2019**, *115*, 48–49.
12. Brown, M.; Chu, G.; Kim, G.H.; Allen-Auerbach, M.; Poon, C.; Bridges, J.; Vidovic, A.; Ramakrishna, B.; Ho, J.; Morris, M.; et al. Computer-aided quantitative bone scan assessment of prostate cancer treatment response. *Nuclear Med. Commun.* **2012**, *33*, 384–394. [CrossRef]
13. Sadik, M.; Jakobsson, D.; Olofsson, F.; Ohlsson, M.; Suurkula, M.; Edenbrandt, L. A new computer-based decision-support system for the interpretation of bone scans. *Nuclear Med. Commun.* **2006**, *27*, 417–423. [CrossRef] [PubMed]
14. Sadik, M.; Hamadeh, I.; Nordblom, P.; Suurkula, M.; Höglund, P.; Ohlsson, M.; Edenbrandt, L. Computer-Assisted Interpretation of Planar Whole-Body Bone Scans. *J. Nuclear Med.* **2008**, *49*, 1958–1965. [CrossRef] [PubMed]
15. Papandrianos, N.; Papageorgiou, E.; Anagnostis, A.; Feleki, A. A Deep-Learning Approach for Diagnosis of Metastatic Breast Cancer in Bones from Whole-Body Scans. *Appl. Sci.* **2020**, *10*, 997. [CrossRef]
16. Papandrianos, N.; Papageorgiou, E.; Anagnostis, A.; Papageorgiou, K. Bone metastasis classification using whole body images from prostate cancer patients based on convolutional neural networks application. *PLoS ONE* **2020**, *15*, e0237213. [CrossRef] [PubMed]
17. Papandrianos, N.; Papageorgiou, E.; Anagnostis, A.; Papageorgiou, K. Efficient Bone Metastasis Diagnosis in Bone Scintigraphy Using a Fast Convolutional Neural Network Architecture. *Diagnostics* **2020**, *10*, 532. [CrossRef]
18. Dang, J. Classification in Bone Scintigraphy Images Using Convolutional Neural Networks. Master's Thesis, Lund University, Lund, Sweden, 2016.
19. Belcher, L. Convolutional Neural Networks for Classification of Prostate Cancer Metastases Using Bone Scan Images. Master's Thesis, Lund University, Lund, Sweden, 2017; Student Paper.
20. EXINI Diagnostics AB. 2020. Available online: https://exini.com/ (accessed on 11 December 2020).
21. ABSI. 510(k) Premarket Submission to U.S. Food & Drug Administration. 2019. Available online: https://www.accessdata.fda.gov/cdrh_docs/pdf19/K191262.pdf (accessed on 12 December 2020).
22. Ulmert, D.; Kaboteh, R.; Fox, J.J.; Savage, C.; Evans, M.J.; Lilja, H.; Abrahamsson, P.A.; Björk, T.; Gerdtsson, A.; Bjartell, A.; et al. A Novel Automated Platform for Quantifying the Extent of Skeletal Tumour Involvement in Prostate Cancer Patients Using the Bone Scan Index. *Eur. Urol.* **2012**, *62*, 78–84. [CrossRef]
23. Domingues, I.; Cardoso, J.S. Using Bayesian surprise to detect calcifications in mammogram images. In Proceedings of the 36th Annual International Conference of the IEEE Engineering in Medicine and Biology Society, Chicago, IL, USA, 24–31 August 2014; pp. 1091–1094. [CrossRef]
24. Otsu, N. A Threshold Selection Method from Gray-Level Histograms. *IEEE Trans. Syst. Man Cybernet.* **1979**, *9*, 62–66. [CrossRef]
25. Alzubaidi, L.; Zhang, J.; Humaidi, A.J.; Al-Dujaili, A.; Duan, Y.; Al-Shamma, O.; Santamaría, J.; Fadhel, M.A.; Al-Amidie, M.; Farhan, L. Review of deep learning: concepts, CNN architectures, challenges, applications, future directions. *J. Big Data* **2021**, *8*, 53. [CrossRef]
26. Schölkopf, B.; Williamson, R.; Smola, A.; Shawe-Taylor, J.; Platt, J. Support vector method for novelty detection. In *Advances in Neural Information Processing Systems 12. Max-Planck-Gesellschaft*; MIT Press: Cambridge, MA, USA, 2000; pp. 582–588.
27. Domingues, I.; Amorim, J.P.; Abreu, P.H.; Duarte, H.; Santos, J. Evaluation of Oversampling Data Balancing Techniques in the Context of Ordinal Classification. In Proceedings of the International Joint Conference on Neural Networks (IJCNN), IEEE, Rio de Janeiro, Brazil, 8–13 July 2018; pp. 1–8. [CrossRef]

28. Marques, F.; Duarte, H.; Santos, J.; Domingues, I.; Amorim, J.P.; Abreu, P.H. An iterative oversampling approach for ordinal classification. In Proceedings of the 34th ACM/SIGAPP Symposium on Applied Computing, Limassol Cyprus, 8–12 April 2019; Volume Part F1477, pp. 771–774. [CrossRef]
29. Lapa, P.; Gonçalves, I.; Rundo, L.; Castelli, M. Semantic learning machine improves the CNN-Based detection of prostate cancer in non-contrast-enhanced MRI. In Proceedings of the Genetic and Evolutionary Computation Conference Companion, New York, NY, USA, 12–16 July 2019; pp. 1837–1845. [CrossRef]
30. Kaboteh, R.; Damber, J.E.; Gjertsson, P.; Höglund, P.; Lomsky, M.; Ohlsson, M.; Edenbrandt, L. Bone Scan Index: A prognostic imaging biomarker for high-risk prostate cancer patients receiving primary hormonal therapy. *EJNMMI Res.* **2013**, *3*, 9. [CrossRef]
31. Poulsen, M.H.; Rasmussen, J.; Edenbrandt, L.; Høilund-Carlsen, P.F.; Gerke, O.; Johansen, A.; Lund, L. Bone Scan Index predicts outcome in patients with metastatic hormone-sensitive prostate cancer. *BJU Int.* **2016**, *117*, 748–753. [CrossRef] [PubMed]
32. Li, D.; Lv, H.; Hao, X.; Dong, Y.; Dai, H.; Song, Y. Prognostic value of bone scan index as an imaging biomarker in metastatic prostate cancer: A meta-analysis. *Oncotarget* **2017**, *8*, 84449–84458. [CrossRef] [PubMed]
33. Mustansar, N. Utility of Bone Scan Quantitative Parameters for the Evaluation of Prostate Cancer Patients. *J. Nuclear Med. Radiat. Ther.* **2018**, *9*. [CrossRef]

Journal of
*Imaging*

*Article*

# 3D Non-Local Neural Network: A Non-Invasive Biomarker for Immunotherapy Treatment Outcome Prediction. Case-Study: Metastatic Urothelial Carcinoma

**Francesco Rundo [1,*], Giuseppe Luigi Banna [2], Luca Prezzavento [3], Francesca Trenta [4], Sabrina Conoci [5] and Sebastiano Battiato [4]**

1 STMicroelectronics—ADG Central R&D Division, 95125 Catania, Italy
2 Medical Oncology Department, United Lincolnshire NHS Hospital Trust, Lincoln LN2, Lincolnshire, UK; giuseppe.banna@nhs.net
3 DIEEI, University of Catania, 95125 Catania, Italy; lucabprezzavento@gmail.com
4 IPLAB, University of Catania, 95125 Catania, Italy; francesca.trenta@unict.it (F.T.); battiato@dmi.unict.it (S.B.)
5 Department of Chemical, Biological, Pharmaceutical and Environmental Sciences, University of Messina, 98100 Messina, Italy; sconoci@unime.it
* Correspondence: francesco.rundo@st.com

Received: 27 October 2020; Accepted: 1 December 2020; Published: 3 December 2020

**Abstract:** Immunotherapy is regarded as one of the most significant breakthroughs in cancer treatment. Unfortunately, only a small percentage of patients respond properly to the treatment. Moreover, to date, there are no efficient bio-markers able to early discriminate the patients eligible for this treatment. In order to help overcome these limitations, an innovative non-invasive deep pipeline, integrating Computed Tomography (CT) imaging, is investigated for the prediction of a response to immunotherapy treatment. We report preliminary results collected as part of a case study in which we validated the implemented method on a clinical dataset of patients affected by Metastatic Urothelial Carcinoma. The proposed pipeline aims to discriminate patients with high chances of response from those with disease progression. Specifically, the authors propose ad-hoc 3D Deep Networks integrating Self-Attention mechanisms in order to estimate the immunotherapy treatment response from CT-scan images and such hemato-chemical data of the patients. The performance evaluation (average accuracy close to 92%) confirms the effectiveness of the proposed approach as an immunotherapy treatment response biomarker.

**Keywords:** 3D-CNN; immunotherapy; radiomics; self-attention

## 1. Introduction

Urothelial carcinoma (also known as bladder cancer) is the most common histological type of urinary tract carcinoma. It accounts for about 3% of all cancers [1]. It is more common between the ages of 60 and 70, which is three times more frequent in men than in women and it is associated with about 165,000 global deaths every year and a five-year survival of approximately 5% in the advanced stage [1]. Metastatic Urothelial Carcinoma (mUC) occurs at disease onset in approximately 10% of patients, mostly arising from the evolution of previous superficial or infiltrating forms [2]. The current standard first-line treatment of mUC is platinum-based chemotherapy. In terms of progression-free survival (PFS), an improvement has been recently reported with the combination of chemotherapy and immunotherapy as a first-line treatment in mUC [3]. However, a significant and mature overall survival (OS) data are still expected. The median OS with cisplatin-based regimens varies between 12 and 15 months [3], while it is approximately 9 months in patients not eligible for cisplatin treatment because of the severity of the side effects treated with carboplatin-based regimens [4]. Immunotherapy has become the standard second-line treatment of mUC based on two phase III studies with Immune Checkpoint Inhibitors (ICIs) immunotherapeutic drugs such as Atezolizumab and Pembrolizumab with a median OS reported of 8.6 [5] and 10.3 months [6], respectively. These results have recently been confirmed in a large population with Atezolizumab [7]. ICIs are also a treatment option in first-line therapy of patients with non-cisplatin-based mUC based on phase Ib-II studies with a median OS between 8.7 and 18.2 months [8–14]. However, no more than about 20% and 30% of patients have a disease response with ICI in post-platinum and first-line treatment, respectively, even though these responses tend to be more durable than those obtained with chemotherapy [12–14]. Therefore, it is a priority to identify and select those patients who can really benefit from immunotherapy, even though, at the moment, there are still no reliable and clinically available biomarkers to properly choose patients who respond or progress with ICIs [15–18]. Since the activity of a high tumor mutational burden and of infiltrating lymphocytes has been associated with a higher probability of response to immunotherapy [8,19], some researchers have tried to characterize the tumor environment by integrating data from instrumental imaging and testing a reliable correlation with patients' outcome, paving the way to the emerging field of radiomics [20]. Radiomics aims to extract a large number of quantitative features from high-throughput medical images by taking advantage of the recent data-characterization learning-based algorithms. Some studies show that these methods have the capability to uncover disease characteristics that, otherwise, cannot be identified by human observers [20]. In this paper, we propose an innovative deep learning pipeline used for the prediction of response to ICIs' immunotherapeutic treatment for patients with advanced metastatic bladder cancer (mUC) who have progressed following a first-line platinum-based chemotherapy. The architecture of the proposed deep model is based on 3D Densely connected Convolutional Neural Network (3D-DCNN) with separable convolutions and self-attention mechanisms through non-local blocks [21]. The model processes computed tomography (CT-scan) imaging data and discriminates patients with high chances of response (complete, partial response, or, at least, stable disease), from those that, instead, are likely to show disease progression. 3D-DCNNs have been widely used in medical imaging for segmentation applications [22] as well as for cancer lesion characterization [23]. We leverage the success of these models and extend them with a self-attention mechanism, based on non-local blocks, for better learning long-range dependencies among the input data (segmented CT scans cancer lesions). As mentioned, it is, thereby, verified that all the patients have undergone a histological exam that confirmed the presence of bladder cancer. Experimental results, carried out on a dataset consisting of 41 patients with bladder cancer (which include, as a whole, 106 cancer lesions to be analyzed), show that the devised self-attention-based model leads to a better characterization of the bladder cancer (i.e., the associated feature maps) and of the radiological visual features for predicting treatment outcome with respect to the state-of-the-art methods. The paper is organized as follows. "Related Works" reviews state-of-the-art pipelines with a focus on deep learning models in the medical imaging field. The section "Methods and Materials" provides an accurate description of the proposed pipeline, together with details about the adopted training and validation dataset procedures. Experimental results of the proposed method as well as a comparison to state-of-the-art models are given in the "Results and Discussion" section. Finally, in the "Conclusions" section, the implications of the proposed approach are discussed, and some ideas for future extensions are briefly outlined.

## 2. Related Works

The feasibility of predicting the response to immunotherapy treatment for patients with neoplastic diseases in the metastatic phase has been recently investigated using standard machine learning and deep learning methods. Traditional machine learning methods based on the analysis of high-dimensional clinical data and CT-based diagnostic imaging have been proposed in order to predict the outcome of bladder cancer treatments [24–26]. Specifically, Reference [24] reports a comparative analysis of different machine learning methods used to process high-dimensional clinical data with the aim to predict mortality after a radical cystectomy in a large dataset of bladder cancer patients. Among the analyzed methods, the Regularized Extreme Learning Machine yields the highest average accuracy. In Reference [25], the authors analyzed the performance of multiple machine learning methods applied to process CT urography imaging data of each subject belonging to the recruited dataset of 84 patients, namely, Linear Discriminant Analysis (LDA), Neural Networks (NNs), Support Vector Machine (SVM), and a Random Forest (RAF) Classifier with SVM slightly outperforming the others. Analogously, Reference [26] compares multiple traditional learning methods (e.g., SVM, Bagged SVM, K-nearest neighbor, AdaBoost, Random Forest, and Gradient Boosted Trees) to automatically determine disease status and prognosis of patients with bladder cancer while also suggesting a recommendation treatment, i.e., neoadjuvant, definitive, and adjuvant therapy. All methods showed a fair accuracy (in terms of sensitivity and specificity), demonstrating the suitability of machine learning in addressing oncological applications. Despite the above machine learning methods showing promising performance in supporting physicians in their clinical investigation, they still rely on an old-fashioned approach, i.e., feature engineering followed by learning methods. With the rediscovery of deep learning in 2011 and the availability of large computation resources (thanks to GPU), the learning paradigm has radically changed in many tasks from computer vision to speech analysis to medical image analysis. About mUC investigation, the Deep Learning breakthrough has solicited the medical image research community to investigate advanced methods mainly for bladder cancer segmentation [27–30], while little effort has been put forward for studying bladder tumor chemotherapy efficacy [31,32]. For instance, in Reference [31], the authors investigate AlexNet [32] (with few architectural variants in terms of kernel size, padding, and stride) for assessing chemotherapy treatment efficacy of bladder cancer, starting from manually-segmented Computed Tomography (CT) scans. Experimental results showed an average accuracy of 86% (Sensitivity 90%, Specificity 89%) on immunotherapy treatment estimation. In Reference [33], the authors propose a two-stage deep-learning analysis pipeline for bladder cancer. The first stage deals with automated CT scan lesion segmentation (Auto-Initialized Cascaded Level Sets system) and the other one handles treatment response prediction (with an AUC of 0.73) using the previously segmented lesions. As for immunotherapy, some of the authors of this paper in Reference [34] proposed a deep network based on auto-encoders for cancer treatment outcome prediction in patients treated with Pembrolizumab (anti PD-1/PD-L1 ICIs checkpoint). To our knowledge, Reference [34] is one of the first works employing artificial intelligence for the prediction of immunotherapy outcomes. The work herein described extends the pipeline proposed in Reference [34] by introducing a 3D deep model enriched with a self-attention mechanism, which improve the learning phase of the joint visual and clinical data features. While it is important to have accurate automated computer methods for cancer treatment efficacy prediction, the interpretability and the explanation for why these methods reach specific decisions are of equal importance for the involved physicians. For these reasons, an ideal model should be accurate as well as explain what features it employs for its decisions. This need has been extensively outlined in both the artificial intelligence and medical imaging domains [35–39]. Our work contributes to the research area in automated immunotherapy outcome prediction through visual investigation of CT scans, as follows.

- We present a generalizable deep model that combines 3D densely connected convolutional layers empowered with self-attention mechanisms for estimating automatically the efficacy of bladder cancer immunotherapy treatment, purely based on CT imaging analysis.
- We investigate, through interpretability methods, such as Grad CAM [40], what are the radiological CT visual features that most likely act as biomarkers for immunotherapy treatment outcome, thus, providing a potentially invaluable support to medical staff in evaluating the progress of bladder cancer. To the best of our knowledge, to date, no method has tackled the task herein proposed, from both the automated treatment outcome prediction and interpretability perspectives.

## 3. Materials and Methods

In this section, we propose our deep learning-based approach for radiomics applied to CT-scan cancer imaging for estimating the outcome of the ICIs based immunotherapy treatment in patients affected by mUC. The proposed framework consists of a combination of 3D Densely Connected Convolutional layers (3D-DCNN) with non-local blocks [21], implementing a self-attention strategy to improve characterization of spatio-temporal dependencies of the neoplastic lesions in CT slices. The use of 3D deep convolutional layers is motivated by the results achieved by our previous work [34] demonstrating the correlation between the biological aggressiveness of bladder tumors with the dynamic morphological evolution of the interested CT lesions. Thus, a 3D deep network provides a very useful tool to characterize not only the static-spatial 2D morphology of the lesions, stratified in the chest-abdomen CT slices, but also the functional dependence between their 3D volumes and the immunotherapy treatment responsiveness [34]. However, as pointed out in bladder treatment cancer guidelines [41], not all the metastatic lesions (generated by the primary neoplasm) play a role in the analysis of the progression of oncological disease. For this reason, a set of guidelines known as Response Evaluation Criteria in Solid Tumors (RECIST) have been developed by the scientific community [34,41]. In this work, a detailed analysis of only the lesions compliant to the RECIST criteria is to be carried out [41]. Thus, to address these indications, we extend 3D convolution architectures with an implicit "attention" to make the models focus—through visual feature learning—only on the most significant parts of the RECIST lesion and their possible correlation. More specifically, the attention mechanism is implemented by concatenating non-local blocks [21] at different layers for capturing long-range dependencies at different scales. We report the problem of treatment outcome estimation as a binary classification task, i.e., the proposed model provides as output of two class probabilities including one in the case of complete/partial regression or stable disease (C1), and the other one for disease progression (C2). The flowchart of the whole approach is shown in Figure 1. The backbone of the proposed model is a sequence of dense blocks, similar to DenseNet Cha et al., 2016, but replacing 2D convolutions with 3D ones. The model processes as input a batch of $16 \times 64 \times 64$ volume (16 slices, among the 64 available, each with a spatial resolution of $64 \times 64$) extracted from CT data and containing the RECIST 1.1 compliant lesion. This input data is first fed to a 3D convolutional layer with a kernel size of $3 \times 3 \times 3$, providing an output of 32 features of depth. These feature maps will be processed by six dense blocks composed by [6, 8, 8, 8, 8, 6] 3D layers, respectively, with the same kernel size as the input layer, followed by ReLU non-linear activations. Each dense block is preceded by [0, 1, 2, 3, 4, 5] Embedded Gaussian Non-Local blocks [21], respectively, and each dense block is followed by a transition-down layer with a $2 \times 2 \times 2$ max pooling. Thus, the input volume is processed by the described blocks (both dense and non-local) generating the feature maps, which will gradually decrease (in dimension) to a one-dimensional feature map. This feature vector is concatenated to additional non-visual features, i.e., blood-stream hemato-chemical indicators. The resulting feature map (size $751 \times 1$) then traverses six fully connected (FC) layers followed by RELU, except the last one that, instead, uses a SoftMax layer for the final binary classification. Negative log-likelihood loss is used during model training.

$$Loss = -\sum_{i=1}^{N} X_i \log(P(X_i)) + (1 - X_i) log(1 - P(X_i)) \tag{1}$$

With $X_i$, we setup the correct class label, while $P\ (X_i)$ represents the model's predicted probability for the correct class. In the next section, more information on the architecture is highlighted, while additional details of the proposed deep model are given in Table 1.

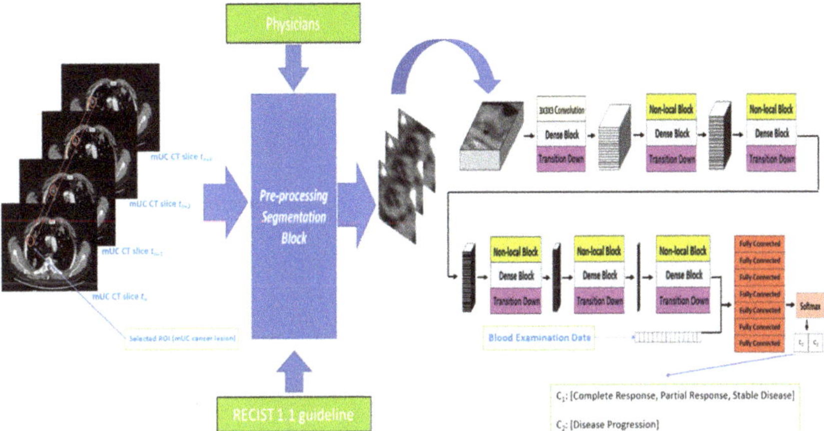


**Figure 1.** The proposed 3D Densely connected Convolutional Neural Network (DCNN) with Non-Local blocks architecture: Input data is a segmented lesion (ROI on CT scan) traverses the deep network (a sequence of dense and non-local blocks) and it is then classified as belonging to class C1 or class C2.

**Table 1.** The layers specification of the proposed deep architecture.

| Block | Output Size | Layer(s) Description | Layers Number |
|---|---|---|---|
| Convolution | $32 \times 16 \times 64 \times 64$ | $3 \times 3 \times 3$ convolution | 1 |
| Dense Block | $128 \times 16 \times 64 \times 64$ | Batch Normalization<br>Rectified Linear Unit (ReLU)<br>$3 \times 3 \times 3$ depth-wise convolution<br>$1 \times 1 \times 1$ point-wise convolution | 6 |
| Transition layer | $128 \times 8 \times 32 \times 32$ | $1 \times 1 \times 1$ convolution, $2 \times 2 \times 2$ maxpool | 1 |
| Dense Block | $256 \times 8 \times 32 \times 32$ | [...] | 8 |
| Transition layer | $256 \times 4 \times 16 \times 16$ | $1 \times 1 \times 1$ convolution, $2 \times 2 \times 2$ maxpool | 1 |
| Dense Block | $384 \times 4 \times 16 \times 16$ | [...] | 8 |
| Transition layer | $384 \times 2 \times 8 \times 8$ | $1 \times 1 \times 1$ convolution, $2 \times 2 \times 2$ maxpool | 1 |
| Dense Block | $512 \times 2 \times 8 \times 8$ | [...] | 8 |
| Transition layer | $512 \times 1 \times 4 \times 4$ | $1 \times 1 \times 1$ convolution, $2 \times 2 \times 2$ maxpool | 1 |
| Dense Block | $640 \times 1 \times 4 \times 4$ | [...] | 8 |
| Transition layer | $640 \times 1 \times 2 \times 2$ | $1 \times 1 \times 1$ convolution, $2 \times 2 \times 2$ maxpool | 1 |
| Dense Block | $736 \times 1 \times 2 \times 2$ | [...] | 6 |
| Transition layer | $736 \times 1 \times 1 \times 1$ | $1 \times 1 \times 1$ convolution, $2 \times 2 \times 2$ maxpool | 1 |
| Concatenation | 751 | Integrates hematochemical patient's data | 1 |
| Fully Connected | 375 | FC, ReLU | 1 |
| Fully Connected | 187 | FC, ReLU | 1 |
| Fully Connected | 93 | FC, ReLU | 1 |
| Fully Connected | 46 | FC, ReLU | 1 |
| Fully Connected | 46 | FC, ReLU | 1 |
| Fully Connected | 46 | FC, ReLU | 1 |
| Classification | 2 | FC, Softmax | 1 |

A visual description of the deep architecture explained in Table 1 is shown in the graph of the model included in Figure 1. Specifically, starting from the pre-processed CT lesions fed as input, the convolutional blocks are highlighted, stacked with dense blocks and the transition layers whose generated feature maps transverse further processing blocks composed of non-local block stacked with a dense block and transition layer up to the final stack of fully connected and SoftMax from which the classification output will be generated. More details in the next sections.

### 3.1. Dense Blocks

The proposed 3D-DCNN includes densely connected blocks (dense blocks) with 3D separable convolution layers (both depth-wise and point-wise). Separable convolutions are adopted to improve efficiency (through a significant reduction of a model's parameters), while not affecting the output performance. Each dense block consists of a sequence of dense layers, including a batch normalization layer, a 3D convolutional layer with a kernel size of $3 \times 3 \times 3$ (depth-wise and point-wise separable) followed by a ReLU. Each dense block is followed by a transition down layer, aiming to half feature map dimension, and composed by a convolutional layer with a kernel size of $1 \times 1 \times 1$ followed by a max pooling layer of kernel $2 \times 2 \times 2$. The output of dense blocks is then passed to non-local blocks.

### 3.2. Self-Attention through Non-Local Blocks

Non-local blocks have been recently introduced [21], as a very promising approach for capturing space-time long-range dependencies and correlation on feature maps, resulting in a sort of "self-attention" mechanism. Non-local blocks take inspiration from the non-local means method, extensively applied in computer vision, and have demonstrated to significantly improve the performance of deep models [21]. Self-attention through non-local blocks aims to enforce the model to extract correlation among feature maps by weighting the averaged sum of the features at all possible positions in the generated feature maps [21]. In our pipeline, non-local blocks operate on almost each convolution layer to extract a feature in dependencies at multiple abstract levels for a holistic morphological modeling of the input RECIST lesions. The mathematical formulation of non-local operation is the following. Given a generic deep network as well as a general input data $x$, the employed non-local operation computes the corresponding response $y_i$ (of the given Deep architecture) at a $i$ location in the input data as a weighted sum of the input data at all positions $j \neq i$.

$$y_i = \frac{1}{\psi(x)} \sum_{\forall j} \zeta(x_i,\ x_j)\beta\left(x_j\right) \tag{2}$$

With $\zeta(\cdot)$ being a pairwise potential describing the affinity or relationship between data positions at index $i$ and $j$, respectively. $\beta(\cdot)$ is, instead, a unary potential modulating $\zeta$ according to the input data. The sum is then normalized by a factor $\psi(x)$. The parameters of $\zeta$, $\beta$, and $\psi$ potentials are learned during the model's training and are defined as follows.

$$\zeta(x_i,\ x_j) \ = \ e^{\Theta(x_i)^T\ \Phi(x_j)} \tag{3}$$

where $\Theta$ and $\Phi$ are two linear transformations of the input data $x$ with learnable weights $W_\Theta$ and $W_\Phi$.

$$\begin{aligned} \Theta(x_i) &= W_\Theta x_i \\ \Phi\left(x_j\right) &= W_\Phi x_j \\ \beta\left(x_j\right) &= \ W_\beta x_j \end{aligned} \tag{4}$$

For the $\beta(\cdot)$ function, a common linear embedding (classical $1 \times 1 \times 1$ convolution) with learnable weights $W\beta$ is employed. The normalization function $\psi$ is:

$$\psi(x) = \sum_{\forall j} \zeta\left(x_i,\ x_j\right) \tag{5}$$

In Equations (2)–(5), an Embedded Gaussian setup is reported [21]. The selection of the Embedded Gaussian based affinity function is compliant with recent self-attention approaches [21,42], specifically recommended for 2D or 3D applications.

### 3.3. Classification Layer: The Stack of Fully Connected

Once features are extracted through the combination of dense and non-local blocks, we obtain a one-dimension visual embedding (736 × 1). These features are then concatenated to an additional side of information consisting of blood tests and other clinical data. The combination of the two sets of features is then fed to a stack of fully connected (FC) layers. The objective of this FC stack is to find additional correlations among the aggregated (at different abstract levels) deep features and clinical data in order to enhance accuracy in assessing the ICI immunotherapy treatment outcome. The proposed full FC stack is composed by seven FC layers, with, respectively, 375, 187, 93, 46, 46, 46, and 2 neurons for each layer. In particular, the sequence of FC layers (whose optimal number of layers was decided during experimental results) is designed to create a hybrid visual-clinical features hierarchy to balance model complexity with learnability.

### 3.4. Dataset: Recruitment and Data Pre-Processing

We recruited a dataset of 43 CT/MRI scans from patients as part of a clinical study performed at a local hospital facility. The recruited patients (details are in Table 2) have histologically confirmed bladder cancer (mUC) progressing after a platinum-based chemotherapy and treated with an anti PD-L1 ICIs agent in the second or beyond line setting. All patients provided their written informed consent to the participation of clinical trials (Nr. D4191C00068 and MO29983 including the use of their clinical information for analysis approved by the Institutional review board (IRB) "Comitato Etico Catania 1", Catania, Italy). The contribution, however, refers to the analysis of CT images for which the dataset will be limited to 41 patients who received an abdominal-chest CT discarding the other two who, instead, received MRI-based imaging. For each recruited patient, a chest-abdomen CT-scan was performed for cancer disease staging. The mentioned CT imaging was performed very close to the start of the ICIs immunotherapy treatment. Such instances of the collected chest-abdomen CT-scans are reported in Figure 2. The used imaging device consists of General Electronics (GE) CT scanner multi-slices (64 slices) with slice thickness of 2.5 mm. The working current is in the range of 10–700 mA. The working voltage is 120 kV and the pitch is 0.98 mm. As known, in a CT scanner, multi-slices of the spatial resolution in the scan plane is influenced by the convolution filter used to reconstruct the image by any other applied post-processing filter. It also depends on the number of projections that make up the image. This number, in turn, depends on the sampling rate and scan time. In this paper, we specify that the software of the previously mentioned GE tomograph allowed us to export CT scan slice images with each having a spatial resolution of 1440 × 810 pixels.

Each CT scan is complemented with the following clinical and personal history data (used in our learning model): primary tumor site, white blood cells (WBC), neutrophils, lymphocytes, eosinophils, platelets, albumin, Lactate DeHydrogenase (LDH), d-dimer, urine pH, proteins and Body Mass Index (BMI), age, gender, and tobacco use. The target of the proposed pipeline is closely related to the predictive estimate of the response to the ICIs immunotherapy treatment based on the analysis of the RECIST compliant lesion (often a metastases) identified by oncologists in the patient's CT imaging. One of the most feared features of malignant cancer is their ability to metastasize to other parts of the body. Metastases can be spread through the blood and/or lymphatic route and clearly follow the anatomy of the interconnection of organs in the human body. The process by which oncologists define the cancer extension is called staging. With a special focus on bladder cancer, the staging requires the use of imaging (usually CT-scan and PET) to characterize the level of disease spread in the subject body [33,34]. For this reason, especially in the advanced mUC stages, CT scans show multiple lesions and radiologists/oncologists select the most significant ones (according to the RECIST guideline) for monitoring cancer evolution over time. The selection is carried out according to the previously mentioned RECIST guidelines that define inclusion criteria, CT scan procedure, patient assessment, lesion features, and how to monitor cancer over time. According to the RECIST 1.1 guideline, lesions of interest (target lesions) are those with the longest diameter (LD) in one dimension ≥20 mm (examples are shown in Figure 2). Lesion dimensions are used to set up a disease baseline for the patient's assessment. According to RECIST 1.1, the sum of the longest diameter (LD) for all selected target lesions is the baseline LD. The baseline LD is used as a reference to evaluate the follow-up and treatment response of analyzed cancer. The LD value is then used for patient assessments after the oncological treatment. In particular:

- A patient shows a complete response (CR) to the medical treatment if all identified target lesions (LD sum) disappear at the end-treatment CT imaging.
- A patient shows a partial response (PR) to drug treatment if the target lesions (LD sum) are reduced by at least 30%.
- A patient shows a progressive disease (PD) if the Longest Diameter(LD) sum increases by, at least, 20% of the LD (LD sum, in case of multiple target lesions).
- A patient instead reports stable disease (SD) if no significant increase or decrease is observed on the target lesions.

**Table 2.** Some statistics of the used clinical dataset. CR: Complete Response; PR: Partial Response; SD: Stable Disease; PD: Progressive Disease.

| Dataset Field Description | Number | % |
|---|---|---|
| **Age** | | |
| ≤60 | 13 | 30 |
| >60 | 30 | 80 |
| **Gender** | | |
| Male | 39 | 91 |
| Female | 4 | 9 |
| **Tobacco Use** | | |
| Never | 5 | 12 |
| Current | 16 | 37 |
| Former | 22 | 51 |
| **Therapy Line** | | |
| 2 | 38 | 88 |
| ≥3 | 5 | 12 |
| **Primary Tumor Site** | | |
| Upper urinary tract | 4 | 9 |
| Lower urinary tract | 36 | 84 |
| Both | 3 | 7 |
| **Metastases Site** | | |
| Lymph-nodes only | 14 | 33 |
| Visceral | 29 | 67 |
| **Treatment Response** | | |
| CR/PR/SD | 16 (43 target lesions) | 37 |
| PD | 27 (63 target lesions) | 63 |
| **Follow-up Median -Months-** | | |
| CR/PR/SD/PD | 13.4 | 11.1–15.6 |
| **Follow-up Imaging** | | |
| CT-scan | 41 | |
| MRI (Magnetic Resonance Imaging) | 2 | |

The 41 patients included in this work were categorized, according to RECIST 1.1. All the CT RECIST 1.1 compliant lesions have been collected for a total amount of 106 RECIST 1.1 findings. Therefore, each RECIST 1.1. compliant lesion was treated individually even if it belonged to the same patient. These are some statistics of the collected clinical dataset. Furthermore, 30% of the patients were under the age of 60.91% of patients were male, with the remaining 9% female. A total of 33% of subjects had lymph node metastases, while the remaining 67% had various visceral metastatic lesions. Forty-three cases (target lesions) are referred to a complete/partial response or a disease stabilization following immunotherapy treatment (CR/PR/SD: Class 1), while 63 lesions are regarded to the disease progression despite anti-PD-L1 drug treatment (PD: Class 2).

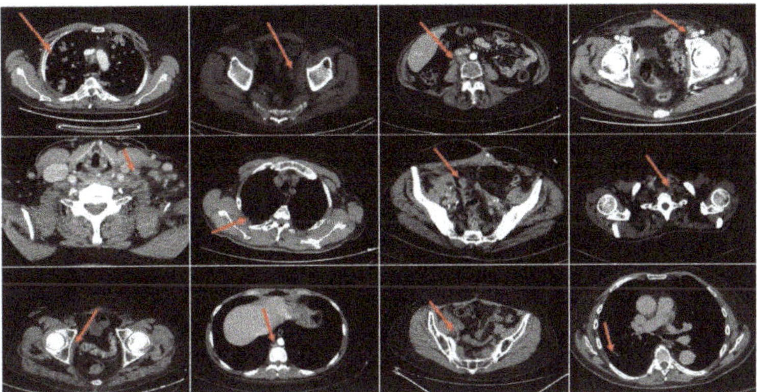

**Figure 2.** Some instances of the selected RECIST (Response Evaluation Criteria in Solid Tumors) compliant cancer lesions in Computerized Tomography (CT) imaging. The red arrow identifies an instance of the selected RECIST 1.1. compliant lesion.

### 3.5. Data Annotation, Training Procedure, and Evaluation Metrics

Data annotation was carried out by an expert oncologist. In particular, starting from the whole CT scan, the oncologist manually selected, according to RECIST 1.1 recommendation, all the target lesions to characterize the disease.

Nowadays, all CT scanner imaging software allows the automatic selection of a Region Of Interest (ROI), according to certain spatial, dimensional, and/or morphological criteria. After this selection, a 64 × 64 bounding box area (ROI) around each selected lesion over 16 consecutive slices is extracted (thus, forming a 16 × 64 × 64 VOI i.e., the Volume of Interest). In order to ensure that the selection manually made by the oncologist includes the whole target lesion to be analyzed, ad-hoc ROI of spatial dimensions include the target lesion in its maximum extension (as it appears in the selected slices) that will be manually applied by the oncologist/physician. If the applied ROI dimension is different from the predetermined input size (64 × 64 in this case), a bi-cubic resizing of the ROI will be applied to bring it to the desired size, i.e., 64 × 64. The oncologist will select the lesion in a slice (the one in which there is the maximum extension of the lesion) and the latter will be propagated to all the selected input slices. The needed bi-cubic resizing will be extended to all the processed slices.

Once the VOI has been selected in the first slice, a software tool we have developed will automatically extract the same VOI in the other slices (for a total of 16) in order to characterize the morophological temporal dynamics of the lesion. As mentioned in the previous section, CT data was complemented with 15 additional clinical and hemato-chemical data that are converted into numeric representation and suitably normalized for being processed by the proposed model. These data are included in the *LabVector (i)* input as the index *i* is used to identify the patient. The temporal depth of processed CT slices (i.e., equal to 16) was identified as the one providing the best trade-off between performance and computational complexity, according to our previous work [34]. We noticed that, depending on the dimension, some RECIST compliant lesions do not appear visible in all 16 CT slices, and, in those cases, we zeroed (input data padding) the slices where the targeted lesion was not visible to keep the input temporal-depth fixed for our deep 3D model. The CT scans (better, VOIs of size 16 × 64 × 64) were properly labelled with reference to the two classes previously identified and described (C1 and C2). However, the selection of the samples in each dataset split was not performed randomly, but, in order to balance suitably, the presence of the patients of the two considered classes (C1 includes patients with some response to immunotherapy, i.e., CR/PR/SD cases, while C2 includes patients with a progressive disease (PD)), and, consequently, to ensure enough variability of the characteristics of the subjects. In particular, the dataset was configured as follows: 76 target lesions

(28 of Class 1 and 48 of Class 2) were used for training and validation sessions, while the remaining 30 CT target image lesions (15 of Class 1 and 15 of Class 2) were used as a test set. Clearly, in configuring the test-set, we avoided using lesions of patients (although different) used in the training or validation set in order to improve the robustness of the validation process of the proposed pipeline. Unfortunately, immunotherapy treatment is still a relatively new strategy and the mechanisms of its functioning and interaction with the human immune system are not widely known and, as such, patients who do not respond to immunotherapy treatment are many more (currently) than those who respond positively to the ICIs' stimulation and, therefore, building a dataset perfectly balanced between the two classes, which is not trivial [4–12]. In any case, to improve the validation reliability, we have implemented a cross-validation mechanism through a k-fold. Specifically, we cross-validated our deep model by configuring k = 5 and reporting the results of this procedure in Table 3 (mean and standard deviation for the main performance indexes). The output of our deep model for each input data (16 × 64 × 64 VOI with additional clinical data) is a (two) class probability vector on which, during training, we compute a negative log-likelihood loss with L2 regularization weighted by a factor $\lambda = 0.0001$. The mini-batch gradient descent was performed for minimizing the model loss, using the Adam optimizer, with an initial learning rate of 0.01, and a mini-batch size of 4. For data augmentation, we perform random translation and rotation (with a random degree value) along the spatial axis, consequently increasing the dataset dimension during the training session. Our deep model is implemented by using the Pytorch framework. Experiments were carried out on a server with 2 Intel Xeon E5620 CPU with 4 cores each, 96GB of RAM equipped with a Nvidia Quadro P6000 GPU with 24 Gbytes of video memory. In order to validate the performance of the proposed architecture with respect to other deep learning-based solutions, the following metrics have been used (FP: False Positive, FN: False Negative, TP: True Positive, TN: True Negative).

$$Accuracy = \frac{TP + TN}{TP + TN + FN + FP}, \tag{6}$$

$$Sensitivity = \frac{TP}{TP + FN}, \tag{7}$$

$$Specificity = \frac{TN}{TN + FP}, \tag{8}$$

$$F1 - Score = \frac{2 \cdot TP}{2 \cdot TP + FN + FP}, \tag{9}$$

**Table 3.** Experimental performance benchmarking (mean ± standard deviation).

| Model | Accuracy | Sensitivity | Specificity | F1-Score |
|---|---|---|---|---|
| 2D ResNet-50 | 0.620 ± 0.052 | 0.604 ± 0.0078 | 0.636 ± 0.061 | 0.613 ± 0.058 |
| 3D DenseNet + H | 0.713 ± 0.047 | 0.711 ± 0.041 | 0.716 ± 0.064 | 0.713 ± 0.043 |
| 3D DenseNet + SepConv + H | 0.733 ± 0.049 | 0.729 ± 0.069 | 0.738 ± 0.047 | 0.731 ± 0.054 |
| 3D DenseNet | 0.640 ± 0.034 | 0.636 ± 0.034 | 0.644 ± 0.048 | 0.638 ± 0.032 |
| 3D DenseNet + NLB + SepConv | 0.878 ± 0.039 | 0.871 ± 0.054 | 0.884 ± 0.075 | 0.877 ± 0.041 |
| Proposed | **0.922 ± 0.037** | **0.929 ± 0.053** | **0.916 ± 0.047** | **0.922 ± 0.038** |
| 2D ResNet-101 | 0.829 ± 0.043 | 0.822 ± 0.054 | 0.836 ± 0.061 | 0.828 ± 0.043 |
| 3D DenseNet-201 | 0.856 ± 0.033 | 0.871 ± 0.047 | 0.840 ± 0.055 | 0.858 ± 0.032 |
| 2D VGG-19 | 0.667 ± 0.033 | 0.662 ± 0.069 | 0.671 ± 0.059 | 0.664 ± 0.041 |
| Previous [34] | 0.861 ± 0.023 | 0.815 ± 0.011 | 0.883 ± 0.048 | 0.810 ± 0.037 |

We considered a "True Positive," which is the right classification of a patient who has shown a certain response to immunotherapy treatment (complete response (CR), partial (PR), or has a stable disease (SD)), and has been previously classified by our pipeline as belonging to class C1. Consequently, we will consider "True Negative," which is a patient who previously classifies as belonging to class C2 and then correctly following the treatment does not show any response to the immunotherapy drug confirming a progression of the disease (PD). The "False Negative" and "False Positive" values are computed accordingly. The accuracy, sensitivity, specificity, and F1-score are evaluated—on the test-set during k-fold cross-validation so that they are reported as mean and standard deviations. The collected experimental results are discussed in the next section.

## 4. Results and Discussion

### *Performance Analysis*

In this section, we report the promising performance results of the proposed approach. As mentioned, the used dataset is composed by 41 patients (106 RECIST compliant input image lesions) recruited as part of a clinical trial including, for each patient, informed consent, chest-abdomen CT scan images, and their blood examinations collected before the start of treatment. Each patient was adequately labeled with histological confirmation of mUC. The cross-validation session is used to select the best model setup of the proposed architecture, i.e., when the maximum k-fold cross-validation accuracy is retrieved (3D deep architecture as reported in Figure 1, network layers structured as per Table 1, fixed learning rate of 0.01, and a weight decay of 0.00001, Adam optimization). As introduced, we compared our architecture with such state-of-the-art deep architectures in order to provide performance benchmarks regarding the proposed application. Specifically, in order to evaluate the improvement in terms of performance compared to similar 2D and 3D deep networks, the authors have validated the performance of the used DenseNet backbone having the same architecture of our pipeline (Table 1), but without the inclusion of Self-Attention (Non-Local Blocks) mechanisms and separable convolutions. In addition, the performance of the proposed method has been compared with respect to such classic architectures: ResNet-50, ResNet-101, VGG-19, and 3D extension of the classical DenseNet-201 [43]. Table 3 reports the collected experimental results and comparisons. The implemented 3D DenseNet backbone baseline (3D DenseNet) showed an accuracy of 0.640 ± 0.034 significantly lower than our full pipeline, which shows a higher accuracy equal to 0.922 ± 0.037. Additionally, in terms of sensitivity, specificity, and the F1-score, our architecture is significantly more performed (0.929 ± 0.053, 0.916 ± 0.047, and 0.922 ± 0.038, respectively) than the simple DenseNet backbone, thus, confirming the improvements that can be obtained in particular through the use of self-attention techniques and separable layers. Specifically, feature maps that suitably weight the spatiotemporal dependencies of the CT imaging selected target lesion (i.e., the result of non-local blocks application) provide more discriminative features to the FC stack. Moreover, the joint contribution of non-local blocks and separable convolution layers allows us to generate feature maps having an informative content that best characterizes the spatiotemporal dependencies between the CT imaging VOIs and treatment response of the associated patient. As well known, the usage of separable convolutional layers significantly reduces the risk of overfitting [30–34]. During this session, we also investigated the performance impact of the jointed link between hemato-chemical data *LabVector (i)* with visual features based on CT imaging. We, therefore, performed a testing session using the same proposed deep architectures but avoiding concatenating the data contained in the *LabVector (i)* ("3D DenseNet+NLB+SepConv" in Table 3). As hypothesized by the oncologists who followed the trial on which this study is based, there is a close correlation between the anamnesis and hematochemical data, the imaging data, and the patient's response to immunotherapy treatment even though this correlation is not perfectly known to date. In fact, our tests revealed a considerable reduction in the overall performance of the tested deep networks if, as inputs, we only use CT visual imaging and do not integrate with blood and medical history (*LabVector (i)*). In more detail, the proposed architecture with visual input but without hemato-chemical data dropped in performance as it showed 0.878 ± 0.039 (Accuracy), 0.871 ± 0.054 (Sensitivity), 0.884 ± 0.075 (Specificity), and 0.877± 0.075 (F1-score) significantly lower with respect to the same proposed pipeline with hemato-chemical data (see Table 3). Therefore, the clinical data included in *LabVector (i)* play a significant role in the discrimination capability of the proposed pipeline so that it is worthy of further study. As reported in Table 3, the accuracy of the classical deep 2D architectures such as ResNet-50, ResNet-101, and VGG-19 is significantly lower than our approach, further confirming the promising performance of the proposed solution. We have compared our implemented pipeline with similar 3D architecture but with more layers (3D DenseNet-201). The performance of our architecture is significantly higher than the compared 3D deep classifier, as reported in Table 3. This confirms that the Self-Attention mechanisms realized through the inclusion of non-local blocks with embedded gaussian setup together with the separable layers allow us to obtain more discriminative and representative features maps (with respect to a deeper network as the 3D-DenseNet-201) of the correlation with the response to immunotherapy treatment. Evidently, the capability of non-local blocks to embed more precise spatial-temporal correlations in the analyzed CT lesions allows the generation of more discriminative feature maps than those obtained by increasing the convolutional layers in the deep classifier. This makes the proposed system particularly performing in the application herein described. All the networks analyzed for a comparison share the same input setup data. We discriminated some setups only in reference to our pipeline to demonstrate that the DenseNet backbone was necessarily enriched with the Separable Convolutional Layers as well as with hemato-chemical data and non-local blocks. The aim of the work is to analyze the impact of imaging in predicting the response to immunotherapy treatment. We tested our model with blood chemistry data only, obtaining poor performances with accuracy below 50% (in cross validation), clearly confirming that the contribution of imaging is fundamental for the overall performance of the proposed model.

In any case, we remark that the proposed deep architecture aims not only to offer valid medical assistance to the physician but rather to highlight the most predictive visual patterns. In doing this, we have tried to investigate and adopt one of the most promising self-deep features explanatory methods already introduced in the scientific literature. The authors propose the usage of GradCAM introduced by Selvaraju et al., 2017 [40]. The GradCAM approach is intuitively very simple. It uses the gradient with respect to the generated convolutional features as a classification score in order to understand which parts of the input image are most significant for classification. By means of a simple combination between the activation saliency map generated by the GradCAM approach [40] with existing correlated input data, we are able to create a combined discriminative image pattern visualization easily understandable and able to guide the physician in the visual analysis of the imaging areas that have a greater weight in the discrimination/classification of the input dataset. More details are present in Reference [40]. Moving into our application, through GradCAM, we tried to understand which parts of the ROI extracted from the chest-abdomen CT images (containing the target RECIST compliant lesion), which were more significant for our 3D pipeline processing. Therefore, having obtained the corresponding GradCam based gradient-weighted activation maps, we combined them with the ROI of the target lesions extracted from the CT slices. The collected results we obtained are reported in Figure 3 and deserves further study in relation to such heuristically hypotheses made by several oncologists. Specifically, oncologists hypothesized that only certain parts of the RECIST compliant target CT lesions are significant in relation to the estimation of an immunotherapy outcome [34,44]. As evident from Figure 3, for some processed lesions (ROI), the GradCAM analysis highlighted such areas with greater salience (red area) than the others (green area). The salient visual areas of such input ROI-lesion are those that most contribute to the performance of the deep network, i.e., those that are best represented in the feature maps. This seems to confirm the hypothesis of some medical researchers.

The researchers hypothesized that the immune-cytochemical hyper-expression of the PD-L1 protein on tissue tumor sections or on cyto-block (together with the analysis of the Tumor Mutational Burder (TMB)) may be predictive of a positive response to the ICIs' immunotherapy treatment [45]. This hyper-expression being found in the surface of such tumor cells is hypothesized to be evident in the morphology of the lesion of the cells of the primary tumor and likely of those of the related metastases [45]. Therefore, such researchers have hypothesized that only the areas of the lesions in which there is a high concentration of expression of the PD-L1 protein can be significant for the estimation of the response to the immunotherapy treatment [34,44,45]. Translating in our case, we investigated the concrete possibility that such micro-areas of the target lesions visible in the GradCam post-processed CT imaging, are significant for the estimation of the ICIs immunotherapy response of patients suffering from bladder cancer (assuming that the selected neoplastic target lesion shows such parts with great surface expression of the ligant protein PD-L1) [46,47]. From Figure 3, it is evident that, with reference to some target RECIST lesions identified in the chest-abdomen CT slice (Figure 3a), a specific sub-area of these lesions appear more significant (red) than others (green) and, therefore, are shown to have a greater weight in the estimation of the response to the immunotherapy treatment (Figure 3b). We have no scientific evidence that this saliency hyper-caption in the feature maps is related to the presence of cancerous cells with high expression of the PD-L1, but it is, however, significant that the explanation of the feature maps is indicative of a self-weighting of such image areas of a target metastatic lesion. Further investigations are underway on this interesting aspect.

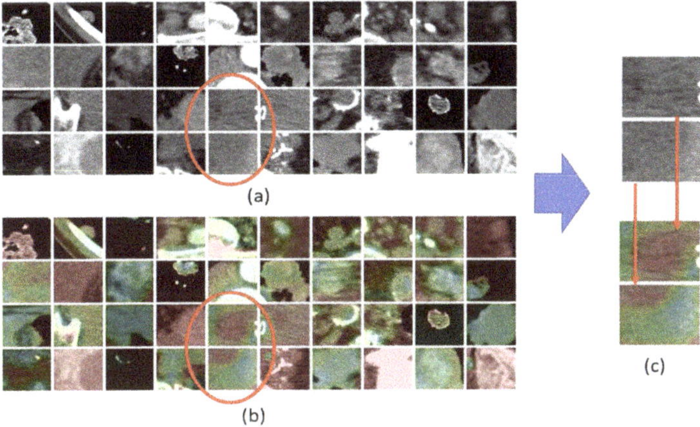


**Figure 3.** (**a**) RECIST (Response Evaluation Criteria in Solid Tumors) 1.1 compliant CT target lesions. (**b**) The corresponding Grad-CAM generated saliency maps. (**c**) A detail of the salient part of the processed RECIST lesion.

## 5. Conclusions

In this paper, a problem of particular interest in the oncological field is tackled, namely the identification of a non-invasive and robust bio-marker that can assist the oncologists in the discrimination of patients potentially eligible for immunotherapy treatment. To the best of our knowledge, although there are several promising lines of research [4–12], there is no method that shows high accuracy in assessing patients who may undergo immunotherapy treatment from those who may not. However, many of the methods that are being studied are based on the research of the expression of the inhibitor ICIs PD-L1, which, however, is not highly discriminating and requires the invasive biopsy of the primary tumor [48,49]. For these reasons, we have investigated recent and innovative deep architectures in order to learn new patterns able to estimate the response to ICIs' immunotherapy treatment based on imaging and clinical data of the cancer patient. Specifically, we investigated the problem of finding a non-invasive image-based biomarker for patients with metastatic bladder cancer. The architecture we have implemented and tested learns image-features from chest-abdomen CT imaging of the patients affected by mUC. As confirmed in Reference [44], a novel mechanism of action of ICIs' treatment with immune and T cell activation leads to unusual patterns of response on CT imaging. For these reasons, an innovative 3D deep architecture with embedded self-attention mechanism and separable convolutions for hyper-parameters model optimization is implemented and analyzed. The proposed pipeline was tested on a sample of 41 recruited patients (clinical trial) for a total of 106 processed input visual RECIST 1.1. compliant lesions. The outcome of the proposed 3D Deep network is a preliminary estimation of the patients who respond or not to ICIs' immunotherapy treatment, i.e., preliminary estimation of the patient belongs to Class C1 (i.e., subjects eligible as they have a high chance to show a complete or partial response or at least stabilization of the disease) with respect to patients belonging to Class C2 (patients who will manifest disease progression). As confirmed by experimental results reported in Table 3, the proposed 3D deep architecture shows very promising performance both in terms of accuracy as well as in terms of sensitivity and specificity, confirming that both adopted a self-attention mechanism. Separable convolutions significantly increase the classification ability of the deep model (as confirmed by a benchmark comparison with the baseline and classical backbones). The integration of blood hemato-chemical numerical data has further improved the classification performance of the proposed pipeline. We remark that the achieved promising results need to be confirmed in a bigger scale dataset (currently an extended clinical trial is under development). We are organizing a large-scale multicenter and multivariate clinical study that can validate with greater robustness. The promising results obtained from this contribution allows us to better understand the correlation between the morphological structure of the tumor lesion found in the CT diagnostic examination with the possible response to immunotherapeutic treatment and, therefore, with the iteration with the used drug.

Future works aim to investigate advanced methods for the automatic segmentation of RECIST target lesion in order to relief physicians from the burden to manually identify the lesion to be processed by our approach. Moreover, we are analyzing such methods based on usage of LSTM and Autoencoder-based architectures for modeling temporal dynamics of each pixel of the segmented CT lesion [50,51]. Interesting results have been collected integrating such bio-signals of the patients during the analysis of the visual pattern of the segmented CT lesion [52].

## 6. Patents

Francesco Rundo, Giuseppe Luigi Banna, Methods and System for Analyzing Skin Lesions, IT Patent Application Number 102016000121060, 29 November 2016—USA Patent Grant Number 10362985, 26 July 2019.

Francesco Rundo, Giuseppe Luigi Banna, Sabrina Conoci, Deep Learning Motion Algorithm for Lung Cancer Early Detection in Embedded Systems, IT Patent Application Number 102018000010833, 05 Dec 2018.

**Author Contributions:** Conceptualization, methodology, software, validation, investigation, F.R. Formal analysis, G.L.B. Writing—original draft preparation, F.T. and L.P. Writing—review and editing, S.C. and S.B. All authors have read and agreed to the published version of the manuscript.

**Funding:** This research received no external funding.

**Conflicts of Interest:** The authors declare no conflict of interest.

## References

1. Ferlay, J.; Soerjomataram, I.; Dikshit, R.; Eser, S.; Mathers, C.; Rebelo, M.; Parkin, D.M.; Forman, D.; Bray, F. Cancer incidence and mortality worldwide: Sources, methods and major patterns in Globocan 2012. *Int. J. Cancer* **2015**, *136*, E359–E386. [CrossRef] [PubMed]
2. Vaidya, A.; Soloway, M.S.; Hawke, C.; Tiguert, R.; Civantos, F. De Novo Muscle Invasive Bladder Cancer: Is There a Change in Trend? *J. Urol.* **2001**, *165*, 47–50. [CrossRef] [PubMed]
3. De Santis, M.; Bellmunt, J.; Mead, G.; Kerst, J.M.; Leahy, M.; Maroto, P.; Gil, T.; Marreaud, S.; Daugaard, G.; Skoneczna, I.; et al. Randomized Phase II/III Trial Assessing Gemcitabine/Carboplatin and Methotrexate/Carboplatin/Vinblastine in Patients With Advanced Urothelial Cancer Who Are Unfit for Cisplatin-Based Chemotherapy: EORTC Study 30986. *J. Clin. Oncol.* **2012**, *30*, 191–199. [CrossRef] [PubMed]

4. Seront, E.; Machiels, J.-P. Molecular biology and targeted therapies for urothelial carcinoma. *Cancer Treat. Rev.* **2015**, *41*, 341–353. [CrossRef] [PubMed]
5. Powles, T.; Durán, I.; van der Heijden, M.S.; Loriot, Y.; Vogelzang, N.J.; De Giorgi, U.; Oudard, S.; Retz, M.M.; Castellano, D.; Bamias, A. Atezolizumab versus chemotherapy in patients with platinum-treated locally advanced or metastatic urothelial carcinoma (IMvigor211): A multicentre, open-label, phase 3 randomised controlled trial. *Lancet* **2018**, *391*, 748–757. [CrossRef]
6. Bellmunt, J.; De Wit, R.; Vaughn, D.J.; Fradet, Y.; Lee, J.-L.; Fong, L.; Vogelzang, N.J.; Climent, M.A.; Petrylak, D.P.; Choueiri, T.K.; et al. Pembrolizumab as Second-Line Therapy for Advanced Urothelial Carcinoma. *N. Engl. J. Med.* **2017**, *376*, 1015–1026. [CrossRef]
7. Sternberg, C.N.; Loriot, Y.; James, N.; Choy, E.; Castellano, D.; Lopez-Rios, F.; Banna, G.L.; De Giorgi, U.; Masini, C.; Bamias, A.; et al. Primary Results from SAUL, a Multinational Single-arm Safety Study of Atezolizumab Therapy for Locally Advanced or Metastatic Urothelial or Nonurothelial Carcinoma of the Urinary Tract. *Eur. Urol.* **2019**, *76*, 73–81. [CrossRef]
8. Balar, A.V.; Galsky, M.D.; E Rosenberg, J.; Powles, T.; Petrylak, D.P.; Bellmunt, J.; Loriot, Y.; Necchi, A.; Hoffman-Censits, J.; Perez-Gracia, J.L.; et al. Atezolizumab as first-line treatment in cisplatin-ineligible patients with locally advanced and metastatic urothelial carcinoma: A single-arm, multicentre, phase 2 trial. *Lancet* **2017**, *389*, 67–76. [CrossRef]
9. Balar, A.V.; Dreicer, R.; Loriot, Y.; Perez-Gracia, J.L.; Hoffman-Censits, J.H.; Petrylak, D.P.; Van Der Heijden, M.S.; Ding, B.; Shen, X.; Rosenberg, J.E. Atezolizumab (atezo) in first-line cisplatin-ineligible or platinum-treated locally advanced or metastatic urothelial cancer (mUC): Long-term efficacy from phase 2 study IMvigor210. *J. Clin. Oncol.* **2018**, *36*, 4523. [CrossRef]
10. Balar, A.V.; Castellano, D.; O'Donnell, P.H.; Grivas, P.; Vuky, J.; Powles, T.; Plimack, E.R.; Hahn, N.M.; De Wit, R.; Pang, L.; et al. First-line pembrolizumab in cisplatin-ineligible patients with locally advanced and unresectable or metastatic urothelial cancer (Keynote-052): A multicentre, single-arm, phase 2 study. *Lancet Oncol.* **2017**, *18*, 1483–1492. [CrossRef]
11. Vuky, J.; Balar, A.V.; Castellano, D.E.; O'Donnell, P.H.; Grivas, P.; Bellmunt, J. (Joaquim); Powles, T.; Bajorin, D.F.; Hahn, N.M.; De Wit, R.; et al. Updated efficacy and safety of KEYNOTE-052: A single-arm phase 2 study investigating first-line pembrolizumab (pembro) in cisplatin-ineligible advanced urothelial cancer (UC). *J. Clin. Oncol.* **2018**, *36*, 4524. [CrossRef]
12. Seliger, B.; Bono, P.; Kim, J.W.; Spiliopoulou, P.; Calvo, E.; Pillai, R.N.; Ott, P.A.; De Braud, F.G.; Morse, M.A.; Le, D.T.; et al. Efficacy and safety of nivolumab monotherapy in metastatic urothelial cancer (mUC): Results from the phase I/II CheckMate 032 study. *J. Clin. Oncol.* **2016**, *34*, 4501. [CrossRef]
13. Massard, C.; Gordon, M.S.; Sharma, S.; Rafii, S.; Wainberg, Z.A.; Luke, J.; Curiel, T.J.; Colon-Otero, G.; Hamid, O.; Sanborn, R.E.; et al. Safety and Efficacy of Durvalumab (MEDI4736), an Anti–Programmed Cell Death Ligand-1 Immune Checkpoint Inhibitor, in Patients With Advanced Urothelial Bladder Cancer. *J. Clin. Oncol.* **2016**, *34*, 3119–3125. [CrossRef] [PubMed]
14. Apolo, A.B.; Infante, J.R.; Balmanoukian, A.; Patel, M.R.; Wang, D.; Kelly, K.; Mega, A.E.; Britten, C.D.; Ravaud, A.; Mita, A.C.; et al. Avelumab, an Anti–Programmed Death-Ligand 1 Antibody, In Patients With Refractory Metastatic Urothelial Carcinoma: Results From a Multicenter, Phase Ib Study. *J. Clin. Oncol.* **2017**, *35*, 2117–2124. [CrossRef] [PubMed]
15. Aggen, D.H.; Drake, C.G. Biomarkers for immunotherapy in bladder cancer: A moving target. *J. Immunother. Cancer* **2017**, *5*, 94. [CrossRef] [PubMed]
16. Paratore, S.; Banna, G.; D'Arrigo, M.; Saita, S.; Iemmolo, R.; Lucenti, L.; Bellia, D.; Lipari, H.; Buscarino, C.I.; Cunsolo, R.; et al. CXCR4 and CXCL12 immunoreactivities differentiate primary non-small-cell lung cancer with or without brain metastases. *Cancer Biomark.* **2012**, *10*, 79–89. [CrossRef]
17. Rundo, F.; Libertino, S.; Banna, G.L.; Ortis, A.; Stanco, F.; Battiato, S. Evaluation of Levenberg–Marquardt neural networks and stacked autoencoders clustering for skin lesion analysis, screening and follow-up. *IET Comput. Vis.* **2018**, *12*, 957–962. [CrossRef]
18. Addeo, A.; Banna, G.L.; Weiss, G.J. Tumor Mutation Burden—From Hopes to Doubts. *JAMA Oncol.* **2019**, *5*, 934–935. [CrossRef]
19. Banna, G.; Olivier, T.; Rundo, F.; Malapelle, U.; Fraggetta, F.; Libra, M.; Addeo, A. The Promise of Digital Biopsy for the Prediction of Tumor Molecular Features and Clinical Outcomes Associated with Immunotherapy. *Front. Med.* **2019**, *6*, 172. [CrossRef]

20. Lambin, P.; Rios-Velazquez, E.; Leijenaar, R.; Carvalho, S.; Van Stiphout, R.G.P.M.; Granton, P.; Zegers, C.M.L.; Gillies, R.; Boellard, R.; Dekker, A.; et al. Radiomics: Extracting more information from medical images using advanced feature analysis. *Eur. J. Cancer* **2012**, *48*, 441–446. [CrossRef]
21. Wang, X.; Girshick, R.; Gupta, A.; He, K. Non-local Neural Networks. In Proceedings of the 2018 IEEE/CVF Conference on Computer Vision and Pattern Recognition, Salt Lake City, UT, USA, 18–23 June 2018; pp. 7794–7803.
22. Williams, H.; Cattani, L.; Li, W.; Tabassian, M.; Vercauteren, T.; Deprest, J.; D'Hooge, J. 3D Convolutional Neural Network for Segmentation of the Urethra in Volumetric Ultrasound of the Pelvic Floor. In Proceedings of the 2019 IEEE International Ultrasonics Symposium (IUS), Glasgow, UK, 6–9 October 2019; pp. 1473–1476.
23. Moradi, P.; Jamzad, M. Detecting Lung Cancer Lesions in CT Images using 3D Convolutional Neural Networks. In Proceedings of the 2019 4th International Conference on Pattern Recognition and Image Analysis (IPRIA), Tehran, Iran, 6–7 March 2019; pp. 114–118.
24. Wang, G.; Lam, K.-M.; Deng, Z.; Choi, K.-S. Prediction of mortality after radical cystectomy for bladder cancer by machine learning techniques. *Comput. Biol. Med.* **2015**, *63*, 124–132. [CrossRef] [PubMed]
25. Garapati, S.S.; Hadjiiski, L.M.; Cha, K.H.; Chan, H.-P.; Caoili, E.M.; Cohan, R.H.; Weizer, A.; Alva, A.; Paramagul, C.; Wei, J.; et al. Urinary bladder cancer staging in CT urography using machine learning. *Med. Phys.* **2017**, *44*, 5814–5823. [CrossRef] [PubMed]
26. Hasnain, Z.; Mason, J.; Gill, K.; Miranda, G.; Gill, I.S.; Kuhn, P.; Newton, P.K. Machine learning models for predicting post-cystectomy recurrence and survival in bladder cancer patients. *PLoS ONE* **2019**, *14*, e0210976. [CrossRef] [PubMed]
27. Cha, K.H.; Hadjiiski, L.; Samala, R.K.; Chan, H.-P.; Caoili, E.M.; Cohan, R.H. Urinary bladder segmentation in CT urography using deep-learning convolutional neural network and level sets. *Med. Phys.* **2016**, *43*, 1882–1896. [CrossRef] [PubMed]
28. Gordon, M.; Hadjiiski, L.; Cha, K.; Chan, H.-P.; Samala, R.; Cohan, R.H.; Caoili, E.M. Segmentation of inner and outer bladder wall using deep-learning convolutional neural network in CT urography. In *Medical Imaging 2017: Computer-Aided Diagnosis, Proceedings of the SPIE Medical Imaging, Orlando, FL, USA, 13–16 February 2017*; International Society for Optics and Photonics: Orlando, FL, USA, 2017.
29. Ma, X.; Hadjiiski, L.M.; Wei, J.; Chan, H.-P.; Cohan, R.H.; Caoili, E.M.; Samala, R.K.; Zhou, C.; Lu, Y.; Cha, K.H. 2D and 3D bladder segmentation using U-Net-based deep-learning. In *Medical Imaging 2019: Computer-Aided Diagnosis, Proceedings of the SPIE Medical Imaging, San Diego, CA, USA, 13 March 2019*; International Society for Optics and Photonics: San Diego, CA, USA, 2019.
30. Shkolyar, E.; Jia, X.; Chang, T.C.; Trivedi, D.; Mach, K.E.; Meng, M.Q.-H.; Xing, L.; Liao, J.C. Augmented Bladder Tumor Detection Using Deep Learning. *Eur. Urol.* **2019**, *76*, 714–718. [CrossRef] [PubMed]
31. Wu, E.; Hadjiiski, L.M.; Samala, R.K.; Chan, H.-P.; Cha, K.H.; Richter, C.; Cohan, R.H.; Caoili, E.M.; Paramagul, C.; Alva, A.; et al. Deep Learning Approach for Assessment of Bladder Cancer Treatment Response. *Tomography* **2019**, *5*, 201–208. [CrossRef] [PubMed]
32. Krizhevsky, A.; Sutskever, I.; Hinton, G.E. Imagenet classification with deep convolutional neural networks. *Commun. ACM* **2017**, *60*, 84–90. [CrossRef]
33. Cha, K.H.; Hadjiiski, L.M.; Chan, H.-P.; Weizer, A.Z.; Alva, A.; Cohan, R.H.; Caoili, E.M.; Paramagul, C.; Samala, R.K. Bladder Cancer Treatment Response Assessment in CT using Radiomics with Deep-Learning. *Sci. Rep.* **2017**, *7*, 1–12. [CrossRef]
34. Rundo, F.; Spampinato, C.; Banna, G.; Conoci, S. Advanced Deep Learning Embedded Motion Radiomics Pipeline for Predicting Anti-PD-1/PD-L1 Immunotherapy Response in the Treatment of Bladder Cancer: Preliminary Results. *Electronics* **2019**, *8*, 1134. [CrossRef]
35. Holzinger, A.; Biemann, C.; Kell, D.; Pattichis, C.S. What do we need to build explainable AI systems for the medical domain? *arXiv* **2017**, arXiv:1712.09923.
36. Samek, W.; Montavon, G.; Vedaldi, A.; Hansen, L.K.; Müller, K.R. (Eds.) *Explainable AI: Interpreting, Explaining and Visualizing Deep Learning*; Springer: Berlin/Heidelberg, Germany, 2019.
37. Lee, H.; Yune, S.; Mansouri, M.; Kim, M.; Tajmir, S.H.; Guerrier, C.E.; Ebert, S.A.; Pomerantz, S.R.; Romero, J.M.; Kamalian, S.; et al. An explainable deep-learning algorithm for the detection of acute intracranial haemorrhage from small datasets. *Nat. Biomed. Eng.* **2018**, *3*, 173–182. [CrossRef]
38. Arrieta, A.B.; Díaz-Rodríguez, N.; Del Ser, J.; Bennetot, A.; Tabik, S.; Barbado, A.; Garcia, S.; Gil-Lopez, S.; Molina, D.; Benjamins, R.; et al. Explainable Artificial Intelligence (XAI): Concepts, taxonomies, opportunities and challenges toward responsible AI. *Inf. Fusion* **2020**, *58*, 82–115. [CrossRef]

39. Tonekaboni, S.; Joshi, S.; McCradden, M.; Goldenberg, A. What clinicians want: Contextualizing explainable machine learning for clinical end use. *Mach. Learn. Res.* **2019**, *106*, 1–21.
40. Selvaraju, R.R.; Cogswell, M.; Das, A.; Vedantam, R.; Parikh, D.; Batra, D. Grad-CAM: Visual Explanations from Deep Networks via Gradient-Based Localization. In Proceedings of the 2017 IEEE International Conference on Computer Vision (ICCV), Venice, Italy, 22–29 October 2017; pp. 618–626.
41. Eisenhauer, E.; Therasse, P.; Bogaerts, J.; Schwartz, L.; Sargent, D.; Ford, R.; Dancey, J.; Arbuck, S.; Gwyther, S.; Mooney, M.; et al. New response evaluation criteria in solid tumours: Revised RECIST guideline (version 1.1). *Eur. J. Cancer* **2009**, *45*, 228–247. [CrossRef]
42. Vaswani, A.; Shazeer, N.; Parmar, N.; Uszkoreit, J.; Jones, L.; Gomez, A.N.; Kaiser, Ł.; Polosukhin, I. Attention is all you need. In *Advances in Neural Information Processing System, Proceedings of the First 12 Conferences (The MIT Press), 21 November 2001*; Jordan, M.I., LeCun, Y., Solla, S.A., Eds.; The MIT Press: Cambridge, MA, USA, 2017.
43. Huang, G.; Liu, Z.; Pleiss, G.; Van Der Maaten, L.; Weinberger, K. Convolutional Networks with Dense Connectivity. *IEEE Trans. Pattern Anal. Mach. Intell.* **2019**, 1. [CrossRef] [PubMed]
44. Chen, H.; Cao, P. Deep Learning Based Data Augmentation and Classification for Limited Medical Data Learning. In Proceedings of the 2019 IEEE International Conference on Power, Intelligent Computing and Systems; Institute of Electrical and Electronics Engineers (IEEE), Shenyang, China, 12–14 July 2019.
45. Dromain, C.; Beigelman, C.; Pozzessere, C.; Duran, R.; Digklia, A. Imaging of tumour response to immunotherapy. *Eur. Radiol. Exp.* **2020**, *4*, 1–15. [CrossRef] [PubMed]
46. Alsaab, H.O.; Sau, S.; Alzhrani, R.; Tatiparti, K.; Bhise, K.; Kashaw, S.K.; Iyer, A. PD-1 and PD-L1 Checkpoint Signaling Inhibition for Cancer Immunotherapy: Mechanism, Combinations, and Clinical Outcome. *Front. Pharmacol.* **2017**, *8*, 561. [CrossRef]
47. Ding, X.; Chen, Q.; Yang, Z.; Li, J.; Zhan, H.; Lu, N.; Chen, M.; Yang, Y.; Wang, J.; Yang, D. Clinicopathological and prognostic value of PD-L1 in urothelial carcinoma: A meta-analysis. *Cancer Manag. Res.* **2019**, *11*, 4171–4184. [CrossRef]
48. Zhou, T.C.; Sankin, A.I.; Porcelli, S.A.; Perlin, D.S.; Schoenberg, M.P.; Zang, X. A review of the PD-1/PD-L1 checkpoint in bladder cancer: From mediator of immune escape to target for treatment 1 1MPS is an investor in and consultant for Urogen. SAP is consultant and advisor for Vaccinex. The remaining authors have nothing to disclose. *Urol. Oncol. Semin. Orig. Investig.* **2017**, *35*, 14–20. [CrossRef]
49. Spencer, K.R.; Wang, J.; Silk, A.W.; Ganesan, S.; Kaufman, H.L.; Mehnert, J.M. Biomarkers for Immunotherapy: Current Developments and Challenges. *Am. Soc. Clin. Oncol. Educ. Book* **2016**, *36*, e493–e503. [CrossRef]
50. Rundo, F.; Trenta, F.; Di Stallo, A.L.; Battiato, S. Advanced Markov-Based Machine Learning Framework for Making Adaptive Trading System. *Computation* **2019**, *7*, 4. [CrossRef]
51. Rundo, F.; Trenta, F.; Di Stallo, A.L.; Battiato, S. Grid Trading System Robot (GTSbot): A Novel Mathematical Algorithm for trading FX Market. *Appl. Sci.* **2019**, *9*, 1796. [CrossRef]
52. Rundo, F.; Petralia, S.; Fallica, G.; Libertino, S. *Lecture Notes in Electrical Engineering*; Springer: Berlin/Heidelberg, Germany, 2019; Volume 539, pp. 473–480.


**Publisher's Note:** MDPI stays neutral with regard to jurisdictional claims in published maps and institutional affiliations.

Journal of
*Imaging*

MDPI

*Article*

# Enhanced Region Growing for Brain Tumor MR Image Segmentation

**Erena Siyoum Biratu [1], Friedhelm Schwenker [2], Taye Girma Debelee [1,3,*], Samuel Rahimeto Kebede [3,4], Worku Gachena Negera [3] and Hasset Tamirat Molla [5]**

1 College of Electrical and Mechanical Engineering, Addis Ababa Science and Technology University, Addis Ababa 120611, Ethiopia; iranasiyoum@gmail.com
2 Institute of Neural Information Processing, Ulm University, 89081 Ulm, Germany; friedhelm.schwenker@uni-ulm.de
3 Artificial Intelligence Center, Addis Ababa 40782, Ethiopia; samuelrahimeto@gmail.com (S.R.K.); worku.gachena2@gmail.com (W.G.N.)
4 Department of Electrical and Computer Engineering, Debreberhan University, Debre Berhan 445, Ethiopia
5 College of Natural and Computational Science, Addis Ababa University, Addis Ababa 1176, Ethiopia; hasset.tamirat@aau.edu.et
* Correspondence: tayegirma@gmail.com

**Abstract:** A brain tumor is one of the foremost reasons for the rise in mortality among children and adults. A brain tumor is a mass of tissue that propagates out of control of the normal forces that regulate growth inside the brain. A brain tumor appears when one type of cell changes from its normal characteristics and grows and multiplies abnormally. The unusual growth of cells within the brain or inside the skull, which can be cancerous or non-cancerous has been the reason for the death of adults in developed countries and children in under developing countries like Ethiopia. The studies have shown that the region growing algorithm initializes the seed point either manually or semi-manually which as a result affects the segmentation result. However, in this paper, we proposed an enhanced region-growing algorithm for the automatic seed point initialization. The proposed approach's performance was compared with the state-of-the-art deep learning algorithms using the common dataset, BRATS2015. In the proposed approach, we applied a thresholding technique to strip the skull from each input brain image. After the skull is stripped the brain image is divided into 8 blocks. Then, for each block, we computed the mean intensities and from which the five blocks with maximum mean intensities were selected out of the eight blocks. Next, the five maximum mean intensities were used as a seed point for the region growing algorithm separately and obtained five different regions of interest (ROIs) for each skull stripped input brain image. The five ROIs generated using the proposed approach were evaluated using dice similarity score (DSS), intersection over union (IoU), and accuracy (Acc) against the ground truth (GT), and the best region of interest is selected as a final ROI. Finally, the final ROI was compared with different state-of-the-art deep learning algorithms and region-based segmentation algorithms in terms of DSS. Our proposed approach was validated in three different experimental setups. In the first experimental setup where 15 randomly selected brain images were used for testing and achieved a DSS value of 0.89. In the second and third experimental setups, the proposed approach scored a DSS value of 0.90 and 0.80 for 12 randomly selected and 800 brain images respectively. The average DSS value for the three experimental setups was 0.86.

**Keywords:** brain MRI image; tumor region; skull stripping; region growing; U-Net; BRATS dataset



**Citation:** Biratu, E.S.; Schwenker, F.; Debelee, T.G.; Kebede, S.R.; Negera, W.G.; Molla, H.T. Enhanced Region Growing for Brain Tumor MR Image Segmentation. *J. Imaging* **2021**, *7*, 22. https://doi.org/10.3390/jimaging7020022

Academic Editor: Leonardo Rundo
Received: 19 November 2020
Accepted: 26 January 2021
Published: 1 February 2021



## 1. Introduction

Cancer is a critical health problem with a very high mortality rate in the world. But we can prevent deaths and illnesses from cancer if we can diagnose it earlier. Globally the mean five-year survival rate of cancer patients has increased from 49% to 67%. The main reason behind this improvement is the rapid growth in diagnostic and treatment techniques [1]. A brain tumor is one of the deadliest cancers among children and adults. A brain tumor is an abnormal mass of brain tissue that grows out of the control of the normal forces that regulate growth inside the skull. These unusual growths can be cancerous or non-cancerous [2]. There are many pieces of research carried out in the past few decades on a brain tumor, but it remained to be one of the major causes among much common type of cancers for the death of people in the entire world [3].

We can classify brain tumors as primary brain tumors and secondary brain tumors depending on the point of origin. Primary brain tumors originate from the brain tissues, whereas secondary tumors originate elsewhere and spread to the brain via hematogenous or lymphatic route. We can categorize brain tumors in terms of severity as benign and malignant [4]:

- Benign brain tumors are those that grow slowly and do not metastasize or spread to other body organs and often can be removed and hence are less destructive or curable. They can still cause problems since they can grow big and press on sensitive areas of the brain (the so-called mass effect). Depending on their location, they can be life-threatening.
- Malignant brain tumors are those with cancerous cells. The rate of growth is fast ranging from months to a few years. Unlike other malignancies, malignant brain tumors rarely spread to other body parts due to the tight junction in the brain and spinal cord.

*Brain Tumor Imaging Technologies*

Medical imaging technologies revolutionized medical diagnosis over the last 40 years allowing doctors to detect tumors earlier and improve the prognosis by visualizing tissue structures [5]. The most common imaging modalities for the detection of brain tumors include computed tomography (CT), magnetic resonance imaging (MRI), and positron emission tomography (PET) [5]. MRI is the most commonly used system to diagnose brain tumors since it presents accurate details about the investigated tumor and has little risk to radiations. Additionally, it is capable of differentiating soft tissue with high resolution and is more sensitive in detecting and visualizing subtle changes in tissue density and the physiological alternations associated with the tumor [6–9]. Usually, one imaging modality is used in the diagnosis of brain tumors. But in some cases, more than one imaging modality might be advantageous in the diagnosis of brain tumors using medical image registration. Rundo et al. [10] explored the use of medical registration, which is a process of combining information from different imaging modalities into single data. These fusions usually require optimization of similarity between the different modality input images. CNN based optimization for medical image registration was performed in [11].

MRI is a non-invasive imaging technique that produces three-dimensional anatomical images by measuring the energy released when a proton changes its polarity after it was altered using a strong magnetic field. MRIs are sain the detection of abnormalities in the soft tissues.

MRI images can be taken in many ways [12]. The most common and widely used modalities include:

- T1-weighted: by measuring the time required for the magnetic vector to return to its resting state(T1-relaxation time)
- T2-weighted: by measuring the time required for the axial spin to return to its resting state (T2-relaxation time).
- Fluid-attenuated inversion recovery(T2-FLAIR): which is T2 weighted by suppressing cerebrospinal fluid(CSF).

T1, especially with the addition of contrast(Gadolinium), is effective in the detection of new lesions, whereas T2 and Flair are effective in defining high-grade glial neoplasm(glioma) and surrounding edema. Flair performs better in defining the actual volume of the neoplasm [13]. In this paper, we considered Flair images since they are effective in the detection of Gliomas (such as glioblastoma, astrocytomas, oligodendrogliomas, and ependymomas), that makeup 81% of malignant brain tumors in adults [14].

## 2. Related Works

Since medical images contain artifacts such as tags, noises, and other body parts that are not the area of interest, they needed to be removed [15]. Then, segmentation tasks are performed to extract the region of interest for the detection and classification step. Recently, deep-learning based methods tried to combine both segmentation and classification of medical images in one process. Brain tumor segmentation can be categorized into region-based and deep-learning-based segmentations. From region-based segmentation algorithms, we will be addressing clustering, region growing, fuzzy means segmentation algorithms. And, from deep learning U-Net has been addressed.

### *2.1. Region-Based Brain Tumor Segmentation*

A lot of researches have been carried out in the area of segmentation for medical images like breast cancer and brain tumor using various segmentation methods [16]. However, the complexity and large variations of the tissue structure and indistinguishable boundaries between regions of the human brain tissues made the brain tumor segmentation a challenging task [17]. In the past few years, different brain image segmentation approaches have been developed for MRI images and evaluated using different evaluation parameters.

One of the most common, easiest, and fastest algorithm for image segmentation is thresholding. The thresholding technique is based on one or more intensity threshold values where these values are compared with pixel intensities. Thresholding performs well when there is homogeneous intensity in the image. However, applying the thresholding segmentation algorithm to brain tumor segmentation is not recommended because of two reasons: optimal threshold selection is not an easy task, and intensity in the brain tumor is not homogeneous [18]. These problems have been tried to be addressed using image enhancement techniques for clearly differentiating between tissue regions on MRI scans. Rundo et al. [19] proposed a novel medical image enhancement technique called medGA, which is a pre-processing technique based on the genetic algorithm. But medGA needs a user input for the ROI from the MRI slices. Acharya and Kumar [20] proposed a particle-swarm-based contrast enhancement technique for brain MRI images. They have compared the proposed algorithm with other contrast enhancement techniques. But they didn't show its performance when it is applied as a pre-processing for segmentation using a thresholding technique.

The other commonly used segmentation algorithm in medical images is the watershed algorithm. The working principle behind the watershed segmentation algorithm is similar to the water flooding in the rigged landscape [18]. The watershed algorithm can accurately segment multiple regions at the same time with complete contour for each section. But, the watershed segmentation algorithm suffers from over-segmentation [21].

The region growing algorithm is one of the most successful approaches for brain tumor segmentation. This approach mainly extracts regions with similar pixels [18]. The region-growing algorithm's performance is highly dependent on the initial seed point selection and the type of similarity measure used between neighboring pixels. However, in most cases selecting an optimal seed point is made manually as presented in Table 1 and a challenging task besides its higher computational cost [18].

Salman et al. [22] and Sarathi et al. [23] stated that region growing segmentation algorithm has shown better performance for brain tumor segmentation to generate ROI. However, Salman et al. [22] in their work manually selected the initial point as the seed for the region growing algorithm-based approach that they proposed to get ROI. Thiruvenkadam [24] explained that manual seed point selection is the most important step for region growing based brain tumor segmentation.

Cui et al. [17] fused two MRI images (MRI-FLAIR and MRI-T2) for generating initial seed points for the region growing algorithm. They automatically select seed points but the overall algorithm is not consistent. The inconsistency comes from the fact that seed points are selected randomly from a set of potential seed points generated by calculating seeds' probability of belonging to a tumor region.

Sarathi et al. [23] proposed a wavelet features based region growing segmentation algorithm for an original 256 × 256 T1-weighted enhanced MRI image. For the selection of seed points, they first convolved the 64 × 64 kernel with the 64 × 64 preprocessed brain images and followed by wavelet feature extraction. Then significant wavelet feature points were used alternatively as a potential initial seed until the best ROI is extracted. In this paper, mean, variance, standard deviation, and entropy were used as similarity properties to include or exclude the neighboring pixels to the seed point. The experimental result showed that the proposed approach gave better performance results with minimum computational time.

In [25] the intensity values of brain tissue from its different regions were considered to decide the selection of the seed points. However, brain map structure and intensity information need to be known in advance. Therefore, to gain detailed information on the brain images, multi-modal images were preferred, and hence in this work Ho et al. [25] used a fusion of multi-modal images to select the initial seed automatically.

Bauer et al. [26] used a soft-margin SVM classifier for the segmentation of brain tumors hierarchically by classifying MRI voxels. 28 features were extracted from the voxel intensity and first-order textures extracted from patches around the voxel. Conditional Random Fields(CRF) regularization was applied to introduces spatial constraints to the SVM classifier since considers each voxel is independent. The proposed algorithm achieved a DSS of 0.84. They didn't specify the size of patches taken around the voxels when extracting texture features. There was no comparison performed with state-of-the-art algorithms.

Rundo et al. [27] used Fuzzy C-Means(FCM) based segmentation algorithms to segment the whole tumor volume using their gross tumor volume (GTV) segmentation in the first step and extract the necrosis volume from the gross tumor volume in the second step. But the proposed algorithm needs human intervention for the GTV algorithm.

**Table 1.** Related Work in region growing seed selection and growth criteria.

| Authors and Citation | Seed Selection | RG Criteria |
|---|---|---|
| Salman et al., 2006 [22] | Manual | Texture |
| Sarathi et al., 2013 [23] | Automatic | variance, Entropy |
| Thiruvenkadam, 2015 [24] | Manual | - |
| Ho et al., 2016 [25] | Automatic | Intensity |
| Cui et al., 2019 [17] | Semi-automatic | Intensity & Spatial Texture |

### 2.2. Deep Learning-Based Brain Tumor Segmentations

Deep learning has been applied for the classification and segmentation of medical images previously [28–32]. Different versions of CNNs were used for the segmentation of brain tumors from MRI scans.

Li et al. [33], applied generative adversarial networks(GANs) to augment brain datasets by generating realistic paired data. The proposed method can augment n data pairs into n 2-n data. Their data augmentation technique was used to train and test different deep learning-based segmentation techniques using the BRATS2017 dataset. The best performer, the U-net algorithm, achieved a DSS of 0.754 when using the original dataset but this performance was improved to 0.765 in the case of whole tumor segmentation. The network architecture of U-Net is symmetric and composed of Encoder and decoder. The encoder is used to extract features from the input images and decoder constructs segmentation from the extracted features in Encoder [34]. U-Net became the most popular semantic segmentation in medical imaging [34]. In this paper, U-Net was implemented for comparing the performances of our proposed model.

Rundo et al. [35] modified the original U-Net architecture by adding squeeze-excitation (SE) blocks in every skip connection. They proposed two architectures, first only the encoder block output was feed to SE blocks at the skip connection. Another architecture was modifying each skip connection by adding SE blocks at every encoder and decoder block and combine the outputs to modify the original skip connection. The SE blocks are designed to model interdependencies between channels and increases the model generalization capabilities when trained using different datasets. The datasets consisted of prostate MRI scans for zonal segmentation collected from various institutions. The SE block's ability to adaptive feature recalibration significantly improves the performances of the U-net architecture, when trained across different datasets.

## 3. Materials and Methods

Figure 1 presents the flowchart of the proposed enhanced region-growing algorithm for brain tumor segmentation. Raw MRI images usually have different artifacts and non-brain parts that affect the segmentation quality and hence a preprocessing step should be applied before segmentation algorithms. The enhanced region-growing algorithm is applied to generate candidate brain tumor regions. The detail methods used in this paper is presented in Section 3.1 through Section 3.4.

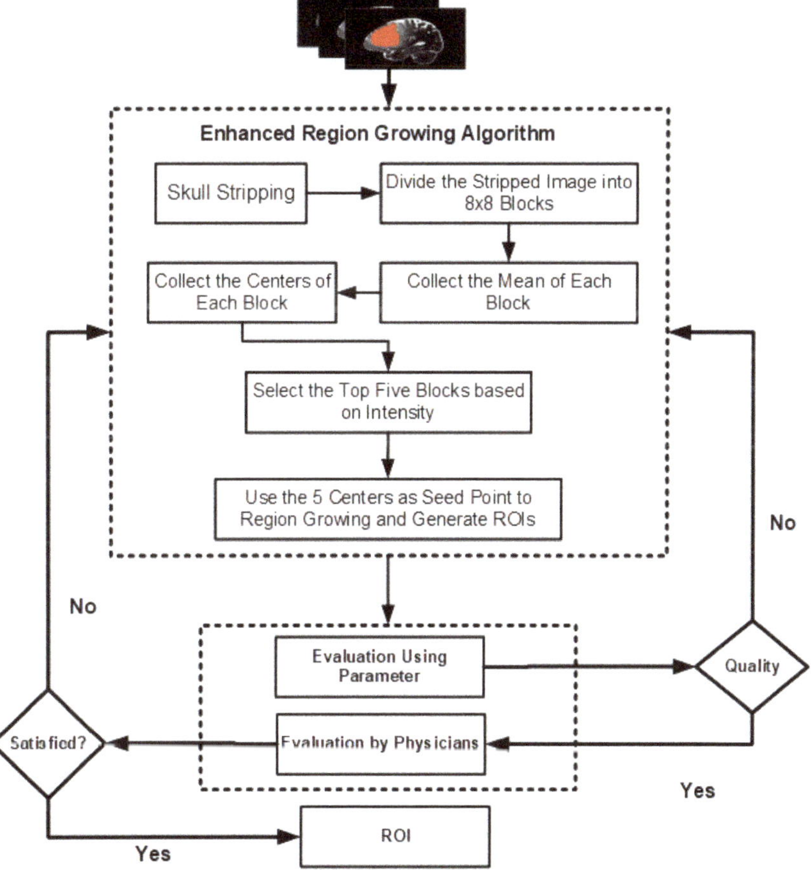


**Figure 1.** Flowchart of the proposed region growing algorithm. In this approach the segmentation result is evaluated both by evaluation parameters and Physicians/Radiologists.

### 3.1. Dataset

The image dataset used in this paper contains multimodal MRI scans of patients with gliomas (54 LGGs and 132 HGGs). It was used for the multimodal Brain Tumor Segmentation (BRATS) 2015 challenge, from the Virtual Skeleton Database (VSD) [36]. Specifically, these image datasets were a combination of the training set (10 LGGs and 20 HGGs) used in the BRATS 2013 challenge [37], as well as 44 LGG and 112 HGG scans provided from the National Institutes of Health (NIH) Cancer Imaging Archive (TCIA). The data of each patient consisted of native and contrast-enhanced (CE) T1-weighted, as well as T2-weighted and T2 Fluid-attenuated inversion recovery (FLAIR) MRI volumes.

In the dataset, the ground truth (GT) was included for training the segmentation model and qualitative evaluation. Specifically, the data from BRATS 2013 were manually annotated, whereas data from TCIA were automatically annotated by fusing the approved by experts results of the segmentation algorithms that ranked high in the BRATS 2012 and 2013 challenges [37]. The GT segmentations comprise the enhancing part of the tumor (ET), the tumor core (TC), which is described by the union of necrotic, non-enhancing, and enhancing parts of the tumor, and the whole tumor (WT), which is the union of the TC and the peritumoral edematous region.

### 3.2. Preprocessing

In digital image processing preprocessing plays an important role in smoothing and normalizing the MRI images [38]. Performing preprocessing suppresses the impact of dark parts in the borders of the brain images [38].

The BRATS2015 dataset is available in a preprocessed format in which unwanted parts are removed. But, preprocessing is essential for raw MRI data. Skull Stripping is one of the popular pre-processing techniques that remove the skull from brain image. The surroundings of a brain are termed as a skull. The skull stripping is the process of eradicating the tissues that are not cerebral. It is difficult to distinguish non-cerebral and the intra-cranial tissues because of their homogeneity in intensities [39]. In brain tumor segmentation, stripping the skull and other non-brain parts is a crucial step to be accomplished but it is a challenging task [40]. The challenge arises from large anatomical variability among brains, different acquisition methods of brain images, and the existence of artifacts on brain images. These are some of the reasons among many that boost the challenge to design a robust algorithm [40]. Segonne et al. [40] proposed a hybrid approach that was used to strip the skull where they combined the watershed algorithm and deformable surface model.

In the proposed approach, we applied thresholding and morphological operation for preprocessing (see Algorithm 1). Since the MRI images in the local dataset are images with three color channels, it was changed into a grayscale image before the preprocessing. Otsu's thresholding technique was employed to determine the threshold between the background and the tissue regions. By thresholding, the largest binary object extracts the brain and removes the skull and other tags from the image. Some examples of skull removal algorithm are presented in Figure 2.

**Algorithm 1** Skull Stripping

1: **input:** gray scale image, *im*
2: Calculate Otsu's Threshold
   $T \leftarrow graythresh(im)$
3: Threshold the image
   $BW \leftarrow im2bw(im, T)$
4: Open the binary image using a disk structuring Seed
   $BW \leftarrow imopen(BW, se)$
5: Dilate the binary image
   $BW \leftarrow imdilate(BW, se)$
6: Select the largest binary image
   $BW \leftarrow largest\_blob(BW)$
7: Dilate the binary image
   $BW \leftarrow imclose(BW, se)$
8: Fill holes on the binary image
   $BW \leftarrow imfill(BW, se)$
9: Remove the skull
   $stripped \leftarrow im(!BW) = 0$
10: **return** *stripped*

### 3.3. Enhanced Region-Growing Approach

The proposed enhanced region-growing based approach that automatically detect the abnormality region and extract the ROI for each brain image is presented in Algorithm 2. This approach is the main contribution of the paper. The role of Algorithm 1 is to strip the skull of the input original brain image. Then, the skull stripped brain image is divided into 32 blocks or patches of size $8 \times 8$. For each $\text{block}_i$, the average (mean) intensity was computed as indicated in Equation (1):

$$AvgI_{i=1:32} = \frac{\sum_{j=1}^{8} \sum_{k=1}^{8} I_{jk}}{64} \tag{1}$$

As presented in Algorithm 2, line 6 and Equation (1), the mean intensities for each of the 32 blocks were computed and selected only the top five brightest pixels as potential candidates to use as seed points for the region-growing segmentation algorithm, refer Figure 3a,c,e,g. Line 12 to 14 of Algorithm 2 presented the five ROIs generated by region-growing segmentation algorithm, and then compared the results against the ground truth using evaluation parameters to select the best ROI as a final segmentation output, see Figure 3b,d,f,h. The region-growing segmentation algorithm's threshold point is determined experimentally to be 0.1 since most of the tumor regions appear homogeneous. However, some of the inhomogeneities parts were accommodated with fill hole operations as shown in Figure 3h. In this particular brain image, the tumor core appears black and our algorithm might detect only the boundaries. But, for such cases, we applied the fill holes operations to include the core of the tumor.

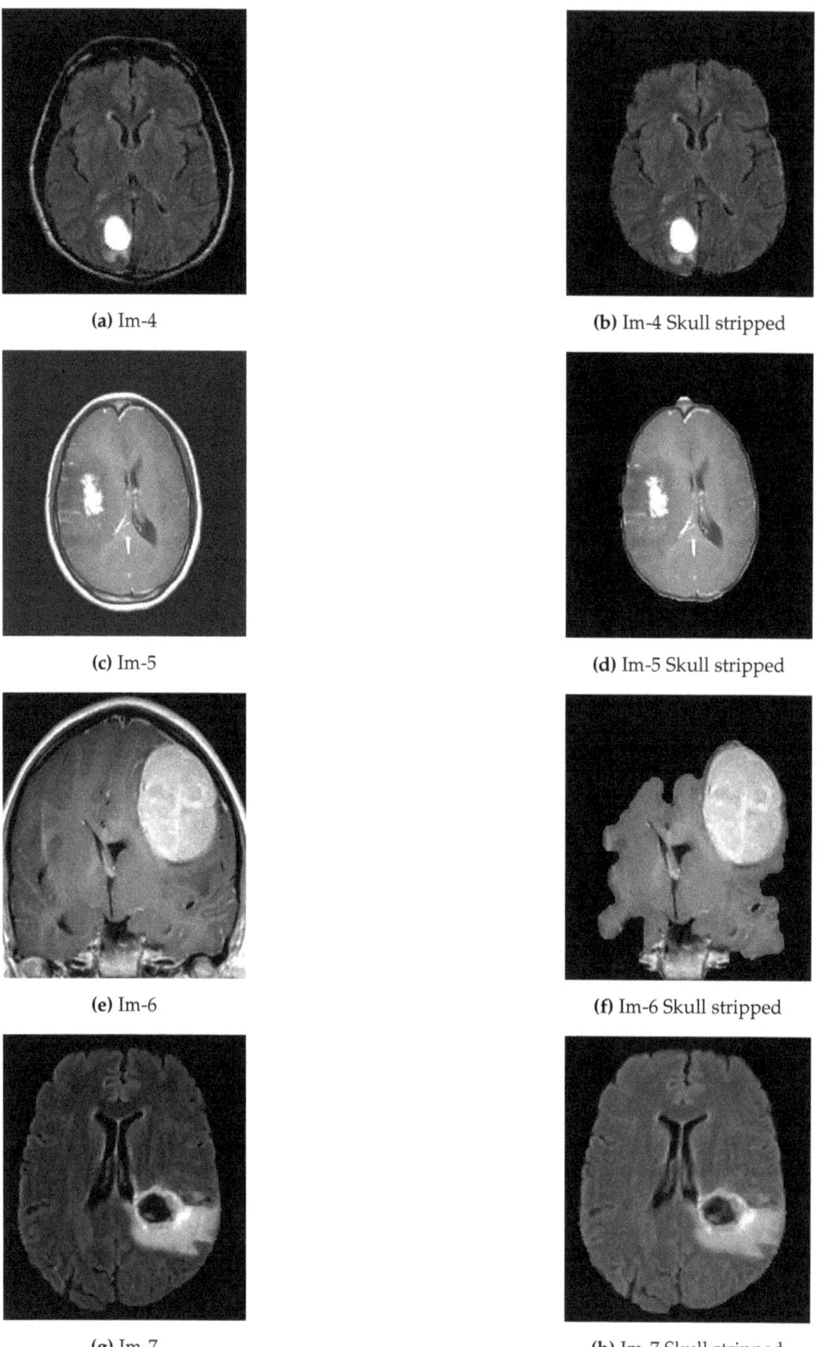


**Figure 2.** Examples of original abnormal brain tumor images before and after skull removed. (**a**,**c**,**e**,**g**) represent original brain images with skull; (**b**,**d**,**f**,**h**) represent the skull removed original brain images.

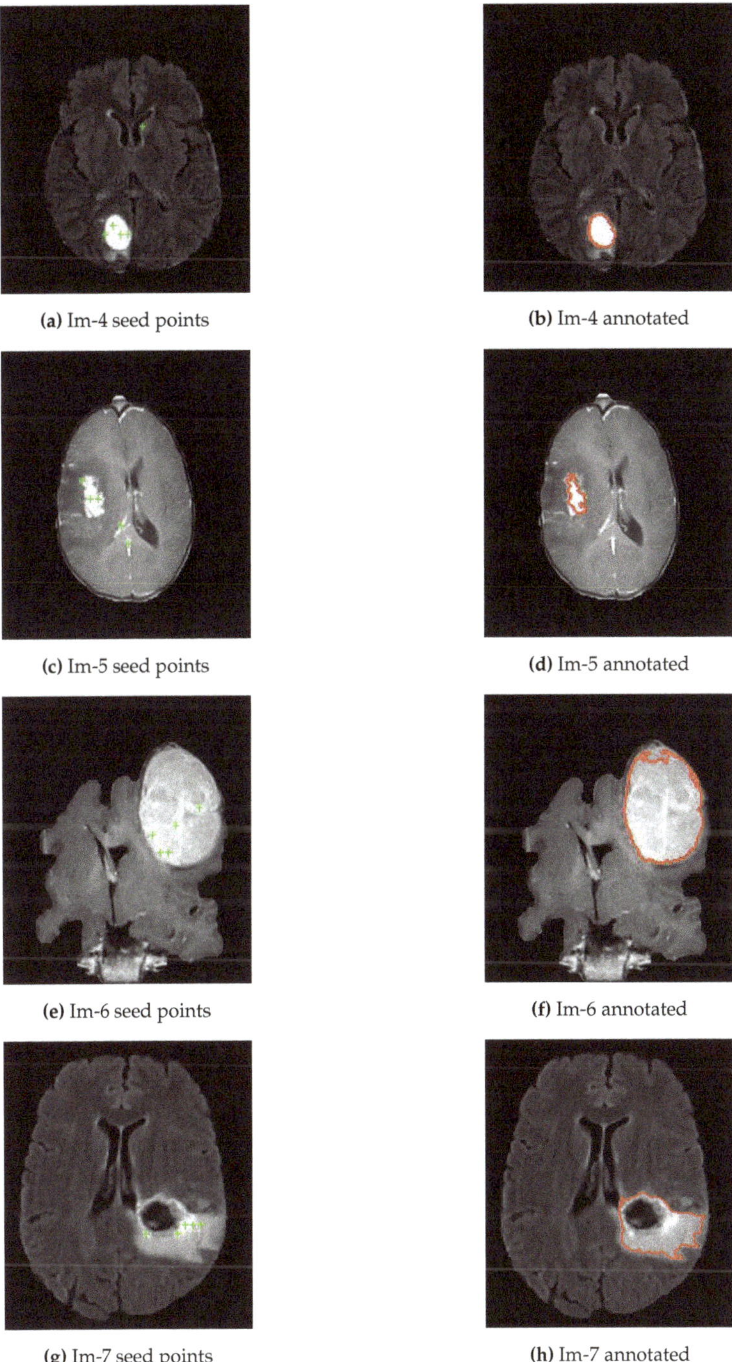


**Figure 3.** Generated possible seed points and annotations using proposed approach. (**a**,**c**,**e**,**g**) represent a skull removed original brain images with five potential seed points for brain images; (**b**,**d**,**f**,**h**) represent the best ROIs of each respective brain images.

**Algorithm 2** Enhanced Region Growing Segmentation for Brain Tumor Segmentation

**input:** skull stripped image, $im$
Resize the Image
  $im \leftarrow imresize(im, [256, 256])$
iterate through each $8 \times 8$ block
**for** $i = 1 : 8 : 256$ **do**
  **for** $j = 1 : 8 : 256$ **do**
    Collect the mean of each block
      $mIs \leftarrow mean(im(i : i + 7, j : j + 7)$
    Collect the centers of each block
    $cBs \leftarrow [i + 3, j + 3]$
  **end for**
**end for**
Select top 5 blocks based on the intensity
  $[ind, vals] = max(mIs, 5)$
  $seeds = cBs(ind)$
**return** $seeds$
**for** $m = 1 : 5$ **do**
  ROI$_m$= Region-growing(seed$_m$)
**end for**
Compare each ROI$_m$ against GT using evaluation parameters for $m = 1 : 5$
Select the best ROI as a final segmentation output.

### 3.4. Evaluation Approach

The most common parameters to be used to evaluate the performance of segmentation algorithms are DSS, Similarity Index (SI), Extra Fraction (EF), Overlap Fraction (OF), Jaccard Similarity (JSI), accuracy (Acc), sensitivity (Sn), specificity (Sp), computation cost, Root Mean Squared Error (RMSE) and intersection over union (IoU). JSI is similar with IoU and Sp is similar with SI.

Consider True Positive (TP) as the number of tumor region pixels correctly identified and classified, False Positive (FP) as the number of normal region pixels in the input image identified as tumor region, False Negative (FN) as the number of tumor region pixels left undetected or misclassified, and True Negative (TN) as the number of normal region pixels in the input region identified as the normal region.

#### 3.4.1. Extra Fraction (EF)

Extra fraction refers to the number of pixels being falsely detected as a tumor region. A minimum extra fraction value means a better segmentation result [41].

$$EF = \frac{FP}{TP + FN} \tag{2}$$

#### 3.4.2. Overlap Fraction (OF)

Overlap fraction or sensitivity value refers to the number of images segmented and classified correctly [41]. Specifically, overlap fraction refers to the tumor region being correctly identified.

$$OF = \frac{TP}{TP + FN} \tag{3}$$

#### 3.4.3. Dice Similarity Score (DSS)

It measures the spatial overlap between the original image and the segmented target region.

$$DSS = \frac{TP}{\frac{1}{2}(2TP + FP + FN)} \tag{4}$$

Besides, we have involved the radiologist to evaluate the final ROIs obtained using the proposed approach for randomly selected brain images to validate our proposed approach qualitatively.

## 4. Experimental Results and Discussion

The first experimental result was the skull stripped brain images as indicated in Figure 2 where Figure 2a,c,e,g were the original brain images of size $256 \times 256$ and Figure 2b,d,f,h were the skull stripped brain images. Then, as presented in Equation (1), we generated 32 average intensities for each skull stripped brain images and selected the five top average intensities for each image and used as potential initial seed points for region growing algorithm as indicated in Figure 3a,c,e,g. Using the five selected initial seed points for each image, we generated five different ROIs and compared against the respective GT and selected the best ROI as presented in Figure 3b,d,f,h.

To validate the proposed approach, we designed three different experimental setups for analysis. In our first experiment, we randomly selected 15 brain images from the BRATS2015 dataset. In the second experiment, we again randomly selected 12 brain images from the same dataset and finally, in the third experimental setup, we used 800 brain images from the same dataset used in the previous two experimental setups.

In all the three experimental setups, the performance of the proposed approach was evaluated in terms of Acc, IoU, DSS, Sn, Sp, EF, OF, and PSNR. In most cases, especially the deep learning algorithms use DSS to evaluate the segmentation algorithms. The highest value of Acc, IoU, DSS, Sn, Sp, OF and PSNR indicate the highest performance whereas the lowest value of EF indicates poor performance.

In the first experimental setup, 15 brain images were used for experimental analysis, and for each image, the corresponding Acc, IoU, DSS, Sn, Sp, OF, EF, and PSNR were computed as indicated in Table 2. The average value of Acc, IoU, DSS, Sn, Sp, OF, EF, and PSNR for the 15 brain images were used to compare the performance of the proposed approach with that of modified adaptive K-means and U-Net.

**Table 2.** Performance comparison of RG with MAKM and U-Net for 15 randomly selected brain images from BRATS2015 Dataset.

| Metric | Algorithm | im01 | im02 | im03 | im04 | im05 | im06 | im07 | im08 | im09 | im10 | im11 | im12 | im13 | im14 | im15 | Avg |
|---|---|---|---|---|---|---|---|---|---|---|---|---|---|---|---|---|---|
| | RG | 100 | 100 | 100 | 100 | 99 | 99 | 99 | 99 | 99 | 99 | 100 | 99 | 99 | 88 | 94 | 98 |
| Acc (%) | MAKM | 99 | 99 | 99 | 82 | 99 | 99 | 99 | 99 | 86 | 86 | 80 | 87 | 99 | 87 | 99 | 93 |
| | U-Net | 100 | 100 | 100 | 100 | 98 | 98 | 74 | 74 | 99 | 99 | 67 | 99 | 100 | 99 | 92 | 93 |
| | RG | 0.94 | 0.94 | 0.94 | 0.93 | 0.88 | 0.88 | 0.85 | 0.85 | 0.85 | 0.85 | 0.84 | 0.83 | 0.81 | 0.31 | 0.04 | 0.78 |
| IoU | MAKM | 0.90 | 0.79 | 0.79 | 0.21 | 0.86 | 0.86 | 0.90 | 0.90 | 0.26 | 0.26 | 0.06 | 0.19 | 0.81 | 0.34 | 0.65 | 0.59 |
| | U-Net | 0.94 | 0.96 | 0.96 | 0.93 | 0.70 | 0.70 | 0.16 | 0.16 | 0.91 | 0.91 | 0.03 | 0.84 | 0.93 | 0.81 | 0.24 | 0.68 |
| | RG | 0.97 | 0.97 | 0.97 | 0.96 | 0.93 | 0.93 | 0.92 | 0.92 | 0.92 | 0.92 | 0.91 | 0.91 | 0.89 | 0.47 | 0.80 | 0.89 |
| DSS | MAKM | 0.95 | 0.88 | 0.88 | 0.35 | 0.92 | 0.92 | 0.95 | 0.95 | 0.42 | 0.42 | 0.11 | 0.33 | 0.90 | 0.51 | 0.79 | 0.68 |
| | U-Net | 0.97 | 0.98 | 0.98 | 0.96 | 0.82 | 0.82 | 0.27 | 0.27 | 0.95 | 0.95 | 0.07 | 0.92 | 0.96 | 0.89 | 0.39 | 0.75 |
| | RG | 97 | 95 | 95 | 98 | 88 | 88 | 87 | 87 | 85 | 85 | 85 | 83 | 81 | 100 | 100 | 90 |
| Sn (%) | MAKM | 91 | 79 | 79 | 100 | 86 | 86 | 96 | 96 | 100 | 100 | 100 | 99 | 85 | 100 | 65 | 91 |
| | U-Net | 100 | 98 | 98 | 93 | 95 | 95 | 99 | 99 | 100 | 100 | 90 | 100 | 96 | 88 | 65 | 94 |
| | RG | 100 | 100 | 100 | 100 | 100 | 100 | 100 | 100 | 100 | 100 | 100 | 100 | 100 | 84 | 00 | 92 |
| Sp(%) | MAKM | 100 | 100 | 100 | 81 | 100 | 100 | 100 | 100 | 85 | 85 | 79 | 86 | 100 | 86 | 100 | 93 |
| | U-Net | 100 | 100 | 100 | 100 | 98 | 98 | 73 | 73 | 99 | 99 | 67 | 99 | 100 | 99 | 93 | 93 |
| | RG | 0.03 | 0.02 | 0.02 | 0.06 | 0.00 | 0.00 | 0.02 | 0.02 | 0.00 | 0.00 | 0.01 | 0.00 | 0.00 | 2.27 | 23.37 | 1.72 |
| EF | MAKM | 0.01 | 0.00 | 0.00 | 3.79 | 0.00 | 0.00 | 0.06 | 0.06 | 2.79 | 2.79 | 16.11 | 4.09 | 0.04 | 1.92 | 0.00 | 2.11 |
| | U-Net | 0.05 | 0.02 | 0.02 | 0.00 | 0.35 | 0.35 | 5.23 | 5.23 | 0.10 | 0.10 | 25.57 | 0.18 | 0.03 | 0.08 | 1.69 | 2.60 |
| | RG | 0.97 | 0.95 | 0.95 | 0.98 | 0.88 | 0.88 | 0.87 | 0.87 | 0.85 | 0.85 | 0.85 | 0.83 | 0.81 | 1.00 | 1.00 | 0.90 |
| OF | MAKM | 0.91 | 0.79 | 0.79 | 1.00 | 0.86 | 0.86 | 0.96 | 0.96 | 1.00 | 1.00 | 1.00 | 0.99 | 0.85 | 1.00 | 0.65 | 0.91 |
| | U-Net | 1.00 | 0.98 | 0.98 | 0.93 | 0.95 | 0.95 | 0.99 | 0.99 | 1.00 | 1.00 | 0.90 | 1.00 | 0.96 | 0.88 | 0.65 | 0.94 |
| | RG | 72.72 | 74.40 | 74.40 | 72.72 | 70.22 | 70.22 | 69.50 | 69.50 | 69.38 | 69.38 | 75.09 | 70.79 | 68.25 | 56.31 | 48.31 | 68.75 |
| PSNR | MAKM | 70.63 | 69.38 | 69.38 | 55.51 | 69.67 | 69.67 | 71.02 | 71.02 | 56.66 | 56.66 | 55.02 | 56.88 | 68.19 | 57.03 | 66.53 | 64.22 |
| | U-Net | 72.99 | 76.49 | 76.49 | 72.64 | 65.12 | 65.12 | 54.06 | 54.06 | 71.02 | 71.02 | 53.00 | 70.37 | 72.64 | 66.70 | 58.92 | 66.71 |

Table 2 indicates that the proposed algorithm outperformed modified adaptive K-means, and U-Net in terms of an average value of Acc, IoU, DSS, EF, and PSNR. However, it achieved a lower average value of Sn, Sp, and OF. The lower average value of Sn, Sp, and OF is achieved because of the least value of respective parameters for images 14 and 15 However, still, the U-Net and MAKM have an insignificant higher performance than the proposed approach. In the case of OF and Sn, U-Net achieved 4% and MAKM achieved 1% higher than the proposed approach. In the case of Sp, both the U-Net and MAKM are 1% higher than the proposed approach.

Table 3 presented the comparison of the proposed approach, MAKAM and U-Net for the 12 randomly selected brain images from BRATS2015. The proposed approach scored a higher value of Acc, IoU, DSS, Sp, EF, and PSNR but a lower value of Sn and OF compared to MAKM and U-Net. The value of Acc, IoU, DSS, Sp, EF, and PSNR were 99.1%, 0.82, 0.90, 99.7%, 0.06, and 163.89 respectively whereas the value of Sn and OF were 89.1% and 0.89 respectively. U-Net achieved a higher value for both Sn and OF compared to MAKM and the proposed approach where performance difference was limited to nearly to 2%.

**Table 3.** Performance comparison of RG with MAKM and U-Net for 12 randomly selected brain images from BRATS2015 Dataset.

| Metric | Algorithm | im081 | im274 | im473 | im551 | im06 | im973 | im689 | im792 | im1507 | im781 | im733 | im1238 | Avg |
|---|---|---|---|---|---|---|---|---|---|---|---|---|---|---|
| | RG | 99.6 | 99.8 | 97.4 | 99.6 | 99.6 | 99.7 | 100.0 | 99.1 | 98.7 | 99.2 | 99.7 | 96.8 | 99.1 |
| Acc(%) | MAKM | 84.9 | 89.1 | 97.2 | 95.9 | 85.4 | 79.7 | 76.9 | 87.7 | 84.3 | 95.6 | 90.4 | 84.6 | 87.6 |
| | U-NET | 99.8 | 99.8 | 93.3 | 99.8 | 99.8 | 98.7 | 99.8 | 89.2 | 99.5 | 99.5 | 99.1 | 86.6 | 97.1 |
| | RG | 0.91 | 0.92 | 0.62 | 0.92 | 0.92 | 0.94 | 0.89 | 0.77 | 0.80 | 0.88 | 0.85 | 0.47 | 0.82 |
| IoU | MAKM | 0.05 | 0.01 | 0.61 | 0.50 | 0.23 | 0.02 | 0.02 | 0.23 | 0.29 | 0.58 | 0.04 | 0.28 | 0.24 |
| | U-NET | 0.95 | 0.93 | 0.39 | 0.94 | 0.95 | 0.76 | 0.61 | 0.25 | 0.92 | 0.93 | 0.45 | 0.31 | 0.70 |
| | RG | 0.95 | 0.96 | 0.76 | 0.96 | 0.96 | 0.97 | 0.94 | 0.87 | 0.89 | 0.94 | 0.92 | 0.64 | 0.90 |
| DSS | MAKM | 0.09 | 0.01 | 0.75 | 0.67 | 0.38 | 0.03 | 0.03 | 0.37 | 0.44 | 0.74 | 0.09 | 0.44 | 0.34 |
| | U-NET | 0.98 | 0.96 | 0.56 | 0.97 | 0.97 | 0.86 | 0.76 | 0.40 | 0.96 | 0.96 | 0.62 | 0.47 | 0.79 |
| | RG | 95.3 | 93.9 | 83.5 | 92.1 | 94.5 | 96.2 | 92.2 | 82.0 | 79.8 | 91.2 | 92.9 | 46.8 | 86.7 |
| Sn(%) | MAKM | 18.2 | 3.5 | 89.1 | 99.3 | 100.0 | 7.4 | 100.0 | 96.9 | 99.4 | 99.9 | 25.8 | 98.5 | 69.8 |
| | U-NET | 98.9 | 97.7 | 85.7 | 98.0 | 98.8 | 97.2 | 60.9 | 98.3 | 92.9 | 95.4 | 45.2 | 99.9 | 89.1 |
| | RG | 99.8 | 100.0 | 98.2 | 100.0 | 99.9 | 99.9 | 100.0 | 99.8 | 100.0 | 99.8 | 99.8 | 100.0 | 99.7 |
| Sp(%) | MAKM | 87.9 | 90.9 | 97.6 | 95.7 | 84.7 | 82.9 | 76.8 | 87.4 | 83.3 | 95.3 | 91.6 | 83.7 | 88.2 |
| | U-NET | 99.8 | 99.9 | 93.7 | 99.8 | 99.8 | 98.8 | 100.0 | 88.9 | 99.9 | 99.8 | 100.0 | 85.8 | 97.2 |
| | RG | 0.05 | 0.02 | 0.35 | 0.01 | 0.03 | 0.02 | 0.03 | 0.06 | 0.00 | 0.03 | 0.09 | 0.00 | 0.06 |
| EF | MAKM | 0.18 | 0.03 | 0.89 | 0.99 | 1.00 | 0.07 | 1.00 | 0.97 | 0.99 | 1.00 | 0.26 | 0.98 | 0.70 |
| | U-NET | 0.99 | 0.98 | 0.86 | 0.98 | 0.99 | 0.97 | 0.61 | 0.98 | 0.93 | 0.95 | 0.45 | 1.00 | 0.89 |
| | RG | 0.95 | 0.94 | 0.83 | 0.92 | 0.94 | 0.96 | 0.92 | 0.82 | 0.80 | 0.91 | 0.93 | 0.47 | 0.87 |
| OF | MAKM | 0.18 | 0.03 | 0.89 | 0.99 | 1.00 | 0.07 | 1.00 | 0.97 | 0.99 | 1.00 | 0.26 | 0.98 | 0.70 |
| | U-NET | 0.99 | 0.98 | 0.86 | 0.98 | 0.99 | 0.97 | 0.61 | 0.98 | 0.93 | 0.95 | 0.45 | 1.00 | 0.89 |
| | RG | 165.52 | 174.19 | 147.51 | 167.26 | 166.43 | 170.25 | 188.41 | 157.89 | 154.39 | 159.74 | 169.80 | 145.23 | 163.89 |
| PNSR | MAKM | 129.72 | 132.99 | 146.42 | 142.69 | 130.06 | 126.75 | 125.49 | 131.81 | 129.36 | 142.02 | 134.30 | 129.55 | 133.43 |
| | U-NET | 172.31 | 175.68 | 137.87 | 170.98 | 171.49 | 154.11 | 175.68 | 133.12 | 163.80 | 164.69 | 157.43 | 130.96 | 159.01 |

Table 4 presented the experimental results of the proposed approach for 800 brain images and compared them with the performance of MAKM and U-Net. The experimental results showed in Table 4 indicated that the proposed approach scored a higher value of Acc, IoU, DSS, Sp, EF, and PSNR but a lower value of Sn and OF compared to MAKM and U-Net. The value of Acc, IoU, DSS, Sp, EF, and PSNR was 98.72%, 0.67, 0.80, 99.8%, 0.06, and 157.0 respectively whereas the value of Sn and OF were 90.7% and 0.91 respectively. The higher value of Sn and OF were scored by U-Net.

**Table 4.** Performance comparison of RG with MAKM and U-Net for 800 brain images from BRATS2015 Dataset.

| Metric | Algorithm | im081 | im274 | im473 | im551 | im06 | im973 | im689 | im792 | im1507 | im781 | im733 | im1238 | im368 | ... | im551 | Ovr_Avg |
|---|---|---|---|---|---|---|---|---|---|---|---|---|---|---|---|---|---|
| | RG | 99.6 | 99.8 | 97.4 | 99.6 | 99.6 | 99.7 | 100.0 | 99.1 | 98.7 | 99.2 | 99.7 | 96.8 | 95.2 | ... | 97.8 | 98.72 |
| Acc(%) | MAKM | 84.9 | 89.1 | 97.2 | 95.9 | 85.4 | 79.7 | 76.9 | 87.7 | 84.3 | 95.6 | 90.4 | 84.6 | 98.8 | ... | 98.7 | 88.60 |
| | U-NET | 99.8 | 99.8 | 93.3 | 99.8 | 99.8 | 98.7 | 99.8 | 89.2 | 99.5 | 99.5 | 77.6 | 86.6 | 83.8 | ... | 99.8 | 98.20 |
| | RG | 0.91 | 0.92 | 0.62 | 0.92 | 0.92 | 0.94 | 0.89 | 0.77 | 0.80 | 0.88 | 0.85 | 0.47 | 0.28 | ... | 0.77 | 0.67 |
| IoU | MAKM | 0.05 | 0.01 | 0.61 | 0.50 | 0.23 | 0.02 | 0.02 | 0.23 | 0.29 | 0.58 | 0.04 | 0.28 | 0.81 | ... | 0.85 | 0.34 |
| | U-NET | 0.95 | 0.93 | 0.39 | 0.94 | 0.95 | 0.76 | 0.61 | 0.25 | 0.92 | 0.93 | 0.45 | 0.31 | 0.26 | ... | 0.27 | 0.60 |
| | RG | 0.95 | 0.96 | 0.76 | 0.96 | 0.96 | 0.97 | 0.94 | 0.87 | 0.89 | 0.94 | 0.92 | 0.87 | 0.43 | ... | 0.96 | 0.80 |
| DSS | MAKM | 0.09 | 0.01 | 0.75 | 0.67 | 0.38 | 0.03 | 0.03 | 0.37 | 0.44 | 0.74 | 0.09 | 0.34 | 0.90 | ... | 0.92 | 0.45 |
| | U-NET | 0.98 | 0.96 | 0.56 | 0.97 | 0.97 | 0.86 | 0.76 | 0.40 | 0.96 | 0.96 | 0.62 | 0.47 | 0.42 | ... | 0.43 | 0.69 |
| | RG | 95.3 | 93.9 | 83.5 | 92.1 | 94.5 | 96.2 | 92.2 | 82.0 | 79.8 | 91.2 | 92.9 | 46.8 | 26.8 | ... | 76.7 | 71.1 |
| Sn(%) | MAKM | 18.2 | 3.5 | 89.1 | 99.3 | 100.0 | 7.4 | 100.0 | 96.9 | 99.4 | 99.9 | 25.8 | 98.5 | 82.4 | ... | 85.5 | 89.6 |
| | U-NET | 98.9 | 97.7 | 85.7 | 98.0 | 98.8 | 97.2 | 60.9 | 98.3 | 92.9 | 95.4 | 45.2 | 99.9 | 89.4 | ... | 97.8 | 90.7 |
| | RG | 99.8 | 100.0 | 98.2 | 100.0 | 99.9 | 99.9 | 100.0 | 99.8 | 100.0 | 99.8 | 99.8 | 100.0 | 100 | ... | 100 | 99.8 |
| Sp(%) | MAKM | 87.9 | 90.9 | 97.6 | 95.7 | 84.7 | 82.9 | 76.8 | 87.4 | 83.3 | 95.3 | 91.6 | 83.7 | 100 | ... | 100 | 88.6 |
| | U-NET | 99.8 | 99.9 | 93.7 | 99.8 | 99.8 | 98.8 | 100.0 | 88.9 | 99.9 | 99.8 | 100.0 | 85.8 | 83.5 | ... | 75.7 | 92.1 |
| | RG | 0.05 | 0.02 | 0.35 | 0.01 | 0.03 | 0.02 | 0.03 | 0.06 | 0.00 | 0.03 | 0.09 | 0.00 | 0 | ... | 0 | 0.06 |
| EF | MAKM | 0.18 | 0.03 | 0.89 | 0.99 | 1.00 | 0.07 | 1.00 | 0.97 | 0.99 | 1.00 | 0.26 | 0.98 | 0.82 | ... | 0.85 | 0.90 |
| | U-NET | 0.99 | 0.98 | 0.86 | 0.98 | 0.99 | 0.97 | 0.61 | 0.98 | 0.93 | 0.95 | 0.45 | 1.00 | 0.89 | ... | 0.98 | 0.91 |
| | RG | 0.95 | 0.94 | 0.83 | 0.92 | 0.94 | 0.96 | 0.92 | 0.82 | 0.80 | 0.91 | 0.93 | 0.47 | 0.27 | ... | 0.77 | 0.71 |
| OF | MAKM | 0.18 | 0.03 | 0.89 | 0.99 | 1.00 | 0.07 | 1.00 | 0.97 | 0.99 | 1.00 | 0.26 | 0.98 | 0.82 | ... | 0.85 | 0.90 |
| | U-NET | 0.99 | 0.98 | 0.86 | 0.98 | 0.99 | 0.97 | 0.61 | 0.98 | 0.93 | 0.95 | 0.45 | 1.00 | 0.89 | ... | 0.98 | 0.91 |
| | RG | 165.52 | 174.19 | 147.51 | 167.26 | 166.43 | 170.25 | 188.41 | 157.89 | 154.39 | 159.74 | 169.80 | 145.23 | 141.3 | ... | 149.8 | 157.0 |
| PNSR | MAKM | 129.72 | 132.99 | 146.42 | 142.69 | 130.06 | 126.75 | 125.49 | 131.81 | 129.36 | 142.02 | 134.30 | 129.55 | 155.0 | ... | 154.1 | 138.6 |
| | U-NET | 172.31 | 175.68 | 137.87 | 170.98 | 171.49 | 154.11 | 175.68 | 133.12 | 163.80 | 164.69 | 157.43 | 130.96 | 129.1 | ... | 125.8 | 152.0 |

Table 5 presented the achieved state-of-the-art deep learning algorithms' results on the BRATS2015 dataset and compared with the scored performance of the proposed approach for three different experimental setups/cases. The experimental results achieved were the DSS value of 0.89, 0.90, and 0.80 for case-1, case-2, and case-3 respectively. The average DSS value of the three experimental setups was 0.86. In this paper, no classifier was applied for final segmentation but the enhanced region growing algorithm was effective in generating candidate regions of interest. We did choose the best ROI against GT from the generated ROIs to compare with the other methods. From the experimental results, we saw that the proposed approach can generate the best ROI in most of the test cases. But still, a classifier should be trained by extracting features from the abnormal ROIs for making the algorithm to detect and determine the tumor type.

Figure 4 presented the segmentation results of the proposed algorithm, MAKM, and U-Net in terms of ROIs and their respective ground truths. For im274, im473, im551, im1507, im781, and im733 the proposed approach achieved ROIs which were almost the same as their respective ground truths (GTs). The proposed approach resulted in under-segmentation for im792 and im1238 as indicated in Figure 4. For the case of U-Net, the good segmentation results were observed only for im274, im551, im1507, and im781 and unable to detect the tumor region for im473, im792, im733, and im1238. In the case of MAKAM, over-segmentation results were achieved in almost all randomly selected brain images except for im274 where it detected the normal brain image part as abnormal.

**Table 5.** Comparison of the proposed approach with U-Net and its variants using BRATS2015 dataset.

| Authors, Year and Citation | Model | Dataset | DSS |
|---|---|---|---|
| Daimary et al. [42] | U-SegNet | BRATS2015 | 0.73 |
| Zhou et al., 2019 | OM-Net + CGAp | BRATS2015 | 0.87 |
| Kayalibay et al., 2017 | CNN + 3D filters | BRATS2015 | 0.85 |
| Isensee et al., 2018 | U-Net + more filters + data augmentation + dice-loss | BRATS2015 | 0.85 |
| Kamnitsas et al., 2016 | 3D CNN + CRF | BRATS2015 | 0.85 |
| Qin et al., 2018 | AFN-6 | BRATS2015 | 0.84 |
| Havaei et al. [43] | CNN(whole) | BRATS2015 | 0.88 |
| Havaei et al. [43] | CNN(core) | BRATS2015 | 0.79 |
| Havaei et al. [43] | CNN(enhanced) | BRATS2015 | 0.73 |
| Pereira et al. [44] | CNN(whole) | BRATS2015 | 0.87 |
| Pereira et al. [44] | CNN(core) | BRATS2015 | 0.73 |
| Pereira et al. [44] | CNN(enhanced) | BRATS2015 | 0.68 |
| Malmi et al. [45] | CNN(whole) | BRATS2015 | 0.80 |
| Malmi et al. [45] | CNN(core) | BRATS2015 | 0.71 |
| Malmi et al. [45] | CNN(enhanced) | BRATS2015 | 0.64 |
| Taye et al., 2018 [46] | MAKM | BRATS2015 | 0.68 |
| Re-implemented | U-Net | BRATS2015 | 0.75 |
| Erena et al., 2020 | Case-1:Proposed Approach (15 randomly selected images) | BRATS2015 | 0.89 |
| Erena et al., 2020 | Case-2:Proposed Approach (12 randomly selected images) | BRATS2015 | 0.90 |
| Erena et al., 2020 | Case-3:Proposed Approach (800 brain images) | BRATS2015 | 0.80 |
| Erena et al., 2020 | Average:Proposed Approach | BRATS2015 | 0.86 |

For comparison purposes, we evaluated the performance of the proposed approach with MAKM and U-Net. MAKM [46] is a modified version of the adaptive k-means algorithm proposed by Debelee et al. The performance of the proposed approach was by far better than the MAKM algorithm that mainly proposed for detection of cancer on mammographic images. For the case of U-Net, we first trained the U-Net architecture from the scratch using 16000 slices extracted from MRI scans of 200 patients obtained from the BRATS2015 datasets, with 80 slices per patient (slice 50 to 130). The 200 patients were affected by the fast-growing and rapidly spreading tumors called High-Grade Glioma. The training was performed for 50 epochs until we got no significant improvements. Since the BRATS2015 datasets consisted of MRI scans with much of the preprocessing (such as tag removal and skull stripping) performed, we just applied intensity normalization before the training. We used DSS as the loss function in the training process, for the training of a nine-layer U-net architecture described in [47]. This architecture has an additional batch normalization after each convolutional layer and for evaluation purposes, we randomly selected 15 brain images for the testing after model validation and the testing DSS score value was less by 14% compared with the proposed approach.

Finally, we compared the performance of the proposed approach with the U-Net and its variants based on the BRATS2015 dataset. Daimary et al. [42] and Zhou et al. proposed a U-Net variant architecture and scored a DSS value of 0.73 and 0.87 respectively which was less than what the proposed approach scored. Havaei et al. [43] have evaluated their approach using the BRATS2015 dataset and achieved 0.88, 0.79, and 0.73 for three modalities, whole, core and enhanced respectively.

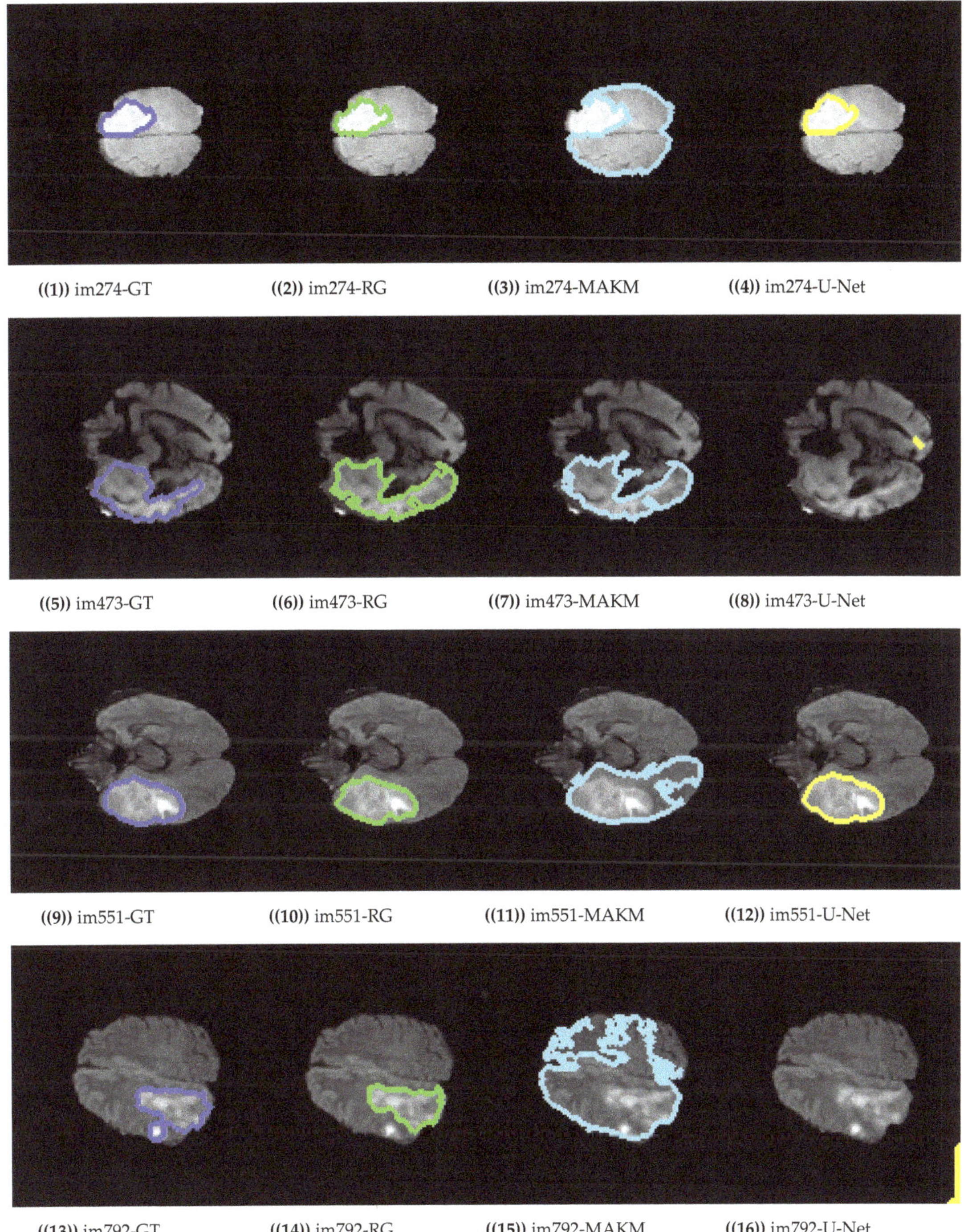

**((1))** im274-GT **((2))** im274-RG **((3))** im274-MAKM **((4))** im274-U-Net

**((5))** im473-GT **((6))** im473-RG **((7))** im473-MAKM **((8))** im473-U-Net

**((9))** im551-GT **((10))** im551-RG **((11))** im551-MAKM **((12))** im551-U-Net

**((13))** im792-GT **((14))** im792-RG **((15))** im792-MAKM **((16))** im792-U-Net

**Figure 4.** *Cont.*

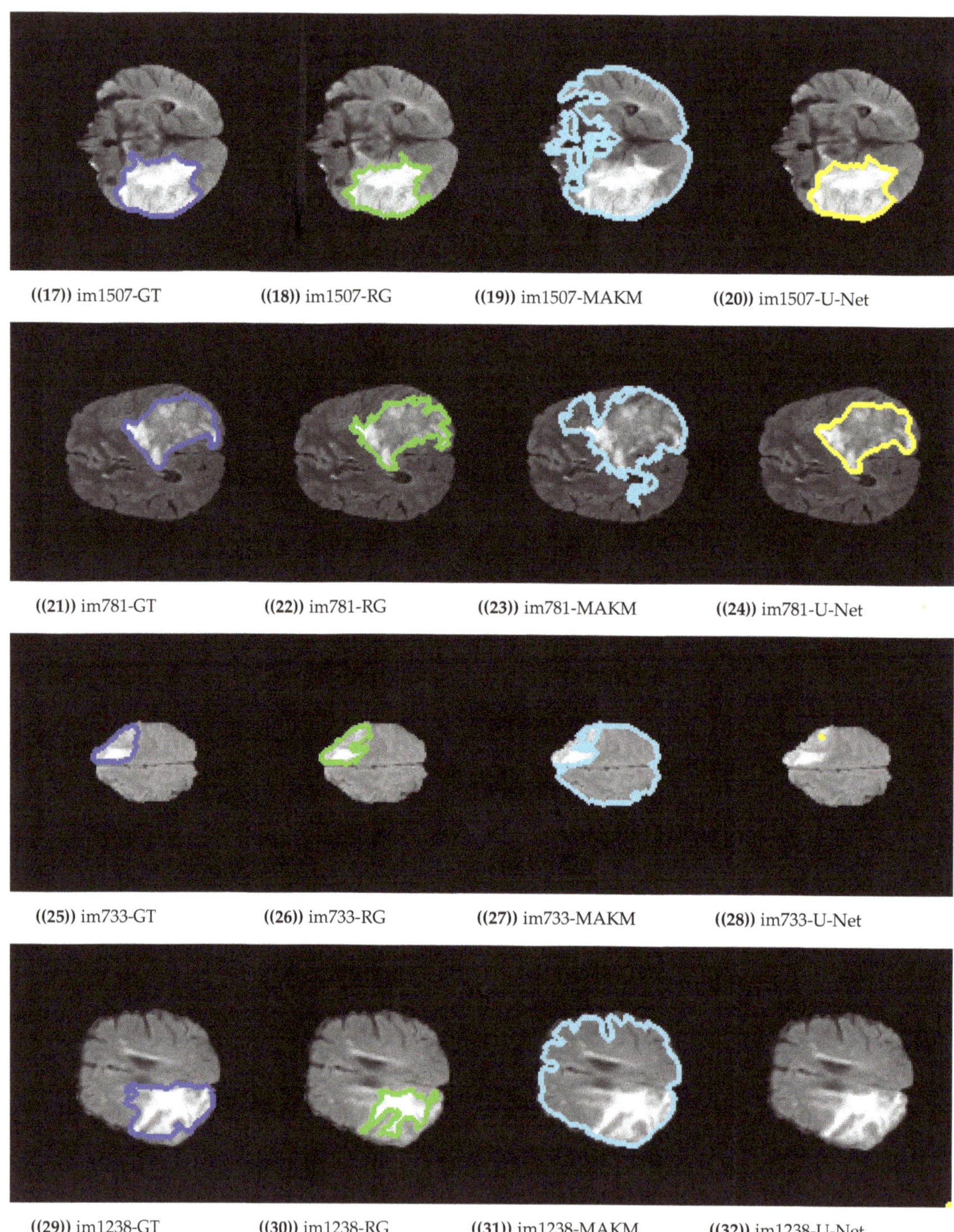

**((17))** im1507-GT **((18))** im1507-RG **((19))** im1507-MAKM **((20))** im1507-U-Net

**((21))** im781-GT **((22))** im781-RG **((23))** im781-MAKM **((24))** im781-U-Net

**((25))** im733-GT **((26))** im733-RG **((27))** im733-MAKM **((28))** im733-U-Net

**((29))** im1238-GT **((30))** im1238-RG **((31))** im1238-MAKM **((32))** im1238-U-Net

**Figure 4.** Segmentation results on BRATS2015 dataset.

## 5. Conclusions

The brain tumor is one of the major cancer types which has been a reason for the higher death rate in the entire world. To combat that a significant number of medical image analysis-based research works have been carried for different types of cancer detection and classification using deep learning and conventional/shallow machine learning approach. Shallow machine learning is usually applied in combination with digital image processing techniques for image-based analysis. In this article, we modified the existing and popular region-growing segmentation algorithm to detect the abnormality region on brain images. The main challenge of the region-growing algorithm is seed point initialization to get the best ROI for any input brain images. In the proposed approach the seed point initialization was made to be automatically generated for any input brain images and tested on the BRATS2015 dataset in three different experimental setups. The experimental result of our approach was compared with MAKM, U-Net architecture, and its variant for brain tumor detection and segmentation. From the experimental result, we have seen that the proposed algorithm can detect brain tumor locations and extract the best ROIs. The results of the proposed method achieved higher performance than modified adaptive k-means. Almost all U-Net architecture and its variants have scored lesser DSS Value for the BRATS2015 brain tumor image dataset. However, in most of the cases, the U-Net either over-segments or missed the tumor region of the brain MRI images. The proposed approach has a problem in thresholding point selection for the region-growing algorithm and was left for future work.

**Author Contributions:** Conceptualization, E.S.B.; Methodology, E.S.B.; Validation, F.S., T.G.D., S.R.K., H.T.M., W.G.N.; Writing—original draft preparation, E.S.B.; Writing—review and editing, E.S.B., F.S., S.R.K., T.G.D. All authors have read and agreed to the published version of the manuscript.

**Funding:** This research received no external funding.

**Institutional Review Board Statement:** Not applicable.

**Informed Consent Statement:** Not applicable.

**Conflicts of Interest:** The authors declare no conflict of interest.

## References

1. Sonar, P.; Bhosle, U.; Choudhury, C. Mammography classification using modified hybrid SVM-KNN. In Proceedings of the 2017 International Conference on Signal Processing and Communication (ICSPC), Coimbatore, India, 28–29 July 2017. [CrossRef]
2. Yasiran, S.S.; Salleh, S.; Mahmud, R. Haralick texture and invariant moments features for breast cancer classification. *AIP Conf. Proc.* **2016**, *1750*, 020022.
3. Aldape, K.; Brindle, K.M.; Chesler, L.; Chopra, R.; Gajjar, A.; Gilbert, M.R.; Gottardo, N.; Gutmann, D.H.; Hargrave, D.; Holland, E.C.; et al. Challenges to curing primary brain tumours. *Nat. Rev. Clin. Oncol.* **2019**, *16*, 509–520. [CrossRef] [PubMed]
4. Zhao, Z.; Yang, G.; Lin, Y.; Pang, H.; Wang, M. Automated glioma detection and segmentation using graphical models. *PLoS ONE* **2018**, *13*, e0200745. [CrossRef] [PubMed]
5. Birry, R.A.K. Automated Classification in Digital Images of Osteogenic Differentiated Stem Cells. Ph.D. Thesis, University of Salford, Salford, UK, 2013.
6. Drevelegas, A.; Papanikolaou, N. Imaging modalities in brain tumors. In *Imaging of Brain Tumors with Histological Correlations*; Springer: Berlin/Heidelberg, Germany, 2011; pp. 13–33.
7. Mechtler, L. Neuroimaging in Neuro-Oncology. *Neurol. Clin.* **2009**, *27*, 171–201. [CrossRef]
8. Strong, M.J.; Garces, J.; Vera, J.C.; Mathkour, M.; Emerson, N.; Ware, M.L. Brain Tumors: Epidemiology and Current Trends in Treatment. *J. Brain Tumors Neurooncol.* **2015**, *1*, 1–21. [CrossRef]
9. Mortazavi, D.; Kouzani, A.Z.; Soltanian-Zadeh, H. Segmentation of multiple sclerosis lesions in MR images: A review. *Neuroradiology* **2011**, *54*, 299–320. [CrossRef]
10. Rundo, L.; Tangherloni, A.; Militello, C.; Gilardi, M.C.; Mauri, G. Multimodal medical image registration using Particle Swarm Optimization: A review. In Proceedings of the 2016 IEEE Symposium Series on Computational Intelligence (SSCI), Athens, Greece, 6–9 December 2016. [CrossRef]
11. Balakrishnan, G.; Zhao, A.; Sabuncu, M.R.; Guttag, J.; Dalca, A.V. VoxelMorph: A Learning Framework for Deformable Medical Image Registration. *IEEE Trans. Med. Imaging* **2019**, *38*, 1788–1800. [CrossRef]
12. MY-MS.org. MRI Basics. 2020. Available online: https://my-ms.org/mri_basics.htm (accessed on 1 October 2020).
13. Stall, B.; Zach, L.; Ning, H.; Ondos, J.; Arora, B.; Shankavaram, U.; Miller, R.W.; Citrin, D.; Camphausen, K. Comparison of T2 and FLAIR imaging for target delineation in high grade gliomas. *Radiat. Oncol.* **2010**, *5*, 5. [CrossRef]

14. Society, N.B.T. Quick Brain Tumor Facts. 2020. Available online: https://braintumor.org/brain-tumor-information/brain-tumor-facts/ (accessed on 3 October 2020).
15. Rahimeto, S.; Debelee, T.; Yohannes, D.; Schwenker, F. Automatic pectoral muscle removal in mammograms. *Evol. Syst.* **2019**. [CrossRef]
16. Kebede, S.R.; Debelee, T.G.; Schwenker, F.; Yohannes, D. Classifier Based Breast Cancer Segmentation. *J. Biomim. Biomater. Biomed. Eng.* **2020**, *47*, 1–21.
17. Cui, S.; Shen, X.; Lyu, Y. Automatic Segmentation of Brain Tumor Image Based on Region Growing with Co-constraint. In *International Conference on Multimedia Modeling, Proceedings of the MMM 2019: MultiMedia Modeling, Thessaloniki, Greece, 8–11 January 2019*; Springer: Berlin/Heidelberg, Germany, 2019; Volume 11295.
18. Angulakshmi, M.; Lakshmi Priya, G.G. Automated Brain Tumour Segmentation Techniques—A Review. *Int. J. Imaging Syst. Technol.* **2017**, *27*, 66–77. [CrossRef]
19. Rundo, L.; Tangherloni, A.; Cazzaniga, P.; Nobile, M.S.; Russo, G.; Gilardi, M.C.; Vitabile, S.; Mauri, G.; Besozzi, D.; Militello, C. A novel framework for MR image segmentation and quantification by using MedGA. *Comput. Methods Programs Biomed.* **2019**, *176*, 159–172. [CrossRef] [PubMed]
20. Acharya, U.K.; Kumar, S. Particle swarm optimized texture based histogram equalization (PSOTHE) for MRI brain image enhancement. *Optik* **2020**, *224*, 165760. [CrossRef]
21. Pandav, S. Brain tumor extraction using marker controlled watershed segmentation. *Int. J. Eng. Res. Technol.* **2014**, *3*, 2020–2022.
22. Salman, Y. Validation techniques for quantitative brain tumors measurements. In Proceedings of the IEEE Engineering in Medicine and Biology 27th annual Conference, Shanghai, China, 17–18 January 2006.
23. Sarathi, M.P.; Ansari, M.G.A.; Uher, V.; Burget, R.; Dutta, M.K. Automated Brain Tumor segmentation using novel feature point detector and seeded region growing. In Proceedings of the 2013 36th International Conference on Telecommunications and Signal Processing (TSP), Rome, Italy, 2–4 July 2013; pp. 648–652.
24. Thiruvenkadam, K.; Perumal, N. Brain Tumor Segmentation of MRI Brain Images through FCM clustering and Seeded Region Growing Technique. *Int. J. Appl. Eng. Res.* **2015**, *10*, 427–432.
25. Ho, Y.L.; Lin, W.Y.; Tsai, C.L.; Lee, C.C.; Lin, C.Y. Automatic Brain Extraction for T1-Weighted Magnetic Resonance Images Using Region Growing. In Proceedings of the 2016 IEEE 16th International Conference on Bioinformatics and Bioengineering (BIBE), Taichung, Taiwan, 31 October–2 November 2016; pp. 250–253.
26. Bauer, S.; Nolte, L.P.; Reyes, M. Fully Automatic Segmentation of Brain Tumor Images Using Support Vector Machine Classification in Combination with Hierarchical Conditional Random Field Regularization. In *Lecture Notes in Computer Science*; Springer: Berlin/Heidelberg, Germany, 2011; pp. 354–361. [CrossRef]
27. Rundo, L.; Militello, C.; Tangherloni, A.; Russo, G.; Vitabile, S.; Gilardi, M.C.; Mauri, G. NeXt for neuro-radiosurgery: A fully automatic approach for necrosis extraction in brain tumor MRI using an unsupervised machine learning technique. *Int. J. Imaging Syst. Technol.* **2017**, *28*, 21–37. [CrossRef]
28. Debelee, T.G.; Schwenker, F.; Ibenthal, A.; Yohannes, D. Survey of deep learning in breast cancer image analysis. *Evol. Syst.* **2019**. [CrossRef]
29. Debelee, T.G.; Gebreselasie, A.; Schwenker, F.; Amirian, M.; Yohannes, D. Classification of Mammograms Using Texture and CNN Based Extracted Features. *J. Biomim. Biomater. Biomed. Eng.* **2019**, *42*, 79–97. [CrossRef]
30. Debelee, T.G.; Kebede, S.R.; Schwenker, F.; Shewarega, Z.M. Deep Learning in Selected Cancers' Image Analysis—A Survey. *J. Imaging* **2020**, *6*, 121. [CrossRef]
31. Afework, Y.K.; Debelee, T.G. Detection of Bacterial Wilt on Enset Crop Using Deep Learning Approach. *Int. J. Eng. Res. Afr.* **2020**, *51*, 131–146. [CrossRef]
32. Debelee, T.G.; Amirian, M.; Ibenthal, A.; Palm, G.; Schwenker, F. Classification of Mammograms Using Convolutional Neural Network Based Feature Extraction. In *Lecture Notes of the Institute for Computer Sciences, Social Informatics and Telecommunications Engineering*; Springer: Cham, Switzerland, 2018; Volume 244, pp. 89–98.
33. Li, Q.; Yu, Z.; Wang, Y.; Zheng, H. TumorGAN: A Multi-Modal Data Augmentation Framework for Brain Tumor Segmentation. *Sensors* **2020**, *20*, 4203. [CrossRef] [PubMed]
34. Ibtehaz, N.; Rahman, M.S. MultiResUNet: Rethinking the U-Net architecture for multimodal biomedical image segmentation. *Neural Netw.* **2020**, *121*, 74–87. [CrossRef] [PubMed]
35. Rundo, L.; Han, C.; Nagano, Y.; Zhang, J.; Hataya, R.; Militello, C.; Tangherloni, A.; Nobile, M.S.; Ferretti, C.; Besozzi, D.; et al. USE-Net: Incorporating Squeeze-and-Excitation blocks into U-Net for prostate zonal segmentation of multi-institutional MRI datasets. *Neurocomputing* **2019**, *365*, 31–43. [CrossRef]
36. Kistler, M.; Bonaretti, S.; Pfahrer, M.; Niklaus, R.; Büchler, P. The Virtual Skeleton Database: An Open Access Repository for Biomedical Research and Collaboration. *J. Med. Internet Res.* **2013**, *15*, e245. [CrossRef]
37. Menze, B.H.; Jakab, A.; Bauer, S.; Kalpathy-Cramer, J.; Farahani, K.; Kirby, J.; Burren, Y.; Porz, N.; Slotboom, J.; Wiest, R.; et al. The Multimodal Brain Tumor Image Segmentation Benchmark (BRATS). *IEEE Trans. Med. Imaging* **2014**, *34*, 1993–2024. [CrossRef]
38. Zhao, J.; Meng, Z.; Wei, L.; Sun, C.; Zou, Q.; Su, R. Supervised Brain Tumor Segmentation Based on Gradient and Context-Sensitive Features. *Front. Neurosci.* **2019**, *13*, 1–11. [CrossRef]
39. Reddy, B.; Reddy, P.B.; Kumar, P.S.; Reddy, S.S. Developing an Approach to Brain MRI Image Preprocessing for Tumor Detection. *Int. J. Res.* **2014**, *1*, 725–731.

40. Ségonne, F.; Dale, A.M.; Busa, E.; Glessner, M.; Salat, D.; Hahn, H.K.; Fischl, B. A Hybrid Approach to the Skull Stripping Problem in MRI. *Neuroimage* **2004**, *22*, 1060–1075. [CrossRef]
41. Vishnuvarthanan, G.; Rajasekaran, M.P.; Vishnuvarthanan, N.A.; Prasath, T.A.; Kannan, M. Tumor Detection in T1, T2, FLAIR and MPR Brain Images Using a Combination of Optimization and Fuzzy Clustering Improved by Seed-Based Region Growing Algorithm. *Int. J. Imaging Syst. Technol.* **2017**, *27*, 33–45. [CrossRef]
42. Daimary, D.; Bora, M.B.; Amitab, K.; Kandar, D. Brain Tumor Segmentation from MRI Images using Hybrid Convolutional Neural Networks. *Procedia Comput. Sci.* **2020**, *167*, 2419–2428. [CrossRef]
43. Havaei, M.; Dutil, F.; Pal, C.; Larochelle, H.; Jodoin, P.M. A Convolutional Neural Network Approach to Brain Tumor Segmentation. In *Brainlesion: Glioma, Multiple Sclerosis, Stroke and Traumatic Brain Injuries*; Springer: Cham, Switzerland, 2016; pp. 195–208. [CrossRef]
44. Pereira, S.; Pinto, A.; Alves, V.; Silva, C.A. Deep Convolutional Neural Networks for the Segmentation of Gliomas in Multi-sequence MRI. In *Brainlesion: Glioma, Multiple Sclerosis, Stroke and Traumatic Brain Injuries*; Springer: Cham, Switzerland, 2016; pp. 131–143. [CrossRef]
45. Malmi, E.; Parambath, S.; Peyrat, J.M.; Abinahed, J.; Chawla, S. CaBS: A Cascaded Brain Tumor Segmentation Approach. *Proc. MICCAI Brain Tumor Segmentation (BRATS)* **2015**, 42–47. Available online: http://www2.imm.dtu.dk/projects/BRATS2012/proceedingsBRATS2012.pdf (accessed on 8 August 2020).
46. Debelee, T.G.; Schwenker, F.; Rahimeto, S.; Yohannes, D. Evaluation of modified adaptive k-means segmentation algorithm. *Comput. Vis. Media* **2019**. [CrossRef]
47. Dong, H.; Yang, G.; Liu, F.; Mo, Y.; Guo, Y. Automatic Brain Tumor Detection and Segmentation Using U-Net Based Fully Convolutional Networks. In *Communications in Computer and Information Science*; Springer: Cham, Switzerland, 2017; pp. 506–517. [CrossRef]

Journal of *Imaging*

MDPI

*Review*

# Deep Learning for Brain Tumor Segmentation: A Survey of State-of-the-Art

**Tirivangani Magadza and Serestina Viriri ***

School of Mathematics, Statistics and Computer Science, University of KwaZulu-Natal, Durban 4000, South Africa; 219098526@stu.ukzn.ac.za
* Correspondence: viriris@ukzn.ac.za

**Abstract:** Quantitative analysis of the brain tumors provides valuable information for understanding the tumor characteristics and treatment planning better. The accurate segmentation of lesions requires more than one image modalities with varying contrasts. As a result, manual segmentation, which is arguably the most accurate segmentation method, would be impractical for more extensive studies. Deep learning has recently emerged as a solution for quantitative analysis due to its record-shattering performance. However, medical image analysis has its unique challenges. This paper presents a review of state-of-the-art deep learning methods for brain tumor segmentation, clearly highlighting their building blocks and various strategies. We end with a critical discussion of open challenges in medical image analysis.

**Keywords:** brain tumor segmentation; deep learning; magnetic resonance imaging; survey



**Citation:** Magadza, T.; Viriri, S. Deep Learning for Brain Tumor Segmentation: A Survey of State-of-the-Art. *J. Imaging* **2021**, *7*, 19. https://doi.org/10.3390/jimaging7020019

Academic Editor: Leonardo Rundo
Received: 23 November 2020
Accepted: 11 January 2021
Published: 29 January 2021



## 1. Introduction

Brain tumors are an abnormal growth of cells in the brain. Their exact causes are not yet known, but there are factors that can increase the risk of brain tumor, such as exposure to radiation and a family history of brain cancer. There has been an increase in incidences of brain tumors in all ages globally over the past few years [1]. In the United States alone, an estimate of 78,980 new cases of primary malignant and non-malignant tumors were expected to be diagonized in 2018. Despite considerable efforts in brain tumor segmentation research, patient diagnosis remains poor [2]. The most common types of tumors in adults are meningiomas (low grade tumors) and gliomas and glioblastomas (high grade tumors). Low grade tumors are less aggressive and they come with a life expectancy of several years. High grade tumors are much more aggressive and they have a median survival rate of less than two years.

Medical imaging techniques, such as Magnetic Resonance Imaging (MRI), CT scans, Positron emission tomography (PET), among others, play a crucial role in the diagnosis of the tumors. These techniques are used to locate and assess the progression of the tumor before and after treatment. MRI is usually the modality of choice for diagnosis and treatment planning for brain tumors [2] due to its high resolution, soft tissue contrast, and non-invasive characteristics. Surgery is the most common form of treatment for brain tumors, but radiation and chemotherapy can also be used to slow the growth of the tumor [1]. More than one MRI slice is required to view different regions of the brain, e.g., T1, T2, T1 contrast and FLAIR images.

Again, in clinical practice, delineation of the tumor is usually done manually. An experienced radiologist will carefully study the scanned medical images of the patient segmenting all of the affected regions. Apart from being time consuming, manual segmentation is dependent on the radiologist and it is subject to large intra and inter rater variability [3]. Consequently, manual segmentation is limited to qualitative assessment or visual inspection only.

Meanwhile, quantitative assessment of the brain tumors provides valuable information for a better understanding of the tumor characteristics and treatment planning [4]. Quantitative analysis of the affected cells reveals clues about the disease progression, its characteristics, and effects on the particular anatomical structure [5]. This task proved to be difficult, because of large variability in shape, size, and location of lesions. Moreover, more than one image modalities with varying contrast need to be considered for accurate segmentation of lesions [4]. As a result, manual segmentation, which provides arguably the most accurate segmentation results, would be impractical for larger studies. Most research endeavors today now focus on using computer algorithms for the automatic segmentation of tumors with the potential to offer objective, reproducible, and scalable approaches to the quantitative assessment of brain tumors.

These methods categorically fall into traditional machine learning and deep learning methods [6]. The application of statistical learning approaches to low-level brain tumor classification features is common in conventional machine learning methods. They mainly focus on the estimation of tumor boundaries and their localization. Additionally, they heavily depend on preprocessing techniques for contrast enhancement, image sharpening, and edge detection/refining, relying on human expertise for feature engineering. Wadhwa et al. [7] provide a concise overview of methods in this category.

On the other hand, deep learning methods rely on large scale dataset availability for training and require minimum preprocessing steps than traditional methods. Over the past few years, convolutional neural networks (CNNs) have dominated the field of brain tumor segmentation [6]. Alom et al. [8] provide a detailed review of deep learning approaches that span across many application domains.

Preliminary investigations [9,10] saw deep learning as a promising technique for automatic brain tumor segmentation. With deep learning, a hierarchy of increasingly complex features is directly learned from in-domain data [1] bypassing the need of feature engineering as with other automatic segmentation techniques. Accordingly, the focus would be on designing network architectures and fine-turning them for task at hand. Deep learning techniques have been popularized by their ground breaking performance in computer vision tasks. Their success can be attributed to advances in high-tech central processing units (CPU) and graphics processing units (GPUs), the availability of huge datasets, and developments in learning algorithms [11]. However, in the medical field, there is hardly enough training samples to train deep models without suffering from over-fitting. Furthermore, ground truth annotation of three-dimensional (3D) MRI is a time consuming and a specialized task that has to be done by experts (typically neurologists). As such, publicly available image datasets are rare and will often have few subjects [12].

In this survey, we highlight state of the art deep learning techniques, as they apply to MRI brain tumor segmentation. Unique challenges and their possible solutions to medical image analysis are also discussed.

## 2. Overview of Brain Tumor Segmentation

This section provides a brief introduction to brain tumor segmentation.

### 2.1. Image Segmentation

A digital image, like an MRI image, can be represented as a two-dimensional function, $f(x, y)$, where $x$ and $y$ are the spatial coordinates and the value of $f$ at any given point $(x, y)$ is the intensity or gray level of the image at that point. Each point in an image represents a picture element, called a pixel. The function $f$ can also be viewed as $M \times N$ matrix , $A$, where $M$ and $N$ represent the number of rows and columns, respectively. Thus,

$$A = f(x, y) = \begin{bmatrix} a_{1,1} & a_{1,2} & \cdots \\ \vdots & \ddots & \\ a_{M,1} & & a_{M,N} \end{bmatrix} \quad (1)$$

In computer vision, image segmentation is the process of partitioning a digital image into multiple disjoint segments, each having certain properties. It is typically used in order to locate objects and their boundaries in images. This is achieved by assigning every pixel. $(x, y)$, in an image $A$, a label depending on some characteristics or computed property, such as color, texture, or intensity.

The goal of brain tumor segmentation as depicted in Figure 1, is to detect the location, and extension of the tumor regions, namely:

- active tumorous tissue;
- necrotic (dead) tissue; and,
- edema (swelling near the tumor).

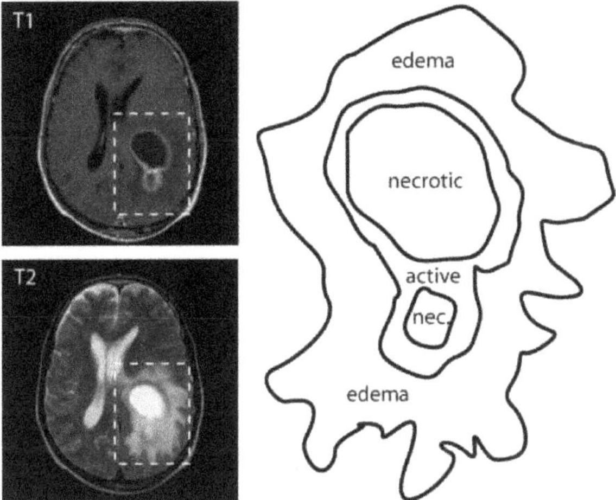


**Figure 1.** Labeled example of a brain tumor illustrating the importance of the different modalities (adapted from [13]).

This is done by identifying abnormal areas when compared to normal tissues [1]. Some tumors, like glioblastomas, are hard to distinguish from normal tissues, because they infiltrate surrounding tissues causing unclear boundaries. As a solution, more than one image modalities with varying contrasts are often employed. In Figure 1, two MRI modalities (T1 with contrast and T2) were used in order to accurately delineate tumor regions.

### 2.2. *Types of Segmentation*

Brain tumor segmentation can be broadly categorised as manual segmentation, semi-automatic segmentation, and fully automatic segmentation, depending on the level of human involvement. Gordillo et al. [14] provide a full description of these methods.

#### 2.2.1. Manual Segmentation

With manual segmentation, a human operator uses specialized tools in order to carefully draw or paint around tumor regions. The accuracy of segmentation results depends heavily on the training and experience of the human operator as well as knowledge of brain anatomy. Apart from being tedious and time consuming, manual segmentation is widely used as a gold standard for semi-automatic and fully automatic segmentation.

#### 2.2.2. Semi-Automatic Segmentation

Semi-automated segmentation combines both computer and human expertise. User interaction is needed for the initialisation of the segmentation process, providing feedback and an evaluation of segmentation results [3]. Although semi-automatic segmentation methods are less time consuming than manual segmentation, their results are still dependent on the operator.

#### 2.2.3. Fully Automatic Segmentation

In fully automatic brain tumor segmentation, no human interaction is required. Artificial intelligence and prior knowledge are combined in order to solve the segmentation problems [3]. Fully automatic segmentation methods are further divided into discriminating and generative methods. Discriminating methods often rely on supervised learning where relationships between input image and manually annotated data are learnt from a huge dataset. Within this group, classical machine learning algorithms, which rely on hand crafted features, have been extensively used with great success over the past years However, these methods may not be able to take full advantage of the training data due to the complexity of medical images [15]. More recently, deep learning methods have gained popularity because of their unprecedented performance in computer vision tasks and their ability to learn features directly from data. On the other hand, generative methods use prior knowledge regarding the appearance and distribution of difference tissue types.

## 3. Deep Learning

Deep learning is a class of machine learning algorithms that uses multiple layers to learn a hierarchy of increasingly complex presentations directly from the raw input. Machine learning models are all about finding appropriate representations for their input data. In this section, we will describe the building blocks, and recent techniques and architectures of deep learning algorithms for brain tumor segmentation that we found in papers surveyed in this work, as summarized in Figure 2.

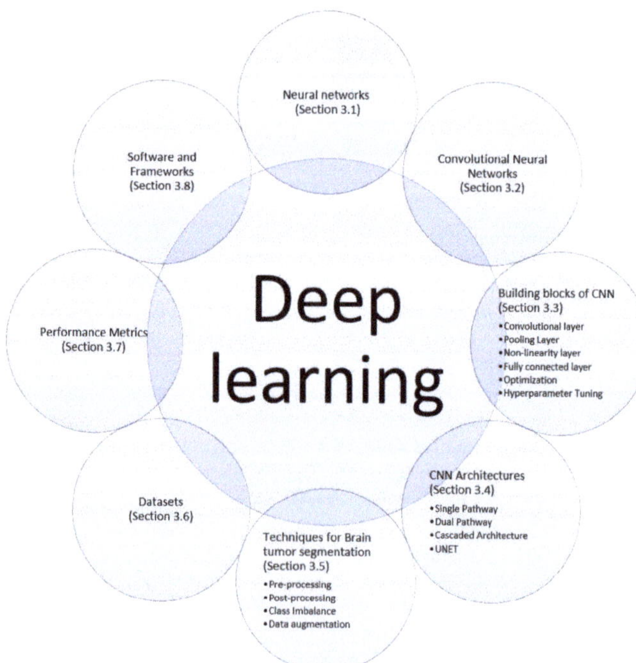


**Figure 2.** Building blocks, architectures and techniques for deep learning algorithms for brain tumor segmentation.

### 3.1. Neural Networks

A neural network is a type of a machine learning algorithm that is able to learn useful representations from data [16,17]. The network is formed by connecting processing units, called neutrons, by directed links. Each link is associated with a weight that adjusts as learning proceeds. When the topology of the network forms a directly acyclic graph, the network is referred to as a feed forward neural network (Figure 3). Associated with each neutron is a function $f(x:\theta)$, which maps an input $x$ to an output $y$ and it learns the value of the parameters $\theta = \{w, b\}$, where $w$ is a weight vector and $b$ is a scalar, through a back-propagation algorithm:

$$f(x:\theta) = \sigma(w \cdot x + b) \tag{2}$$

where $\sigma(\cdot)$ is element-wise non-linearity activation function.

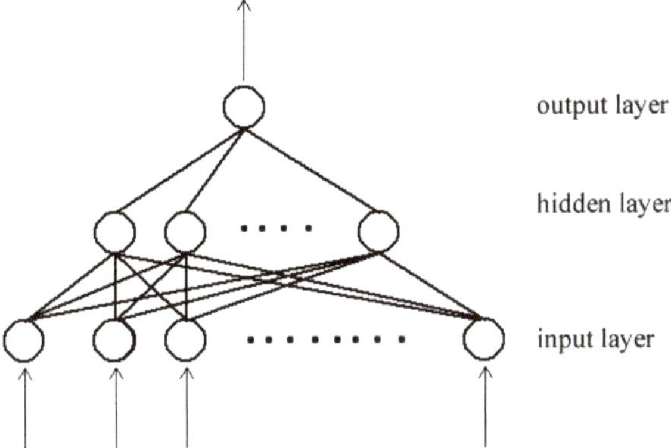


**Figure 3.** Typical feed-forward neural network composed of three layers. (adapted from [18]).

In a typical neural network, neurons are organized in layers. The input of each neuron in a layer is connected to all or some of the output of neurons in the up-stream layer. Likewise, the output of each neuron is connected to all or some of the input of neurons in the downstream layer. The first layer in the network is the input layer, and the final layer is the output layer. Layers in the middle are referred to as hidden layers. When each neuron in a layer is connected to all of the neurons in the next layer, the network is called fully connected network. A deep neural network is formed when there are many hidden layers, hence the term *deep learning*.

### 3.2. Convolutional Neural Network (CNN)

A convolutional neural network is a type of a neural network that performs a convolutional operation in some of its layers. The convolutional layer is able to learn local features from the input data. By stacking many convolutional layers one after the other, the network is able to learn a hierarchy of increasingly complex features. A polling layer is usually added in-between successive convolutional layers to summarize important features. This will reduce the number parameters that are passed to downstream layers and, at the same time, introducing translation invariant (able to recognize learned patterns, regardless of their geometric transformations) to the network.

Recently, CNN has become the de factor model for brain tumor segmentation because of its record shattering performance in classical computer vision problems as well as in medical image analysis as compared to other models. CNN models are able to learn spatial hierarchies of features within data, for example, the first convolutional layer will learn small local patterns, like edges, the second layer will learn larger patterns made up of features of the preceding layer and so on. This ability make them a better fit for image analysis task. Furthermore, units in convolutional layers share weights, thereby reducing the number of parameter to learn and improve the efficiency of the network.

### 3.3. Building Blocks CNN

#### 3.3.1. Convolutional Layer

This layer consists of a set of learnable filters or kernels (the typical size is usually $3 \times 3$ or $3 \times 3 \times 3$, depending whether the input is a two-dimensional (2D) or three-dimensional (3D) image, respectively) that are used to slide over the entire input volume, performing a dot product between entries of the filter and the input at that point. Thus, the convolutional operation first extracts patches from its input in a sliding window fashion, and then applies the same linear transformation to all of these patches. The output of the convolution operation is sometimes referred to as the feature map. The network will learn filters that recognize certain visual patterns present in the input data. When convolutional layers are stacked one after the other, the network is able to learn a hierarchy of increasing complex features, from simple edges to being able to recognize the presence of a face for example.

Over the past few years, there were various attempts meant to improve the performance of deep learning models by replacing the conventional convolutional layer with blocks that increase the network's capacity while using less computational resources. For example, Szegedy et al. [19] introduced the inception block that captured sparse correlation patterns whlie using multi-scale receptive fields. Their network architecture, the GoogleNet, a winner of ILSVRC 2014, had fewer network parameters and required less computational resources than its predecessors AlexNet [20] or VGG [21]. The residual block was another notable improvement [22], which facilitated very deep networks that do not suffer from the vanishing gradient problem. Hu et al. [23] introduced the Squeeze-and-Excitation (SE) block that captured the interdependencies between the network's feature maps.

#### 3.3.2. Pooling Layer

A pooling layer usually follow a convolutional layer or a set of convolutional layers. The goal is to reduce the dimensions of the feature maps, and at the same time, keep important features. A pooling operation is applied to a rectangular neighbourhood in a sliding window fashion. For example, the max pooling is used in order to produce a maximum of a rectangular neighbourhood. Other popular pooling operations include average and weighted average pooling.

#### 3.3.3. Non-Linearity Layer

Typical convolutional layers involves three steps [16]. In the first step, the layer performs convolutional operation on input feature maps to produce a set of linear activations. Second, a non-linear transformation is performed on the output feature maps. Third, a pooling layer is used in order to modify the output further. Non-linear transformations can be obtained by using special class of functions, called activation functions. Non-linearity gives the network the ability to learn nontrivial representations that are sparse. Hence, making the network resilient to slight modifications or noise in the input data as well as improving computational efficiency of the representations.

In the past, sigmoid and hyperbolic tangent functions were commonly used for the non-linearity layer. Today, the most popular activation function is the rectified linear unit (ReLU), which is expressed as $f(z) = max(z, 0)$. It was observed in [20,24], where ReLU typically learns faster in network with many layers and does not suffer from vanishing/exploding gradients, as with the sigmoidal activations. However, ReLU presents some potential drawbacks when the network saturates with a constant zero gradient causing the network to converge slowly. As a solution, Maas et al. [25] proposed a Leaky ReLU (LReLU) that allows for small, non-zero gradient to flow when the network is saturated. This function is defined as

$$f(z) = max(z, 0) + \alpha min(0, z) \quad (3)$$

where $\alpha$ is a constant leakiness parameter(typically 0.01). Another common variant of ReLU is Parametric Rectified Linear Unit (PReLU) [26]. This activation function adaptively learns the parameter $\alpha$ in Equation (3), thus improving the accuracy with less computational cost.

#### 3.3.4. Fully Connected Layer

The convolutional layers are used as feature extractors. The features that they produce are then passed to the fully connected (FC) layers for classification. Each unit in the FC layer is connected to all of the units in the previous layer, as shown in Figure 3. The final layer is usually a softmax classifier, which produces a probability vector map over the different classes. All of the features are converted in to a one-dimensional feature vector before being passed to a FC layer. By doing so, spatial information inherent in image data is lost. Another issue with the FC layers is that they have a larger number of parameters as compared to other layers that increase the computational costs and require input images to be of the same size.

As a solution to above problems, Long et al. [27] proposed converting FC layers to $1 \times 1$ convolutional layers, thus transforming the the network into a fully convolutional network (FCN). The network takes the input of any arbitrary sizes and outputs a grid of classification maps.

#### 3.3.5. Optimization

The performance of the deep CNN can be improved (or optimized) by training the network on a large dataset. Training involves finding the parameters $\theta$ of the model that significantly reduce a cost function $J(\theta)$. Gradient descent is the widely used method for updating network parameters through a back-propagation algorithm. Optimization can be done per single sample, subset, or full set of the training samples. Thus, stochastic, mini-batch, or batch gradient descent, respectively. Today, many optimization algorithms for deep learning use mini-batches and it is now common to just call them stochastic methods [16].

Stochastic gradient descent (SDG) comes with few notable challenges. Choosing an appropriate learning rate can be difficult. A learning rate that is too small leads to very slow convergence (tiny updates to the model parameters) and, at the same time, too large will result in undesired divergence behavior in the loss function. All of the parameter updates are based on the same learning rate, disregarding the fact that some of the features might have higher frequency than other. Another key challenge is that optimization can be trapped in sub-optimal local minima or saddle points, especially for non-convex optimization [28].

Various variants of SDG have been proposed in the literature that address the aforementation challenges. Memontum-based SDG methods [29] can help in accelerating SDG in relevant direction, dampening undesirable oscillations in local optima. Adagrad [30] addressed the issue of manually turning the learning by adapting the learning rate to the parameters, performing larger updates for infrequent parameters as compared to frequent ones. However, Adagrad suffers from monotonically decreasing learning rate to a point at which the algorithm stops learning. Adadelta [31], RMSprop [32], and Adam [33] addressed the shortcomings of Adagrad by dividing the learning rate by an exponentially decaying average of past gradients.

#### 3.3.6. Loss Function

In machine learning, a loss function is used in order to evaluate how well a specific algorithm models the given data. When the output is far from the true value, loss will be very high and low when the predictions are close to the true values. The primary goal of training a neural network is to minimize the loss (or cost) function of the network as much as possible and, at the same time, ensuring that the network generalizes well with unseen data.

The choice of the cost function depends on the problem area, whether it is a classification or regression problem and the choice of the output unit [16]. The majority of the image classification algorithms use softmax loss, withhs a combination of softmax and CE loss or log-loss [28]. The softmax function produces a probability distribution over a number of given output classes, while the CE loss takes the probability of predictions and penalizes predictions that are confident but wrong. Class imbalance is one major issue in medical image analysis, where one class will have fewer instances than the other. For example, a brain tumor occupies a small portion when compared to healthy tissues. As a result, the classifier will tend to be biased to the majority class. One way of addressing such a problem is to adapt loss functions for class imbalance. Some works [34–36] proposed a loss function that is based on the Dice coefficient. Ronneberger et al. [37] proposed a weighted CE loss, which gives more importance to some pixels in the training data.

#### 3.3.7. Parameter Initialization

Deep learning optimization algorithms are iterative in nature, thus requiring the user to specify initial starting point of the algorithms [16]. The choice of initialization will influence how quickly learning can converge if it can converge at all. Empirical studies have shown that a carefully chosen initialization scheme dramatically improves the rate of convergence [38], while gradient-based optimization starting from random initialization may get stuck near poor solutions [39].

Ref. [38] proposed a normalized initialization scheme (Xavier initialization), which guarantees that weight initialization should not obtain values that are too small or too large, thus reducing saturation and vanishing gradients, thereby improving convergence. This approach was later improved in [26] to perform much better on Relu or PRelu activations and extreme deep models.

#### 3.3.8. Hyperparameter Tuning

Hyperparameters are parameters that are supplied by the user to control the algorithm's behavior before training commences, such as learning rate, batch size, image size, number of epochs, kernel size etc. While the learning algorithms do not adapt these parameters, their choice has varying effects on the resulting model and its performance. The majority of the works studied in this review set their hyperparameters manually or perform a grid search while using the validation set. However, these approaches will become impractical when the number of hyperparameters is large [40] and they rely on human expertise, intuition, or guessing. As a solution to these challenges, automated approaches, like AutoML (http://www.automl.org) and Keras Tuner, (https://keras-team.github.io/keras-tuner/) are beginning to gain much attention.

#### 3.3.9. Regularization

Regularization is a technique for improving the performance of a machine learning algorithm on unseen data. It is a way of reducing over-fitting on training set. Over-fitting occurs when the gap between the training error and test error is too large [16]. When that happens, the model performs well on training data, but poorly on previously unseen data. There are various techniques that can be employed in order to reduce the generalization error, such as reducing the model capacity, which is, reducing the number of learnable parameters in the model; adding $L^2$ or $L^1$ weight decay regularization term to the cost function to force the model to only take small weight values; introducing early stopping whenever the model performance stops improving on validation dataset; randomly dropping out (skipping) the output of some units during training [41]. The last approach is one of the most effective and most commonly used technique [17], mainly because it is computationally inexpensive and prevents interdependent learning amongst units. Batch Normalization [42] can also be used as a regularizer by ensuring that the distribution of non-linearity inputs remains more stable as the model trains, thereby improving the training of the model.

Training a machine learning model with more data is the best way to reduce the generalization error. However, in the medical domain, acquiring a training dataset is time-consuming, more expensive, and requires highly trained personnel to annotate ground truth labels. Data augmentation can increase the dataset and reduce over-fitting by flipping, applying small rotations, warping, and using the non-rigid deformation transformation of images. However, great care must be taken when performing transformations of the medical image dataset since the patch's label is determined by the center of pixel [43]. Some recent works used generative models that include variational autoencoders [44] and generative adversarial networks [45] to act as additional regularization that deals with data scarcity.

### 3.4. Deep CNN Architectures

#### 3.4.1. Single Pathway

A single pathway architecture is a basic network that resembles a feed-forward deep neural network. Data flows from the input layer to the classification layer using a single path. Urban et al. [10] proposed a 3D single path CNN which has fully connected convolutional layer as the classification layer. This gave the network the ability to classify multiple 3D pixel in one go. In [46], each image's modality was fed to a different two-dimensional (2D) CNN. The result of each CNN was then used as features to train a random forest classier. Extracts from XY, XZ, and YZ planes around each center pixel were used as the neighborhood information. Pereira et al. [43] used small kernels in their convolutional layers. As a result, a very deep network, DeepMedic, was obtained, which can learn more feature hierarchies. Their architecture obtained first and second positions in BRATS 2013 and 2015 challenge,respectively.

#### 3.4.2. Dual Pathway

Many segmentation algorithms perform pixel-wise classification, where an input patch is extracted from an MRI image and then predicts the label of the central pixel without considering global neighborhood information. This can be risky because of infiltrating nature of brain tumors, which produces unclear boundaries. Hence, local information cannot be enough to accurately produce good segmentation results. As a solution, other researchers [1,47] introduced neighbourhood information to the mix by using CNN with two data streams (dual pathway) that are combined in order to influence label predictions of each pixel. One of the streams will represent local information, the visual details of the region around the center pixel. The other stream represents the global context, which takes the location of the extracted patch in the brain into account.

#### 3.4.3. Cascaded Architecture

In a cascaded architecture, the output one CNN is concatenated with the other. There many variations with this architecture in the literature, but the most prominent is the input cascade [1,48]. In this architecture the output of one CNN becomes a direct input of another CNN. The Input cascade is used in order to concatenate contextual information to the second CNN as additional image channels. This is an improvement to the dual-path way that performs multi-scale label predictions separately from each other. Another variation of cascaded architecture is the local pathway concatenation [1]. In this architecture, the output of the first CNN is concatenated with the output of the first hidden layer of the second CNN instead of its input.

Hierarchical segmentation [34,49] is another form of a cascaded architecture. In this architecture, the segmentation of brain tumor regions is sequentially done by reducing the multi-class segmentation problem into the multi-stage binary segmentation problem. This architecture takes full advantage of the hierarchical nature of tumor sub-regions and helps in reducing false positives as well as mitigating the inherent class imbalance problem. The first stage of architecture segments the whole tumor from the input MRI modalities, which is then used as a bounding box for the next stage. For the second stage, the output of the first stage is used as an input to perform either a multi-class intra-tumoral segmentation, as in [49], or perform successive binary segmentation of the remain tumor sub-regions [34]. Wang et al. [34] observed an increase in the training and inference time of a multi-stage binary segmentation as compared to a single multi-class network approach.

#### 3.4.4. UNET

The UNET architecture [37] is an improvement of FCN [27], which resembles an encoder and decoder network designed specifically for biomedical image segmentation. The network consists of a contracting path (encoder) and an expansive path (decoder), which gives it the u-shaped architecture. The contracting path consists of the repeated application of two convolutional layers, followed by a rectified linear unit (ReLU) and max pooling layer. Along the path, the spacial information is reduced, while feature information is increased. The expansive path consists of a series of up-sampling operations combined with high-resolution features from the contracting path through skip connections.

### *3.5. Techniques for Brain Tumor Segmentation*

#### 3.5.1. Pre-Processing

Data preprocessing is a very crucial step of preparing raw input data to be more amenable to neural networks. MRI images contains various artifacts that are caused by the acquisition protocol and the hardware used. These artifacts need to be corrected before the images are fed into the network for better performance. One of the notable artifacts is the presence of smooth intensity variations within the image, which is also known as bias field. Among various techniques for bias field correction, the non-parametric nonuniform normalization (N3) [50] approach has become the technique of choice for bias field correction due to its ease of use and its availability as an open source project [51]. This technique was later improved in [51] and it is also well known as N4ITK. These techniques are limited to a single image. Accordingly, for uniform intensity distribution across patients and acquisitions, the intensity normalization proposed by Nyul et al. [52] can be applied.

Another popular preprocessing technique is to normalize image dataset to have a mean zero and a standard deviation of one. This technique assists in removing the bias from features. Image cropping can also be applied to remove as much background pixels as possible.

#### 3.5.2. Post-Processing

The post-processing step is performed to further refine the segmentation results. It helps in reducing the number of misclassifications or false positives in the segmentation results while using algorithms, like conditional random fields (CRF) [4,34,53], markov random fields (MRF) [54], connected component analysis [1,53,55], and morphological operators [48,56]. CRF and MRF based techniques effectively remove false positives by combining model predictions with low-level image information, like local interations of pixels and edges when making finer adjustments. However, these techniques are computationaly expensive [14]. Connected compents analysis involves finding and extracting connected components and then applying a simple thresholding technique to remove unwanted blobs. Another technique of removing false positive around edges of the segmentation image is to apply morphological operations, erosion, and dilation in succession.

#### 3.5.3. Class Imbalance

The performance of the segmentation task is affected by the class imbalance problem, where there is an unequal distribution of voxel classes in the training dataset. For example, in brain tumor segmentation, healthy voxels constitute 98% of the total voxels [1]. Training the model on this distribution will cause the model to be more biased towards the majority class. Whereas, training with equal distribution results in bias towards tumor classes [57]. Several techniques have been explored in the literature in order to address this problem.

Many works incorporated loss-based methods of addressing the class-imbalance problem. Lin et al. [58] proposed a loss function that addresses the problem by dynamically scaling the loss based on the model's confidence in classifying samples. The scaling factor was reduced when the model's accuracy in classifying classed increases. As a result, the model pays more attention to misclassified samples. In [59], dice loss was used as a means of addressing the problem. Some works [60,61] incorporated a weighted-loss function, where voxels (or pixels) belonging to different classes are assigned weights according to their distribution in the training data. This ensures that each class in the segmentation problem has an equal contribution to the model's loss. Kuzima et al. [62] combined the CE loss with Dice based loss as means of addressing class imbalance problem. Other works explored hard negative mining [63,64] as a solution to the class-imbalance problem. Voxels with largest negative losses and positive voxels are used in order to update the model's weights.

Two-phase training [1,5,57] is also another way of dealing with the class imbalance problem. In the first phase, the network is trained with patches that have equal class distribution and then trained with true class distribution in the second phase. Hussain et al. [57] reported that two-phased training helped in removing most of the false positives.

In [34], Wang et al. pointed out that hierarchical segmentation also assists in addressing the class-imbalance problem.

#### 3.5.4. Data Augmentation

Data augmentation is a technique for reducing the generalization error of a machine learning algorithm. As indicated earlier, one way of effectively increasing the machine learning model's generalization capabilities is to train it on more data. However, acquiring a considerable amount of high-quality training data is nearly impossible in practice, especially for the medical domain. Data augmentation has emerged in order to increase the training data by creating more synthetic data and adding (augment) it to the training set.

Data augmentation can be broadly divided into two categories [65]: the transformation of original data and artificial data generation. With the transformation of original data, new data are generated by applying various transformations on the original data, which include affine transformations (which involves rotation, zooming, cropping, flipping, and translations), elastic transformations (shape variations), and pixel-level transformation (intensity variations). While these transformations assist in mitigating insufficient data challenges, they fundamentally produce very correlated images [66], which results in very little performance improvement [66,67] and sometimes generates anatomically incorrect examples (e.g. using rotation) [65]. However, their use in the literature is widespread, due to the ease of implementation.

On the other hand, artificial data generation [67,68] exploits the Generative adversarial networks (GANs) [69] to generate realistic data that are indistinguishable from the real data and also serves as a effective method for data anonymization [66]. GANs are able to generate a wide variety of realistic samples that can bring invariance and robustness. However, there are scenarios where they can generate samples that are very similar to the real ones, resulting in poor performance [65].

### *3.6. Datasets*

Over the past few years, there have been considerable research interests in automatic brain tumor segmentation. As research output continued to grow, the objective evaluation of different algorithms became a challenge because researchers used private datasets with varying attributes. As a result, benchmarking challenges, such as Multi-modal Brain Tumor Image Segmentation (BRATS), emerged to standardize performance evaluation while using publicly accessible datasets. Table 1 show a summary of the mostly used datasets for brain tumor segmentation.

Since 2012, the BRATS Challenge [2], in conjunction with the International Conference on Medical Image Computing and Computer-Assisted Interventions (MICCAI), has been the primary bench-marking resource for brain tumor segmentation. It offers the medical research community publicly accessible datasets for training and validation and standardized metrics in order to objectively evaluate model performance against an online evaluation platform. The dataset initially contained as small as 30 clinically acquired scans of glioma patience, and the number has continued to grow over the subsequent years.

**Table 1.** Summary of commonly used public datasets for brain tumor segmentation.

| Name | Total | Training Data | Validation Data | Testing Data |
|---|---|---|---|---|
| BRATS 2012 [2] | 50 | 35 | - | 15 |
| BRATS 2013 [2] | 60 | 35 | - | 25 |
| BRATS 2014 [2] | 238 | 200 | - | 38 |
| BRATS 2015 [2] | 253 | 200 | - | 53 |
| BRATS 2016 [2] | 391 | 200 | - | 191 |
| BRATS 2017 [2] | 477 | 285 | 46 | 146 |
| BRATS 2018 [2] | 542 | 285 | 66 | 191 |
| BRATS 2019 [2] | 653 | 335 | 127 | 191 |
| Decathlon [70] | 750 | 484 | - | 266 |

Medical Segmentation Decathlon Challenge offers a relatively large dataset that supports a wide range of segmentation task. The Challenge aims to facilitate research in general-purpose segmentation algorithms that solve various functions without any human intervention. For brain tumor segmentation, the dataset comprises a subset of the 2016 and 2017 BRATS Challenge data.

### 3.7. *Performance Evaluation Metrics*

In order to objectively measure the performance of segmentation algorithms, researchers have to group different tumor structures into three mutually inclusive regions:

- the *whole* tumor (includes all tumor structures);
- the *tumor* core (exclusive of edema); and,
- the *active* tumor (only consists of the "enhancing core").

Subsequently, they measure the algorithm's performance on each region against several metrics that include the Dice score, Sensitivity, Specificity, and Hausdorff measure.

### 3.8. *Software and Frameworks*

Researchers and engineers have always relied on open-source software frameworks from idea generation to experimentation to production deployments in order to accelerate the deep learning workflow. This section described some of the popular machine learning frameworks that were used in the reviewed papers.

Theano [71] is a free and open-source python framework for the fast computation of large-scale dataflow mathematical expressions compiled and executed naively on both CPUs and GPUs. Moreover, the research community has been utilizing the platform in order to conduct machine learning research. However, it is not a purely a machine learning framework, but rather a compiler for mathematical expressions that are defined in NumPy-like syntax. Several high-level software packages like Pylearn2, Keras, blocks, and Lasagne have been built on top of Theano, leveraging its strengths as an efficient mathematical powerhouse.

Pylearn2 [72] is a free and open-source machine learning library that is built on top of the Theano framework. It started gaining popularity after being used to win a transfer learning challenge and implementing various state of the art computer vision benchmarks. The library focuses on flexibility and extensibility, allowing for researchers to implement arbitrary machine learning models at ease. Unfortunately, the library no longer has an active developer and has, ever since, fallen behind other actively maintained frameworks, like Keras.

Caffe [73] is a C++ deep learning framework that was initially developed for computer vision applications and later spread to other domains like robotics, neuroscience, and astronomy. It offers a complete toolkit for a deep learning pipeline, from training to production deployment. Each processing stage is supplemented with well-documented examples. Moreover, the framework is shipped with implementations of popular deep learning building block and reference models allowing for quick experimentation with state-of-the-art deep learning methods. The definition of models is done in config files, rather than being hard-coded, ensuring the separation of representation from implementation.

Pytorch [74] is yet another fully-fledged open-source deep learning framework. Its design philosophy moved away from the define and execute style, as in many frameworks that create a static computational graph before running the model. While this approach is powerful, it sacrifices usability, the ease of debugging, and flexibility. Instead, Pytorch took an imperative approach by dynamically constructing the computational graph, allowing for the models to be idiomatically defined following the python programming model. The framework also offers a seamless transition from research to production, distributed training, and the seamless execution of models on edge devices.

Tensorflow [75] is an end-to-end distributed deep learning platform for large scale machine learning applications. The platform supports the execution of dataflow graphs across a span of heterogeneous devices, such as mobile devices and large-scale distributed systems, with little or no change. Its design philosophy has been used to simplify model parallelism within a single machine and across thousands of distributed systems. It has a complete toolbox for quick experimentation with state-of-the-art deep learning models, seamless transition from research to heterogeneous deployments, and the visualization and debugging of large-scale models.

Keras [76] is a fast-growing high-level API for deep learning applications. Although it initially supported multiple data-flow graph back-ends, like Theano, it is now deeply woven into the Tensorflow 2 ecosystem. It provides consistent and simple APIs to quickly experiment with new models and leverage Tensorflow in order to export the models to run in browsers and mobile devices. Moreover, it comes bundled with building blocks and pre-trained state-of-the-art models for various machine learning domains. The industry and the research community have adopted the platform, because of its ease of use, user-centric approach, and extensive documentation.

## 4. Discussion

Deep learning methods to medical image analysis have received tremendous attention over the past few years. This is evident in the considerable increase in the number of published works each year [2]. Deep learning techniques are able to learn a hierarchy of increasingly complex features directly from data, as stated earlier. For example, in brain tumor segmentation, deep learning algorithms can learn to segment MRI images by being trained on a sufficiently large dataset. For this reason, CNN based models have been widely adopted in medical image analysis, following their success in solving many problems in computer vision, speech recognition, and natural language processing. Table 2 shows a summary of deep learning methods that were reviewed in this work. Many techniques differ considerably in terms of architectural design, with recent works following the Unet [37] architecture and ensemble methods as shown in Table 3. Moreover, several techniques have been developed in order to address inherent problems in automated brain MRI analysis.

**Table 2.** Overview of Deep learning methods for brain tumor segmentation. BN = Batch normalization, GN = Group normalization, outliers = remove top 1%, hist-norms = Histogram normalization,RN = Range normalization, HS = Histogram standardization, slice-norm = Slice-based normalization, PLN = Piece-wise linear normalization, IN = Instant normalization, CE = Cross entropy,BS = Bootstrapping, SS = Sensitivity-specification, NM = Negative Mining, WCE = Weighted cross-entropy, neg-mining = Hard negative mining.

| Reference | Input | Preprocessing | Regulization | Loss | Optimizer | Activation |
|---|---|---|---|---|---|---|
| *Unet Architecture* | | | | | | |
| [47] | 3D | Z-score | | | | ReLu |
| [77] | 2D | | BN | Dice, WCE, BS, SS | Adam | ReLU |
| [34] | 2D | Z-score, hist-norms | dropout | CE | SDG | LReLU |
| [78] | 3D | cropping | BN | Jaccard loss, CE | | PReLU |
| [79] | | Z-score, N4ITK,lin-norm | | | | |
| [80] | 2D | | | Dice | Adam | |
| [81] | 2D | Z-score, HM | BN | CE | Adam | ReLU |
| [82] | 3D | bounding box | dropout | Dice | Adam | |
| [83] | 3D | Z-score, rescaling, outliers | IN, L2 | Dice | Adam | LReLU |
| [84] | 2D | slice-norm | | CE | Adam | |
| [85] | 3D | | BN | Dice | Adam | |
| [15] | 2D | Z-score | BN | CE | Adam | ReLU |
| [63] | 3D | Z-score | GN | CE, neg-mining | SGD | |
| [36] | 2D | bounding-box, cropping, Z-score, intensity-windowing | BN | Dice | Adam | Relu |
| [86] | 2D | N4ITK, Nyúl | BN, spatial-dropout | CE | Adam | ReLU |
| [60] | 2D | | BN | CE | | ReLU |
| [87] | 2D | Z-score, remove outliers | BN | WCE, Dice | SGD | PReLU |
| [88] | 3D | Z-score | IN, L2 | CE, Dice | Adam | LReLU |
| [5] | | N4ITK, remove outliers | | WCE | Adam | |
| [35] | 2D | Z-score | BN | Dice | Adam | Relu |
| [59] | 3D | Z-score | BN | Dice | Adam | PReLU |
| [89] | 3D | Z-score | BN, L2 | CE, Dice, focal | Adam | ReLU |
| [90] | | Z-score | | | Adam | RelU |
| [91] | 3D | Z-score | GN, L2, Dropout | Dice | Adam | ReLU |
| [92] | 3D | RN, random axis mirror | | CE, Dice | SDG | |
| [64] | 3D | Z-score, N4ITK | BN, L2 | CE, NM | Adam | ReLU |
| *Dual-pathay Architecture* | | | | | | |
| [10] | 2D | | L1, L2 Dropout | | SDG | |
| [1] | 2D | Z-score, N4ITK, outliers | L1, L2, Dropout | log-loss | Maxout | ReLU |
| [47] | 2D | Z-score | | | Adam | ReLU |
| [57] | 2D | Z-score, N4ITK | BN, Dropout | log-loss | SDG | ReLU |
| [63] | 3D | | GN | CE, NM | SDG | |
| [53] | 2D | N4ITK | | | | PReLU |
| [5] | | N4ITK, outliers | | WCE | SGD | |
| [93] | 3D | N4ITK, LIN | | | | ReLU |
| [94] | 3D | | Dropout | log-loss | SDG | PReLU |
| [95] | 2D | N4ITK | Dropout | | SGD | ReLU |
| [4] | 3D | Z-score | | log-loss | | ReLU |
| [79] | | Z-score, N4ITK, PLN | | | | |
| [96] | 3D | Z-score | BN, L2, Dropout | | | ReLU |

**Table 2.** *Cont.*

| Reference | Input | Preprocessing | Regularization | Loss | Optimizer | Activation |
| --- | --- | --- | --- | --- | --- | --- |
| *Single-pathway Architecture* | | | | | | |
| [9] | 2D | | | log-loss | SGD | ReLU |
| [46] | 2D | | Dropout | CE | SGD | ReLU |
| [43] | 2D | | | | SGD | ReLU |
| [64] | 3D | Z-score, N4ITK | BN | CE, NM | Adam | ReLU |
| [97] | 2D | | | CE | Nesterov, RMSProp | ReLu |
| [98] | 2D | Z-score, outliers | | | Adam, SGD, RMSProp | ReLu |
| [99] | 3D | | | | | ReLU |
| [43] | 3d | Z-score, N4ITK, Nyúl | Dropout | CE | Nesterov | LReLU |
| *Ensemble Architecture* | | | | | | |
| [59] | 3D | Z-score | BN | dice | Adam | PReLU |
| [64] | 3D | Z-score, N4ITK | BN | CE, NM | Adam | ReLU |
| [63] | 3D | | GN | CE, NM | SDG | |
| [61] | 2D | Z-score, N4ITK, HN, | Dropout | CE | Adam | |
| [98] | 2D | Z-score, outliers | | | Adam, SGD, RMSProp | ReLu |
| [44] | 3D | Z-score | GN, L2, spatial dropout | Dice | Adam | ReLU |
| [79] | | Z-score, N4ITK, PLN | | | | |
| *Cascaded Architecture* | | | | | | |
| [34] | 2D | HS, Z-score | dropout | CE | SGD | LReLU |
| [1] | 2D | Z-score, N4ITK, remove outliers | Dropout L2, L1 | log-loss | Maxout | |
| [48] | 2D | | | | Maxout | RelU |
| [85] | 3D | | BN | Dice | Adam | LReLU |
| [100] | 2D | Z-score, BN,outliers | L2, dropout | CE | SGD | ReLU |
| [34] | 2.5D | Z-score | BN | Dice | Adam | PReLU |
| [59] | 3D | Z-score | BN | Dice | Adam | PReLU |
| [89] | 3D | Z-score | | | Adam | ReLU |
| [86] | 2D | Z-score, N4ITK | BN, spatial dropout | CE | SDG | ReLU |
| [34] | 3D | Z-score | BN | Dice | Adam | PReLU |
| [86] | | N4ITK, Nyúl | BN, dropout | CE | Adam | ReLU |
| [35] | 2D | Z-score | BN | Dice | Adam | ReLU |
| [91] | 3D | Z-score | GN, L2, dropout | Dice | Adam | ReLU |

**Table 3.** A summary of top performing methods on BraTS 2017, 2018, and 2019 validation data as reported by the online evaluation platform. ET—Enhancing tumor, WT—Whole tumor, and TC—Tumor core.

| Rank | Reference | Architecture | Dice | | | Sensitivity | | | Specificity | | | Hausdorff 95 | | |
|---|---|---|---|---|---|---|---|---|---|---|---|---|---|---|
| | | | ET | WT | TC | ET | WT | TC | ET | WT | TC | ET | WT | TC |
| *BraTS 2017* | | | | | | | | | | | | | | |
| 1 | [79] | Ensemble | 0.738 | 0.901 | 0.797 | 0.783 | 0.895 | 0.762 | 0.998 | 0.995 | 0.998 | 4.499 | 4.229 | 6.562 |
| 2 | [34] | Cascaded | 0.786 | 0.905 | 0.838 | 0.771 | 0.915 | 0.822 | 0.999 | 0.995 | 0.998 | 3.282 | 3.890 | 6.479 |
| 3 | [83] | Unet | 0.776 | 0.903 | 0.819 | 0.803 | 0.902 | 0.786 | 0.998 | 0.996 | 0.999 | 3.163 | 6.767 | 8.642 |
| 3 | [101] | SegNet | 0.706 | 0.857 | 0.716 | 0.687 | 0.811 | 0.660 | 0.999 | 0.997 | 0.999 | 6.835 | 5.872 | 10.925 |
| *BraTS 2018* | | | | | | | | | | | | | | |
| 1 | [44] | Ensemble | 0.825 | 0.912 | 0.870 | 0.845 | 0.923 | 0.864 | 0.998 | 0.995 | 0.998 | 3.997 | 4.537 | 6.761 |
| 2 | [88] | Unet | 0.809 | 0.913 | 0.863 | 0.831 | 0.919 | 0.844 | 0.998 | 0.995 | 0.999 | 2.413 | 4.268 | 6.518 |
| 3 | [102] | Ensemble | 0.792 | 0.901 | 0.847 | 0.829 | 0.911 | 0.836 | 0.998 | 0.994 | 0.998 | 3.603 | 4.063 | 4.988 |
| 3 | [103] | Ensemble | 0.814 | 0.909 | 0.865 | 0.813 | 0.914 | 0.868 | 0.998 | 0.995 | 0.997 | 2.716 | 4.172 | 6.545 |
| *BraTS 2019* | | | | | | | | | | | | | | |
| 1 | [91] | two-stage Unet | 0.802 | 0.909 | 0.865 | 0.804 | 0.924 | 0.862 | 0.998 | 0.994 | 0.997 | 3.146 | 4.264 | 5.439 |
| 2 | [92] | Unet | 0.746 | 0.904 | 0.840 | 0.780 | 0.901 | 0.811 | 0.990 | 0.987 | 0.990 | 27.403 | 7.485 | 9.029 |
| 3 | [104] | Ensemble | 0.634 | 0.790 | 0.661 | 0.604 | 0.727 | 0.587 | 0.983 | 0.980 | 0.983 | 47.059 | 14.256 | 26.504 |

Deep learning algorithms require a relatively large amount of training data to generalize well on unseen data. However, this poses many challenge in the medical domain. Firstly, it takes a well trained radiologist a considerable amount of time to annotate even a single MRI volume. Moreover, the work is subject to an intra-rater and inter-rater variability. Therefore, all of the annotations are approved by one to many experienced neuro-radiologists [105], before they can be used in supervised training, which makes the process of creating training and testing datasets not only time consuming, but expensive. Secondly, medical data is protected by data protection laws that restrict the usage and sharing of this kind of data to other parties. Consequently, a lot of time is spent seeking approvals and removing personal identifiable information from medical data. Fortunately, Table 1 shows a consistent increase of training and testing data for the BraTS Challenge. Hopefully, this trend will continue in the coming years. Thus, facilitating training relative deep networks and reducing over-fitting.

Because the lack of large-scale datasets restricts deep learning models' full potential, researchers have adopted data augmentation as an immediate solution to the data challenges that are mentioned above. Other works have recently explored weakly-supervised learning [106–108] as a promising solution to address the need for fully annotated pixel-wise labels. Instead of performing pixel-level annotations, known to be tedious and time-consuming, weakly-supervised annotation uses bounding box or image-level annotations in order to signify the presence or absence of lesions in images. This approach has the benefit of being cheap, contains less labeling noise [107], far larger volumes of data can be generated than pixel-level annotation, and training of deep learning models can leverage both kinds of datasets.

Moreover, deep learning techniques require a huge amount of computational and memory resources [28]. Very deep networks, which are becoming a widespread, have millions of parameters that result in many costly mathematical computations that are restrictive on the kind of computational hardware that can be used by researchers. Furthermore, the use of 3D deep learning models increases the computational and memory requirements by large margins. All of the reviewed literature use deep learning software libraries to provide an infrastructure to define and train deep neural networks in parallel or distributed manner while leveraging multi-core or multi-GPU environments. Currently, researchers are being limited by the amount of GPU memory at their disposal (typically 12 gigabytes). For this reason, batch sizes and model complexities are being limited to what can fit into the available memory.

The performance of brain tumor segmentation algorithms have continued to increase over the past few years due to the availability of more training data and use of more sophisticated CNN architectures and training schemes. However, their robustness is still lagging behind expert performance [105]. Recently, researchers have used the ensemble methods to achieve state-of-the-art performance (see Table 3). Precisely, the ensemble methods fuse the segmentation results of several models to improve the robustness of individual approach, resulting in superior performance as compared to inter-rater agreements [105]. Interestingly, single Unet [37] based models [91] continue to produce exceptional performance, supporting the argument that: *"a well trained Unet is hard to beat"* [88]. The reviewed literature have shown that careful initialization of hyper-parameters, a selection of pre-processing techniques, employing advanced training schemes, as well as dealing with the class imbalance problem will immensely improve the accuracy and robustness of segmentation algorithms.

## 5. Summary

This paper has discussed several building blocks, state-of-the-art techniques, and tools for implementing automatic brain tumor segmentation algorithms. Despite the tremendous advance in the field, the robustness of deep learning methods are still inferior to expert performance. Some notable architectures, including ensemble methods and UNet based models, have shown great potential for improving the state-of-the-art with careful pre-processing, weight initialization, advanced training schemes, and techniques in order to address inherent class imbalance problems. The lack of a large-scale medical training dataset is the leading factor in many segmentation algorithms' poor performance.

**Author Contributions:** Conceptualization, T.M. and S.V.; methodology, T.M. and S.V; formal analysis, S.V.; investigation, T.M.; resources, S.V.; writing original draft preparation, T.M.; writing review and editing, S.V.; supervision, S.V. Both authors have read and agreed to the published version of the manuscript.

**Funding:** This research received no external funding.

**Institutional Review Board Statement:** Not applicable.

**Informed Consent Statement:** Informed consent was obtained from all subjects involved in the study.

**Data Availability Statement:** Data available in publicly accessible repositories.

**Conflicts of Interest:** The authors declare no conflict of interest.

## References

1. Havaei, M.; Davy, A.; Warde-Farley, D.; Biard, A.; Courville, A.; Bengio, Y.; Pal, C.; Jodoin, P.M.; Larochelle, H. Brain tumor segmentation with Deep Neural Networks. *Med. Image Anal.* **2017**, *35*, 18–31. [CrossRef] [PubMed]
2. Menze, B.H.; Jakab, A.; Bauer, S.; Kalpathy-Cramer, J.; Farahani, K.; Kirby, J.; Burren, Y.; Porz, N.; Slotboom, J.; Wiest, R.; et al. The Multimodal Brain Tumor Image Segmentation Benchmark (BRATS). *IEEE Trans. Med. Imaging* **2015**, *34*, 1993–2024. [CrossRef] [PubMed]
3. Işın, A.; Direkoğlu, C.; Şah, M. Review of MRI-Based Brain Tumor Image Segmentation Using Deep Learning Methods. *Procedia Comput. Sci.* **2016**, *102*, 317–324. [CrossRef]
4. Kamnitsas, K.; Ledig, C.; Newcombe, V.F.J.; Simpson, J.P.; Kane, A.D.; Menon, D.K.; Rueckert, D.; Glocker, B. Efficient multi-scale 3D CNN with fully connected CRF for accurate brain lesion segmentation. *Med. Image Anal.* **2017**, *36*, 61–78. [CrossRef] [PubMed]
5. Razzak, M.I.; Imran, M.; Xu, G. Efficient Brain Tumor Segmentation With Multiscale Two-Pathway-Group Conventional Neural Networks. *IEEE J. Biomed. Health Inform.* **2019**, *23*, 1911–1919. [CrossRef]
6. Muhammad, K.; Khan, S.; Ser, J.D.; de Albuquerque, V.H.C. Deep Learning for Multigrade Brain Tumor Classification in Smart Healthcare Systems: A Prospective Survey. *IEEE Trans. Neural Netw. Learn. Syst.* **2020**, 1–16. [CrossRef]
7. Wadhwa, A.; Bhardwaj, A.; Singh Verma, V. A review on brain tumor segmentation of MRI images. *Magn. Reson. Imaging* **2019**, *61*, 247–259. [CrossRef]
8. Alom, M.Z.; Taha, T.M.; Yakopcic, C.; Westberg, S.; Sidike, P.; Nasrin, M.S.; Hasan, M.; Van Essen, B.C.; Awwal, A.A.S.; Asari, V.K. A State-of-the-Art Survey on Deep Learning Theory andArchitectures. *Electronics* **2019**, *8*, 292. [CrossRef]
9. Zikic, D.; Ioannou, Y.; Brown, M.; Criminisi, A. Segmentation of Brain Tumor Tissues with Convolutional Neural Networks. In Proceedings of the BRATS-MICCAI, Boston, MA, USA, 14 September 2014; pp. 36–39.
10. Urban, G.; Bendszus, M.; Hamprecht, F.A.; Kleesiek, J. Multi-Modal Brain Tumor Segmentation Using Deep Convolutional Neural Networks. In Proceedings of the BRATS-MICCAI, Boston, MA, USA, 14 September 2014; pp. 31–35.
11. Shen, D.; Wu, G.; Suk, H.I. Deep learning in medical image analysis. *Annu. Rev. Biomed.* **2017**, *19*, 221–248. [CrossRef]
12. Havaei, M.; Guizard, N.; Larochelle, H.; Jodoin, P.M. Deep Learning Trends for Focal Brain Pathology Segmentation in MRI. In *Machine Learning for Health Informatics*; Holzinger, A., Ed.; Springer International Publishing: Berlin/Heidelberg, Germany, 2016; Volume 9605, pp. 125–148. [CrossRef]
13. Corso, J.J.; Sharon, E.; Dube, S.; El-Saden, S.; Sinha, U.; Yuille, A. Efficient Multilevel Brain Tumor Segmentation With Integrated Bayesian Model Classification. *IEEE Trans. Med. Imaging* **2008**, *27*, 629–640. [CrossRef]
14. Gordillo, N.; Montseny, E.; Sobrevilla, P. State of the Art Survey on MRI Brain Tumor Segmentation. *Magn. Reson. Imaging* **2013**, *31*, 1426–1438. [CrossRef] [PubMed]
15. Chen, L.; Bentley, P.; Mori, K.; Misawa, K.; Fujiwara, M.; Rueckert, D. DRINet for Medical Image Segmentation. *IEEE Trans. Med. Imaging* **2018**, *37*, 2453–2462. [CrossRef] [PubMed]
16. Goodfellow, I.; Bengio, Y.; Courville, A. *Deep Learning*; Adaptive Computation and Machine Learning; The MIT Press: Cambridge, MA, USA, 2016.
17. Chollet, F. *Deep Learning with Python*; Manning Publications Co.: Shelter Island, NY, USA, 2018.
18. Svozil, D.; Kvasnicka, V.; Pospichal, J. Introduction to Multi-Layer Feed-Forward Neural Networks. *Chemom. Intell. Lab. Syst.* **1997**, *39*, 43–62. [CrossRef]
19. Szegedy, C.; Liu, W.; Jia, Y.; Sermanet, P.; Reed, S.; Anguelov, D.; Erhan, D.; Vanhoucke, V.; Rabinovich, A. Going Deeper with Convolutions. In Proceedings of the IEEE Conference on Computer Vision and Pattern Recognition (CVPR), Boston, MA, USA, 7–12 June 2015.

20. Krizhevsky, A.; Sutskever, I.; Hinton, G.E. ImageNet Classification with Deep Convolutional Neural Networks. In *Proceedings of the Advances in Neural Information Processing Systems*; Pereira, F., Burges, C.J.C., Bottou, L., Weinberger, K.Q., Eds.; Curran Associates, Inc.: Red Hook, NY, USA, 2012; Volume 25, pp. 1097–1105.
21. Simonyan, K.; Zisserman, A. Very Deep Convolutional Networks for Large-Scale Image Recognition. *arXiv* **2015**, arXiv:1409.1556.
22. He, K.; Zhang, X.; Ren, S.; Sun, J. Deep Residual Learning for Image Recognition. *arXiv* **2015**, arXiv:1512.03385.
23. Hu, J.; Shen, L.; Albanie, S.; Sun, G.; Wu, E. Squeeze-and-Excitation Networks. *arXiv* **2019**, arXiv:1709.01507.
24. Glorot, X.; Bordes, A.; Bengio, Y. Deep Sparse Rectifier Neural Networks. In Proceedings of the Fourteenth International Conference on Artificial Intelligence and Statistics, JMLR Workshop and Conference Proceedings, Ft. Lauderdale, FL, USA, 11–13 April 2011; pp. 315–323.
25. Maas, A.L.; Hannun, A.Y.; Ng, A.Y. Rectifier Nonlinearities Improve Neural Network Acoustic Models. In Proceedings of the ICML Workshop on Deep Learning for Audio, Speech and Language Processing, Atlanta, GA, USA, 16 June 2013.
26. He, K.; Zhang, X.; Ren, S.; Sun, J. Delving Deep into Rectifiers: Surpassing Human-Level Performance on ImageNet Classification. *arXiv* **2015**, arXiv:1502.01852.
27. Long, J.; Shelhamer, E.; Darrell, T. Fully Convolutional Networks for Semantic Segmentation. *arXiv* **2015**, arXiv:1411.4038.
28. Bernal, J.; Kushibar, K.; Asfaw, D.S.; Valverde, S.; Oliver, A.; Martí, R.; Lladó, X. Deep Convolutional Neural Networks for Brain Image Analysis on Magnetic Resonance Imaging: A Review. *Artif. Intell. Med.* **2019**, *95*, 64–81. [CrossRef]
29. Sutskever, I.; Martens, J.; Dahl, G.; Hinton, G. On the Importance of Initialization and Momentum in Deep Learning. In *Proceedings of the 30th International Conference on Machine Learning*; Dasgupta, S., McAllester, D., Eds.; PMLR: Atlanta, GA, USA, 2013; Volume 28, pp. 1139–1147.
30. Duchi, J.; Hazan, E.; Singer, Y. Adaptive subgradient methods for online learning and stochastic optimization. *J. Mach. Learn. Res.* **2011**, *12*, 2121–2159.
31. Zeiler, M.D. ADADELTA: An Adaptive Learning Rate Method. *arXiv* **2012**, arXiv:1212.5701.
32. Tieleman, T.; Hinton, G. Lecture 6.5-rmsprop: Divide the gradient by a running average of its recent magnitude. *COURSERA Neural Netw. Mach. Learn.* **2012**, *4*, 26–31.
33. Kingma, D.P.; Ba, J. Adam: A method for stochastic optimization. *arXiv* **2014**, arXiv:1412.6980.
34. Wang, G.; Li, W.; Ourselin, S.; Vercauteren, T. Automatic Brain Tumor Segmentation Based on Cascaded Convolutional Neural Networks With Uncertainty Estimation. *Front. Comput. Neurosci.* **2019**, *13*, 56. [CrossRef] [PubMed]
35. Li, H.; Li, A.; Wang, M. A novel end-to-end brain tumor segmentation method using improved fully convolutional networks. *Comput. Biol. Med.* **2019**, *108*, 150–160. [CrossRef]
36. Cahall, D.E.; Rasool, G.; Bouaynaya, N.C.; Fathallah-Shaykh, H.M. Inception Modules Enhance Brain Tumor Segmentation. *Front. Comput. Neurosci.* **2019**, *13*, 44. [CrossRef]
37. Ronneberger, O.; Fischer, P.; Brox, T. U-Net: Convolutional Networks for Biomedical Image Segmentation. *arXiv* **2015**, arXiv:1505.04597.
38. Glorot, X.; Bengio, Y. Understanding the Difficulty of Training Deep Feedforward Neural Networks. In Proceedings of the Thirteenth International Conference on Artificial Intelligence and Statistics, Chia Laguna Resort, Sardinia, Italy, 13–15 May 2010; Volume 9, pp. 249–256.
39. Bengio, Y.; Lamblin, P.; Popovici, D.; Larochelle, H. Greedy Layer-Wise Training of Deep Networks. In *Advances in Neural Information Processing Systems 19*; Schölkopf, B., Platt, J.C., Hoffman, T., Eds.; MIT Press: Cambridge, MA, USA, 2007; pp. 153–160.
40. Claesen, M.; De Moor, B. Hyperparameter Search in Machine Learning. *arXiv* **2015**, arXiv:1502.02127.
41. Srivastava, N.; Hinton, G.; Krizhevsky, A.; Sutskever, I.; Salakhutdinov, R. Dropout: A Simple Way to Prevent Neural Networks from Overfitting. *J. Mach. Learn. Res.* **2014**, *15*, 1929–1958.
42. Ioffe, S.; Szegedy, C. Batch Normalization: Accelerating Deep Network Training by Reducing Internal Covariate Shift. *arXiv* **2015**, arXiv:1502.03167.
43. Pereira, S.; Pinto, A.; Alves, V.; Silva, C.A. Brain Tumor Segmentation Using Convolutional Neural Networks in MRI Images. *IEEE Trans. Med. Imaging* **2016**, *35*, 1240–1251. [CrossRef] [PubMed]
44. Myronenko, A. 3D MRI Brain Tumor Segmentation Using Autoencoder Regularization. *arXiv* **2018**, arXiv:1810.11654.
45. Rezaei, M.; Harmuth, K.; Gierke, W.; Kellermeier, T.; Fischer, M.; Yang, H.; Meinel, C. Conditional Adversarial Network for Semantic Segmentation of Brain Tumor. *arXiv* **2017**, arXiv:1708.05227.
46. Rao, V.; Sarabi, M.S.; Jaiswal, A. Brain tumor segmentation with deep learning. In Proceedings of the MICCAI Multimodal Brain Tumor Segmentation Challenge (BraTS), 2015; pp. 56–59. Available online: https://www.researchgate.net/profile/Mona_Sharifi2/publication/309456897_Brain_tumor_segmentation_with_deep_learning/links/5b444445458515f71cb8a65d/Brain-tumor-segmentation-with-deep-learning.pdf (accessed on 1 June 2020).
47. Casamitjana, A.; Puch, S.; Aduriz, A.; Sayrol, E.; Vilaplana, V. 3D Convolutional Networks for Brain Tumor Segmentation. In Proceedings of the MICCAI Challenge on Multimodal Brain Tumor Image Segmentation (BRATS), 2016; pp. 65–68. Available online: https://imatge.upc.edu/web/sites/default/files/pub/cCasamitjana16.pdf (accessed on 1 June 2020).
48. Hussain, S.; Anwar, S.M.; Majid, M. Brain Tumor Segmentation Using Cascaded Deep Convolutional Neural Network. In Proceedings of the 2017 39th Annual International Conference of the IEEE Engineering in Medicine and Biology Society (EMBC), Seogwipo, Korea, 11–15 July 2017; pp. 1998–2001. [CrossRef]

49. Pereira, S.; Oliveira, A.; Alves, V.; Silva, C.A. On hierarchical brain tumor segmentation in MRI using fully convolutional neural networks: A preliminary study. In Proceedings of the 2017 IEEE 5th Portuguese Meeting on Bioengineering (ENBENG), Coimbra, Portugal, 16–18 February 2017; pp. 1–4. [CrossRef]
50. Sled, J.; Zijdenbos, A.; Evans, A. A nonparametric method for automatic correction of intensity nonuniformity in MRI data. *IEEE Trans. Med. Imaging* **1998**, *17*, 87–97. [CrossRef]
51. Tustison, N.J.; Avants, B.B.; Cook, P.A.; Zheng, Y.; Egan, A.; Yushkevich, P.A.; Gee, J.C. N4ITK: Improved N3 Bias Correction. *IEEE Trans. Med. Imaging* **2010**, *29*, 1310–1320. [CrossRef]
52. Nyul, L.; Udupa, J.; Zhang, X. New variants of a method of MRI scale standardization. *IEEE Trans. Med. Imaging* **2000**, *19*, 143–150. [CrossRef]
53. Zhao, X.; Wu, Y.; Song, G.; Li, Z.; Zhang, Y.; Fan, Y. A deep learning model integrating FCNNs and CRFs for brain tumor segmentation. *Med. Image Anal.* **2018**, *43*, 98–111. [CrossRef]
54. Milletari, F.; Navab, N.; Ahmadi, S.A. V-Net: Fully Convolutional Neural Networks for Volumetric Medical Image Segmentation. *arXiv* **2016**, arXiv:1606.04797.
55. Vaidhya, K.; Thirunavukkarasu, S.; Alex, V.; Krishnamurthi, G. Multi-Modal Brain Tumor Segmentation Using Stacked Denoising Autoencoders. In *Proceedings of the Brainlesion: Glioma, Multiple Sclerosis, Stroke and Traumatic Brain Injuries*; Crimi, A., Menze, B., Maier, O., Reyes, M., Handels, H., Eds.; Lecture Notes in Computer Science; Springer International Publishing: Berlin/Heidelberg, Germany, 2016; pp. 181–194. [CrossRef]
56. Pereira, S.; Pinto, A.; Alves, V.; Silva, C.A. Deep Convolutional Neural Networks for the Segmentation of Gliomas in Multi-Sequence MRI. In *Proceedings of the Brainlesion: Glioma, Multiple Sclerosis, Stroke and Traumatic Brain Injuries*; Crimi, A., Menze, B., Maier, O., Reyes, M., Handels, H., Eds.; Lecture Notes in Computer Science; Springer International Publishing: Berlin/Heidelberg, Germany, 2016; pp. 131–143. [CrossRef]
57. Hussain, S.; Anwar, S.M.; Majid, M. Segmentation of glioma tumors in brain using deep convolutional neural network. *Neurocomputing* **2018**, *282*, 248–261. [CrossRef]
58. Lin, T.Y.; Goyal, P.; Girshick, R.; He, K.; Dollár, P. Focal Loss for Dense Object Detection. *arXiv* **2018**, arXiv:1708.02002.
59. Sun, L.; Zhang, S.; Chen, H.; Luo, L. Brain Tumor Segmentation and Survival Prediction Using Multimodal MRI Scans with Deep Learning. *Front. Neurosci.* **2019**, *13*, 810. [CrossRef] [PubMed]
60. Mlynarski, P.; Delingette, H.; Criminisi, A.; Ayache, N. Deep learning with mixed supervision for brain tumor segmentation. *J. Med. Imaging* **2019**, *6*, 034002. [CrossRef] [PubMed]
61. Iqbal, S.; Ghani Khan, M.U.; Saba, T.; Mehmood, Z.; Javaid, N.; Rehman, A.; Abbasi, R. Deep learning model integrating features and novel classifiers fusion for brain tumor segmentation. *Microsc. Res. Tech.* **2019**, *82*, 1302–1315. [CrossRef] [PubMed]
62. Kuzina, A.; Egorov, E.; Burnaev, E. Bayesian Generative Models for Knowledge Transfer in MRI Semantic Segmentation Problems. *Front. Neurosci.* **2019**, *13*, 844. [CrossRef]
63. Kao, P.Y.; Ngo, T.; Zhang, A.; Chen, J.W.; Manjunath, B.S., Brain Tumor Segmentation and Tractographic Feature Extraction from Structural MR Images for Overall Survival Prediction. In *Brainlesion: Glioma, Multiple Sclerosis, Stroke and Traumatic Brain Injuries*; Crimi, A., Bakas, S., Kuijf, H., Keyvan, F., Reyes, M., van Walsum, T., Eds.; Springer International Publishing: Berlin/Heidelberg, Germany, 2019; Volume 11384, pp. 128–141. [CrossRef]
64. Kao, P.Y.; Shailja, F.; Jiang, J.; Zhang, A.; Khan, A.; Chen, J.W.; Manjunath, B.S. Improving Patch-Based Convolutional Neural Networks for MRI Brain Tumor Segmentation by Leveraging Location Information. *Front. Neurosci.* **2020**, *13*. [CrossRef]
65. Nalepa, J.; Marcinkiewicz, M.; Kawulok, M. Data Augmentation for Brain-Tumor Segmentation: A Review. *Front. Comput. Neurosci.* **2019**, *13*, 83. [CrossRef]
66. Shin, H.C.; Tenenholtz, N.A.; Rogers, J.K.; Schwarz, C.G.; Senjem, M.L.; Gunter, J.L.; Andriole, K.; Michalski, M. Medical Image Synthesis for Data Augmentation and Anonymization Using Generative Adversarial Networks. *arXiv* **2018**, arXiv:1807.10225.
67. Han, C.; Rundo, L.; Araki, R.; Nagano, Y.; Furukawa, Y.; Mauri, G.; Nakayama, H.; Hayashi, H. Combining Noise-to-Image and Image-to-Image GANs: Brain MR Image Augmentation for Tumor Detection. *IEEE Access* **2019**, *7*, 156966–156977. [CrossRef]
68. Han, C.; Murao, K.; Noguchi, T.; Kawata, Y.; Uchiyama, F.; Rundo, L.; Nakayama, H.; Satoh, S. Learning More with Less: Conditional PGGAN-Based Data Augmentation for Brain Metastases Detection Using Highly-Rough Annotation on MR Images. In Proceedings of the 28th ACM International Conference on Information and Knowledge Management, Beijing China, 3–7 November 2019; pp. 119–127. [CrossRef]
69. Goodfellow, I.J.; Pouget-Abadie, J.; Mirza, M.; Xu, B.; Warde-Farley, D.; Ozair, S.; Courville, A.; Bengio, Y. Generative Adversarial Networks. *arXiv* **2014**, arXiv:1406.2661.
70. Simpson, A.L.; Antonelli, M.; Bakas, S.; Bilello, M.; Farahani, K.; van Ginneken, B.; Kopp-Schneider, A.; Landman, B.A.; Litjens, G.; Menze, B.; et al. A Large Annotated Medical Image Dataset for the Development and Evaluation of Segmentation Algorithms. *arXiv* **2019**, arXiv:1902.09063.
71. Team, T.T.D.; Al-Rfou, R.; Alain, G.; Almahairi, A.; Angermueller, C.; Bahdanau, D.; Bastien, F.; Bayer, J.; Belikov, A.; Belopolsky, A.; et al. Theano: A Python Framework for Fast Computation of Mathematical Expressions. *arXiv* **2016**, arXiv:1605.02688.
72. Goodfellow, I.J.; Warde-Farley, D.; Lamblin, P.; Dumoulin, V.; Mirza, M.; Pascanu, R.; Bergstra, J.; Bastien, F.; Bengio, Y. Pylearn2: A Machine Learning Research Library. *arXiv* **2013**, arXiv:1308.4214.
73. Jia, Y.; Shelhamer, E.; Donahue, J.; Karayev, S.; Long, J.; Girshick, R.; Guadarrama, S.; Darrell, T. Caffe: Convolutional Architecture for Fast Feature Embedding. *arXiv* **2014**, arXiv:1408.5093.

74. Paszke, A.; Gross, S.; Massa, F.; Lerer, A.; Bradbury, J.; Chanan, G.; Killeen, T.; Lin, Z.; Gimelshein, N.; Antiga, L.; et al. PyTorch: An Imperative Style, High-Performance Deep Learning Library. *arXiv* **2019**, arXiv:1912.01703.
75. Abadi, M.; Agarwal, A.; Barham, P.; Brevdo, E.; Chen, Z.; Citro, C.; Corrado, G.S.; Davis, A.; Dean, J.; Devin, M.; et al. Tensorflow: Large-Scale Machine Learning on Heterogeneous Distributed Systems. *arXiv* **2016**, arXiv:1603.04467.
76. Chollet, F. Keras: The Python Deep Learning API. 2020. Available online: https://keras.io/ (accessed on 1 June 2020).
77. Zhang, J.; Shen, X.; Zhuo, T.; Zhou, H. Brain tumor segmentation based on refined fully convolutional neural networks with a hierarchical dice loss. *arXiv* **2017**, arXiv:1712.09093.
78. Kayalibay, B.; Jensen, G.; Smagt, P.V.D. CNN-based segmentation of medical imaging data. *arXiv* **2017**, arXiv:1701.03056.
79. Kamnitsas, K.; Bai, W.; Ferrante, E.; McDonagh, S.; Sinclair, M.; Pawlowski, N.; Rajchl, M.; Lee, M.; Kainz, B.; Rueckert, D.; et al. Ensembles of Multiple Models and Architectures for Robust Brain Tumour Segmentation. *arXiv* **2017**, arXiv:1711.01468.
80. Dong, H.; Yang, G.; Liu, F.; Mo, Y.; Guo, Y. Automatic Brain Tumor Detection and Segmentation Using U-Net Based Fully Convolutional Networks. In *Proceedings of the Medical Image Understanding and Analysis*; Valdés Hernández, M., González-Castro, V., Eds.; Communications in Computer and Information Science; Springer International Publishing: Berlin/Heidelberg, Germany, 2017; pp. 506–517. [CrossRef]
81. Alex, V.; Safwan, M.; Krishnamurthi, G. Automatic Segmentation and Overall Survival Prediction in Gliomas Using Fully Convolutional Neural Network and Texture Analysis. *arXiv* **2017**, arXiv:1712.02066.
82. Erden, B.; Gamboa, N.; Wood, S. *3D Convolutional Neural Network for Brain Tumor Segmentation*; Technical Report; Computer Science, Stanford University: Stanford, CA, USA, 2017.
83. Isensee, F.; Kickingereder, P.; Wick, W.; Bendszus, M.; Maier-Hein, K.H. Brain Tumor Segmentation and Radiomics Survival Prediction: Contribution to the BRATS 2017 Challenge. *arXiv* **2018**, arXiv:1802.10508.
84. Meng, Z.; Fan, Z.; Zhao, Z.; Su, F. ENS-Unet: End-to-End Noise Suppression U-Net for Brain Tumor Segmentation. In Proceedings of the 2018 40th Annual International Conference of the IEEE Engineering in Medicine and Biology Society (EMBC), Honolulu, HI, USA, 18–21 July 2018; pp. 5886–5889. [CrossRef]
85. Liu, J.; Chen, F.; Pan, C.; Zhu, M.; Zhang, X.; Zhang, L.; Liao, H. A Cascaded Deep Convolutional Neural Network for Joint Segmentation and Genotype Prediction of Brainstem Gliomas. *IEEE Trans. Bio-Med. Eng.* **2018**, *65*, 1943–1952. [CrossRef] [PubMed]
86. Pereira, S.; Pinto, A.; Amorim, J.; Ribeiro, A.; Alves, V.; Silva, C.A. Adaptive feature recombination and recalibration for semantic segmentation with Fully Convolutional Networks. *IEEE Trans. Med. Imaging* **2019**. [CrossRef] [PubMed]
87. Kermi, A.; Mahmoudi, I.; Khadir, M.T. Deep Convolutional Neural Networks Using U-Net for Automatic Brain Tumor Segmentation in Multimodal MRI Volumes. In *Brainlesion: Glioma, Multiple Sclerosis, Stroke and Traumatic Brain Injuries*; Crimi, A., Bakas, S., Kuijf, H., Keyvan, F., Reyes, M., van Walsum, T., Eds.; Lecture Notes in Computer Science; Springer International Publishing: Berlin/Heidelberg, Germany, 2019; pp. 37–48. [CrossRef]
88. Isensee, F.; Kickingereder, P.; Wick, W.; Bendszus, M.; Maier-Hein, K.H. No New-Net. *arXiv* **2019**, arXiv:1809.10483.
89. Wang, L.; Wang, S.; Chen, R.; Qu, X.; Chen, Y.; Huang, S.; Liu, C. Nested Dilation Networks for Brain Tumor Segmentation Based on Magnetic Resonance Imaging. *Front. Neurosci.* **2019**, *13*, 285. [CrossRef]
90. Ribalta Lorenzo, P.; Nalepa, J.; Bobek-Billewicz, B.; Wawrzyniak, P.; Mrukwa, G.; Kawulok, M.; Ulrych, P.; Hayball, M.P. Segmenting brain tumors from FLAIR MRI using fully convolutional neural networks. *Comput. Methods Programs Biomed.* **2019**, *176*, 135–148. [CrossRef]
91. Jiang, Z.; Ding, C.; Liu, M.; Tao, D. Two-Stage Cascaded U-Net: 1st Place Solution to BraTS Challenge 2019 Segmentation Task. In *Proceedings of the Brainlesion: Glioma, Multiple Sclerosis, Stroke and Traumatic Brain Injuries*; Crimi, A., Bakas, S., Eds.; Lecture Notes in Computer Science; Springer International Publishing: Berlin/Heidelberg, Germany, 2020; pp. 231–241. [CrossRef]
92. Zhao, Y.X.; Zhang, Y.M.; Liu, C.L. Bag of Tricks for 3D MRI Brain Tumor Segmentation. In *Brainlesion: Glioma, Multiple Sclerosis, Stroke and Traumatic Brain Injuries*; Lecture Notes in Computer Science; Crimi, A., Bakas, S., Eds.; Springer International Publishing: Berlin/Heidelberg, Germany, 2020; pp. 210–220. [CrossRef]
93. Zhuge, Y.; Krauze, A.V.; Ning, H.; Cheng, J.Y.; Arora, B.C.; Camphausen, K.; Miller, R.W. Brain tumor segmentation using holistically nested neural networks in MRI images. *Med. Phys.* **2017**, *44*, 5234–5243. [CrossRef]
94. Liu, Y.; Stojadinovic, S.; Hrycushko, B.; Wardak, Z.; Lau, S.; Lu, W.; Yan, Y.; Jiang, S.B.; Zhen, X.; Timmerman, R.; et al. A deep convolutional neural network-based automatic delineation strategy for multiple brain metastases stereotactic radiosurgery. *PLoS ONE* **2017**, *12*, e0185844. [CrossRef]
95. Li, Z.; Wang, Y.; Yu, J.; Guo, Y.; Cao, W. Deep Learning based Radiomics (DLR) and its usage in noninvasive IDH1 prediction for low grade glioma. *Sci. Rep.* **2017**, *7*, 5467. [CrossRef]
96. Kamnitsas, K.; Chen, L.; Ledig, C.; Rueckert, D.; Glocker, B. Multi-Scale 3D Convolutional Neural Networks for Lesion Segmentation in Brain MRI. *Ischemic Stroke Lesion Segm.* **2015**, *13*, 46.
97. Hoseini, F.; Shahbahrami, A.; Bayat, P. AdaptAhead Optimization Algorithm for Learning Deep CNN Applied to MRI Segmentation. *J. Digit. Imaging* **2019**, *32*, 105–115. [CrossRef]
98. Naceur, M.B.; Saouli, R.; Akil, M.; Kachouri, R. Fully Automatic Brain Tumor Segmentation using End-To-End Incremental Deep Neural Networks in MRI images. *Comput. Methods Programs Biomed.* **2018**, *166*, 39–49. [CrossRef] [PubMed]
99. Yi, D.; Zhou, M.; Chen, Z.; Gevaert, O. 3-D convolutional neural networks for glioblastoma segmentation. *arXiv* **2016**, arXiv:1611.04534.

100. Cui, S.; Mao, L.; Jiang, J.; Liu, C.; Xiong, S. Automatic Semantic Segmentation of Brain Gliomas from MRI Images Using a Deep Cascaded Neural Network. *J. Healthc. Eng.* **2018**, *2018*, 4940593. [CrossRef] [PubMed]
101. Yang, T.; Ou, Y.; Huang, T. Automatic Segmentation of Brain Tumor from MR Images Using SegNet: Selection of Training Data Sets. In Proceedings of the 6th MICCAI BraTS Challenge, Quebec City, QC, Canada, 14 September 2017; pp. 309–312.
102. McKinley, R.; Meier, R.; Wiest, R. Ensembles of Densely-Connected CNNs with Label-Uncertainty for Brain Tumor Segmentation. In *Proceedings of the Brainlesion: Glioma, Multiple Sclerosis, Stroke and Traumatic Brain Injuries;* Crimi, A., Bakas, S., Kuijf, H., Keyvan, F., Reyes, M., van Walsum, T., Eds.; Lecture Notes in Computer Science; Springer International Publishing: Berlin/Heidelberg, Germany, 2019; pp. 456–465. [CrossRef]
103. Zhou, C.; Chen, S.; Ding, C.; Tao, D. Learning Contextual and Attentive Information for Brain Tumor Segmentation. In *Proceedings of the Brainlesion: Glioma, Multiple Sclerosis, Stroke and Traumatic Brain Injuries;* Crimi, A., Bakas, S., Kuijf, H., Keyvan, F., Reyes, M., van Walsum, T., Eds.; Lecture Notes in Computer Science; Springer International Publishing: Berlin/Heidelberg, Germany, 2019; pp. 497–507. [CrossRef]
104. McKinley, R.; Rebsamen, M.; Meier, R.; Wiest, R. Triplanar Ensemble of 3D-to-2D CNNs with Label-Uncertainty for Brain Tumor Segmentation. In *Proceedings of the Brainlesion: Glioma, Multiple Sclerosis, Stroke and Traumatic Brain Injuries;* Crimi, A., Bakas, S., Eds.; Lecture Notes in Computer Science; Springer International Publishing: Berlin/Heidelberg, Germany, 2020; pp. 379–387. [CrossRef]
105. Bakas, S.; Reyes, M.; Jakab, A.; Bauer, S.; Rempfler, M.; Crimi, A.; Shinohara, R.T.; Berger, C.; Rozycki, M.; Prastawa, M.; et al. Identifying the Best Machine Learning Algorithms for Brain Tumor Segmentation, Progression Assessment, and Overall Survival Prediction in the BRATS Challenge. *arXiv* **2019**, arXiv:1811.02629.
106. Ji, Z.; Shen, Y.; Ma, C.; Gao, M. Scribble-Based Hierarchical Weakly Supervised Learning for Brain Tumor Segmentation. *arXiv* **2019**, arXiv:1911.02014.
107. Pavlov, S.; Artemov, A.; Sharaev, M.; Bernstein, A.; Burnaev, E. Weakly Supervised Fine Tuning Approach for Brain Tumor Segmentation Problem. *arXiv* **2019**, arXiv:1911.01738.
108. Wu, K.; Du, B.; Luo, M.; Wen, H.; Shen, Y.; Feng, J. Weakly Supervised Brain Lesion Segmentation via Attentional Representation Learning. In *Proceedings of the Medical Image Computing and Computer Assisted Intervention—MICCAI 2019;* Shen, D., Liu, T., Peters, T.M., Staib, L.H., Essert, C., Zhou, S., Yap, P.T., Khan, A., Eds.; Lecture Notes in Computer Science; Springer International Publishing: Berlin/Heidelberg, Germany, 2019; pp. 211–219. [CrossRef]

Journal of *Imaging*

MDPI

*Article*

# A Survey of Brain Tumor Segmentation and Classification Algorithms

**Erena Siyoum Biratu [1], Friedhelm Schwenker [2,*], Yehualashet Megersa Ayano [3], Taye Girma Debelee [1,3]**

1 College of Electrical and Mechanical Engineering, Addis Ababa Science and Technology University, Addis Ababa 120611, Ethiopia; iranasiyoum@gmail.com (E.S.B.); tayegirma@gmail.com (T.G.D.)
2 Institute of Neural Information Processing, Ulm University, 89081 Ulm, Germany
3 Ethiopian Artificial Intelligence Center, Addis Ababa 40782, Ethiopia; yehualeuven@gmail.com
* Correspondence: friedhelm.schwenker@uni-ulm.de

**Abstract:** A brain Magnetic resonance imaging (MRI) scan of a single individual consists of several slices across the 3D anatomical view. Therefore, manual segmentation of brain tumors from magnetic resonance (MR) images is a challenging and time-consuming task. In addition, an automated brain tumor classification from an MRI scan is non-invasive so that it avoids biopsy and make the diagnosis process safer. Since the beginning of this millennia and late nineties, the effort of the research community to come-up with automatic brain tumor segmentation and classification method has been tremendous. As a result, there are ample literature on the area focusing on segmentation using region growing, traditional machine learning and deep learning methods. Similarly, a number of tasks have been performed in the area of brain tumor classification into their respective histological type, and an impressive performance results have been obtained. Considering state of-the-art methods and their performance, the purpose of this paper is to provide a comprehensive survey of three, recently proposed, major brain tumor segmentation and classification model techniques, namely, region growing, shallow machine learning and deep learning. The established works included in this survey also covers technical aspects such as the strengths and weaknesses of different approaches, pre- and post-processing techniques, feature extraction, datasets, and models' performance evaluation metrics.

**Keywords:** brain tumor; classification; segmentation; region growing; shallow machine learning; deep learning



**Citation:** Biratu, E.S.; Schwenker, F.; Ayano, Y.M.; Debelee, T.G. A Survey of Brain Tumor Segmentation and Classification Algorithms. *J. Imaging* **2021**, *7*, 179. https://doi.org/10.3390/jimaging7090179

Academic Editors: Leonardo Rundo, Carmelo Militello, Vincenzo Conti, Fulvio Zaccagna and Changhee Han

Received: 29 June 2021
Accepted: 28 August 2021
Published: 6 September 2021



## 1. Introduction

Machine learning has been applied in different sectors, the majority of the studies indicate that it was applied in agriculture [1], and health sectors [2,3] for disease detection, prediction, and classifications. In health sectors the most researched areas are breast cancer segmentation and classification [4–7], brain tumor detection and segmentation [8], and lung and colon cancer segmentation and classification [3].

The gold standard in brain tumor diagnosis is biopsy which includes resection and pathological examination using various cellular (histologic) examination techniques. However, the diagnosis using biopsy is invasive that may result in bleeding and even injury that results in functional loss [9]. As a result, non-invasive brain tumor diagnosis using magnetic resonance imaging is the mainstay of modern neuroimaging that enables physician to characterize structural, cellular, metabolic, and functional properties of brain tumor [9,10].

In a conventional structural MRI scan, a healthy brain contains white mater (WM), gray matter (GM), cerebrospinal fluid (CSF) [11]. The main variation of these tissues in a structural MRI scan depends on their water content. The white matter (WM), which is 70% water, is a myelinated axon that connects the cerebral cortex with other brain regions. Furthermore, it carries information between neurons and connects the right and left hemispheres of the brain. The gray matter, which is 80% water, contains neuronal and glial cells that control brain activity, and the basal nuclei which are located deep within the white matter. Whereas, the cerebrospinal fluid is almost 100% water, and fills the space between the infoldings of the brain, between the brain and skull, and between the ventricular system in the brain[11,12].

Clinically, due to the variability in size, locality, rate of growth, and pathology, it is difficult to understand the manifestation of a brain tumor. However, a brain tumor is an abnormal mass of tissue, in which some cells grow and multiply uncontrollably. This uncontrollable growth takes up space within the skull and interferes with normal brain activity and damages the brain cells. The damage may be caused through increasing pressure in the brain, by shifting the brain or pushing against the skull, and by invading nerves and healthy brain tissues [13,14]. Different criteria can be used to classify brain tumor. A layered based tumor classification schema that has been proposed by WHO provides a detailed classification techniques that is more pertinent to radiological use. In this schema the hierarchy from top to bottom four layers, that are, final integrated diagnosis, histologic classification, WHO grade, molecular information [15]. However, brain tumors can be more generally grouped into primary and secondary (metastatic) tumors depending on their place of origin [16]. Primary brain tumors originates in the brain itself and are named for the cell types from which they originated. These primary tumors can be benign (non-cancerous) and malignant (cancerous). Benign tumors grow slowly and do not spread elsewhere or invade the surrounding tissues. However, they can put pressure on the brain and compromise its function. On the contrary, the malignant tumors grow rapidly and spread to surrounding tissues. On the other hand, secondary brain tumors originate from another part of the body. These tumors mainly occur due to cancer cells from somewhere else in the patient's body that spread to the brain. The most common causes of secondary brain tumors are lung cancer, breast cancer, melanoma, kidney cancer, bladder cancer, certain sarcomas, and testicular and germ cell tumors [13,16,17]. Each of these tumors has unique clinical, radiographic, and biological characteristics [13].

In MRI scanning, brain examination can be normal or abnormal. The normal brain tissues in MRI are characterized by gray matter (GM), white matter (WM), and cerebrospinal fluid (CSF) tissues. Apart from the normal tissues listed earlier the tumorous brain scan often contains core tumor, necrosis, and edema. Necrosis is a dead cell located inside a core tumor, while edema is located near active tumor borders. Edema is a swelling that exists due to trapped fluids around a tumor. It can be vasogenic in non-infiltrative extra-axial tumors, such as meningioma, or it can be infiltrative that invades WM tracts of a brain in tumors, such as glioma [10,18]. Furthermore, these tissues often have indistinguishable intensity features in structural MRI sequences, such as T1-w, T2-w, FLAIR. For instance, the difficulty in differentiating between the core tumor and associated inflammation was discussed [19]. In addition to that, Alves et al. [19] demonstrated the difficulty in differentiating tumors using signal intensities alone. They demonstrated using a case where two patients were diagnosed with two different brain tumor types due to both tumors have similar intensity features and both are surrounded by extensive edema.

### 1.1. *Brain Tumor Imaging Modalities*

There are a variety of imaging techniques used to study brain tumors, such as magnetic resonance imaging (MRI), computed tomography (CT), positron emission tomography (PET), and single-photon emission computed tomography (SPECT) imaging. However, CT and MR imaging are the most widely used techniques, because of their widespread availability and their ability to produce high-resolution images of normal anatomic structures and pathologies [20].

#### 1.1.1. Magnetic Resource Imaging

Magnetic resonance imaging (MRI) of a brain generates several 3-dimensional image data that comprise the three anatomical views of a brain (axial, sagittal, and coronal) at different depths of a brain. Depending on the strength of the magnetic field and the sampling protocols, the image quality, slice thickness, and inter-slice gap vary [21,22].

During MR imaging, a patient lay in a strong magnetic field, almost 10,000 times stronger than the earth's magnetic field, that forces the protons in the water molecule of the body to align in either a parallel (low energy) or anti-parallel (high energy) orientation with the magnetic field. Then, a radiofrequency pulse is introduced that forces the spinning protons to move out of the equilibrium state. When a radiofrequency pulse pauses, the protons return to an equilibrium state and produce a sinusoidal signal at a frequency dependent on the local magnetic field. Finally, a radio antenna within the scanner detects the sinusoidal signal and creates the image [22,23]. The amount of signal produced by specific tissue types is determined by their number of mobile hydrogen protons, the speed at which they are moving, the time needed for the protons within the tissue to return to their original state of magnetization (T1), and the time required for the protons perturbed into coherent oscillation by the radiofrequency pulse to lose their coherence (T2) relaxation times. As T1 (spin-lattice, also known as longitudinal relaxation) and T2 (spin-spin, also known as traversal relaxation) times are time-dependent, the timing of the radio frequency pulse and the reading of the radiated RF energy change the appearance of the image. In addition, the repetition time (TR) describes the time between successive applications of RF pulse sequences, and the echo time (TE) tells the delay before the RF energy radiated by the tissue in question is measured. The variation of T1 and T2 relaxation times between tissues gives image contrast on T1- and T2-weighted (T1-w and T2-w) images. The T1-w sequence is characterized by short TR and short TE while the T2-w sequence is characterized by long TR and short TE. Tissues with shorter T1 (for example, white matter) appear brighter when compared to tissues with a longer T1 (for example, gray matter) in magnetic resonance images. The other intermediate sequence that adopts long TR from T2-w and short TE from T1-w is a proton density-weighted (PD-w). In PD-w, the number of protons per unit volume in tissues is the main factor in determining the formation of image [23,24].

In the current neuroimaging techniques different MRI brain scan procedures can be performed, these include, the conventional structural MRI, functional MRI, diffusion-weighted imaging (DWI), and diffusion tensor imaging (DTI) [10]. In structural MRI procedure which mainly differentiates healthy and abnormal brain tissues based on their water molecule content is the most commonly employed standard imaging technique. This procedure helps to visualize healthy brain tissues and to map gross brain anatomy, tumoral vascularity, calcification, and radiation-induced micro hemorrhage [10,11]. The structural sequences include T1-w, T2-w, FLAIR, and contrast-enhanced T1-w [10]. The functional MRI (fMRI) on the other hand is used to capture the neural activity inside a brain through the ratio of oxygenated to the deoxygenated level of blood in the neighboring vasculature while performing a cognitive or motor task. The fMRI is used to localize eloquent cortex and differentiate between tumor grades [10]. The DWI captures the random motion of water molecules in a brain and it is used to characterize a tumor through identification of its cellularity and hypoxia, peritumoral edema, the integrity of WM tracts, and to differentiate between posterior fossa tumors [10,25]. Whereas, diffusion tensor imaging (DTI) is used to analyze the 3D diffusion direction, also known as diffusion tensor, of the water molecule. The DTI helps to determine local effects of the tumor on white matter tract integrity including tract displacement, the existence of vasogenic edema, tumor infiltration, and tract destruction [26].

#### 1.1.2. Computed Tomography Imaging

A computed tomography (CT) scan was used in neuroimaging to help understand the functional and structural status of clinically significant signs of diseases. However, it provides less information than an MRI in brain tumor diagnosis. For instance, CT is inferior to MRI in the characterization of soft tissues like a brain and its use of ionizing radiation. However, a computed tomography (CT) scan can provide more detailed images of the bone structures near a brain tumor, such as the skull or spine. A CT scan may also be used to diagnose a brain tumor if the patient has implants like a pacemaker and when an MRI is not available. Currently, a CT is commonly used in the diagnosis of diseases like acute hemorrhage Parkinson's, head trauma, and in determining age [27,28]. Therefore, in this survey work, brain tumor segmentation and classification techniques that use the brain scan image of MRI are only explored.

The remaining part of the paper is organized as follows, Section 2 illustrates related works to this survey work and shows their strengths and limitations. In Section 3, the literature search strategy, including the chronological span, journal databases, the keywords used for search, and the inclusion and exclusion criteria, is presented. In Section 4, the commonly used model performance metrics in evaluating the performance of brain tumor segmentation and classification algorithms are highlighted. In Section 5, different region growing, conventional shallow supervised machine learning, and deep learning-based brain tumor segmentation techniques are discussed. Furthermore, the reported performances are presented. The techniques used in conventional machine learning-based brain tumor classification and their classification performance are elaborated in Section 6. In addition, different deep learning models based brain tumor classification techniques with their reported performance are presented. Finally, the paper presents a discussion on Section 7 and a conclusion in Section 8.

## 2. Related Works

The quest to find a better autonomous brain tumor segmentation and classification technique that can aid physicians in brain tumor diagnosis have been an active research area. As a result, several survey works have been completed to foster the research in the field and recap techniques used in brain tumor segmentation and classification. In Table 1, only some of the recent pieces of literature that are related to our survey work are listed. Furthermore, their strengths and limitations are clearly discussed.

**Table 1.** Survey literature on brain tumor segmentation and classification techniques.

| Author and Publication Year | Strength | Limitation |
|---|---|---|
| Sharma and Shukla [29] 2021 | Thresholding, conventional supervised and unsupervised based segmentation techniques are briefly described. | • A very shallow discussion on deep learning based brain tumor segementation and classification.<br>• The performances of the surveyed literature are not inculded. |
| Rao and Karunakara [30] 2021 | • Differnt brain tumor segmentation techniques that includes thresholding, region growing, atlas, deep learning, and conventional supervised and unsupervised machine learning based have been discussed.<br>• The performances of tumor classification techniques were clearly presented. | • Chronologically majority of the reviewed papers on brain tumor classification are from 2019 and earlier. Except two literature that are published on 2020.<br>• The segmentation and classification techniques are not clearly distingushed while presenting their performce metrices. |
| Magadza and Viriri [31] 2021 | • Deep learning based brain tumor segmentation techniques are presented in detail; including, their building blocks | • The survey does not include brain tumor classification techniques and conventional machine learning based tumor classification and segmentation techniques.<br>• Segmentation performce of top performing models on BRATs dataset is provided. |

Table 1. *Cont.*

| Author and Publication Year | Strength | Limitation |
|---|---|---|
| Tiwari et al. [32] 2020 | • A detailed hierarchical classification of brain tumor presented.<br>• A brain tumor segmentation techniques, including: those based on thresholding, conventional supervised and unsupervised machine learning, and deep learning are discussed.<br>• Conventional machine learning and deep learning based brain tumor classification techniques are surveyed. | • Chronologically, literature earlier than and including 2019 are reviewed.<br>• A small number of deep learning based brain tumor segmentation and classification literature are reviewed. |
| Kumari and Saxena [33] 2018 | • A limited literature that encompases different segmentation techniques including thresholding, deep learning, and supervised and unsupervised machine learning techniques were reviewed. | • Rather than reviewing literature on brain tumor classification, the paper only discusses the pros and cons of the classification algorithms.<br>• Aside from the limited discussion on brain tumor segmentation techniques, the review did not include the performance of proposed techniques.<br>• Furthermore, the review work incorporates literature before 2018. |

Our work is tailored to provide a comprehensive survey of recently proposed different brain tumor segmentation and classification techniques, including region growing, shallow machine learning, and deep learning. The established work in this survey also covers technical aspects, such as the strengths and weaknesses of different approaches, together with their performance.

## 3. Method

In this survey work, peer reviewed research papers from 2015 to 2021 that were published on Scopus and Web of Science indexed journals are surveyed to investigate the region growing, deep learning based brain tumor segmentation techniques, and machine learning and deep learning based brain tumor classification techniques. The databases that are extensively searched for this survey work were: (1) IEEE Xplore Digital Library, (2) Science Direct, (3) PubMed, (4) Google Scholar, and (5) MDPI. The search criterion includes ("Brain Tumor") AND ("Region Growing") AND ("Segmentation") AND ("Deep Learning") AND ("Machine Learning") AND ("Classification"). The methodology used for selecting literature is clearly shown in Algorithm 1. In addition, the paper inclusion criteria (IC) and exclusion criteria (EC) is indicated on Table 2.

Table 2. Inclusion and exclusion criteria for paper selection.

| IC | EC |
|---|---|
| IC1: Paper must be peer reviewed. | EC1: Duplicate studies in different databases. |
| IC2: Journals on which papers published must be either scopus or web of science indexed | EC2: Study that uses imaging techniques other than MRI. |
| IC3: The paper should use only MRI brain images | EC3: Study which is less cited by other peer reviewed papers. |
| | EC4: MSc and PhD papers. |
| | EC5: Case study papers. |

**Algorithm 1** Paper search strategy from different search databases.

```
procedure TOPIC(Application of Machine Learning and Region Growing Techniques in Brain Tumor Segmentation and Classification)
    SearchDatabases ← IEEEXplore, GoogleScholar, ScienceDirect, PubMed, MDPI
    SearchYear ← 2015 − 2021 AND Few papers from older years asexceptional to enrich Section 1
    i ← 1                                                    ▷ Initialize counter
    N ← 5                                                    ▷ N is the number of search databases

    for i ≤ N do
        Keyword ← braintumor, deeplearning, machinelearning, regiongrowing, segmentation, classification
        if SearchLink ∈ SearchDatabases and Year ∈ SearchYear then
            Search (Brain Tumor AND Region Growing AND Segmentation AND Deep Learning AND Machine Learning AND Classification)
        end if
    end for
    if NumberofPapers ≥ 0 then
        Refine Papers
        ApplyInclusionCriteria ← IC1, IC2, IC3
        ApplyExclusionCriteria ← EC1, EC2, EC3, EC4, EC5
    end if
end procedure
```

## 4. Performance Measuring Metrics

Evaluating the segmentation and classification performance of a machine learning algorithm is an essential part of a research project. A machine learning model may give a satisfying result when evaluated using a metric, for instance, accuracy score but may give poor results when evaluated against other metrics such as precision or any other metric. Therefore, most of the time various evaluation metrics are applied to measure and compare the model performance.

In a segmentation task, true positive (TP) represents a pixel that is correctly predicted to belong to the given class according to the ground truth, whereas a true negative (TN) represents a pixel that is correctly identified as not belonging to the given class. On the other hand, a false positive (FP) is an outcome where the model incorrectly predicts a pixel not belonging to a given class. A false negative (FN) is an outcome where the model incorrectly predicts the pixel belonging to a given class. Similarly, for tumor classification task, TP represents a tumor class that is correctly predicted to belong to the given class according to the ground truth whereas a TN represents a tumor class that is correctly identified as not belonging to the given class. By the same token, false positive (FP) is an outcome where the model incorrectly predicts a tumor class not belonging to a given class. A false negative (FN) is an outcome where the model incorrectly predicts the class belonging to a given class. Therefore, keeping different performance metrics used in brain tumor segmentation and classification literature are listed as follows.

Accuracy (ACC) measures the ability of a model in correctly identifying all class or pixels, no matter if it is positive or negative.

$$ACC = \frac{TP + TN}{TP + TN + FP + FN} \tag{1}$$

Sensitivity (SEN) indicates the frequency of correctly predicted positive samples/pixels among all real positive/samples. It measures the models ability in identifying positive samples/pixels.

$$SEN = \frac{TP}{TP + FN} \tag{2}$$

Specificity (SPE) is the proportion of actual negatives, which was predicted as the negative (or true negative). It tells the percentage of classes/pixels could not correctly identified.

$$SPE = \frac{TN}{TN + FP} \tag{3}$$

Recall (RE) describes the completeness of the machine learning model's positive predictions relative to the ground truth. It tells the percentage of classes/pixels annotated in our ground truth, are also included in model's prediction.

$$RE = \frac{TN}{TP + FN} \tag{4}$$

Precision (PR) also known as positive predictive value (PPV) describes how often the model predicting correct class/pixel. It tells the the correct proportion of models predicted positives.

$$PR = \frac{TP}{TP + FP} \tag{5}$$

F1-Score is the most popular metric that combines both precision and recall. It represents harmonic mean of the two.

$$F1score = 2\frac{PR * RE}{(PR + RE)} \tag{6}$$

Intersection over union (IoU) also known as Jaccard index (JI) measures the percent overlap between the annotated ground truth mask and the model's prediction output.

$$IoU = \frac{TP}{TP + FP + FN} \tag{7}$$

Dice similarity coefficient (DSC) measures the spatial overlap between the ground truth tumor region and the model segmented region. A zero DSC value indicates no spatial overlap between the ground truth tumor region and model annotated result whereas a value indicates a indicating complete overlap between the two.

$$DSC = \frac{TP}{\frac{1}{2}(2TP + FP + FN)} \tag{8}$$

Area under the curve (AUC) measure of the ability of a classifier to distinguish between classes and is used as a summary of the receiver characteristics curve and it is an area under true positive rate vs. false positive rate.

Similarity index (SI) refers to the similarity between the expert annotated ground truth and the model's segmentation. It describes the similar identity between the input image and the detected tumor region.

$$SI = \frac{2TP}{2TP + FP + FN} \tag{9}$$

## 5. Brain Tumor Segmentation Methods

Brain tumor imaging using techniques, such as MRI and CT, generate a significantly large number of images. Brain MRI scan of a single individual consists of several slices across the 3D anatomical view. Therefore, manual segmentation of brain tumors from magnetic resonance (MR) images is a challenging and time-consuming task. In addition, the artifacts introduced in the imaging process results in low-quality images that make the interpretation difficult. As a result, the manual brain MRI segment is susceptible for inter and intra observable variability. To alleviate these challenges and help radiologist, different automatic brain tumor segmentation techniques have been proposed in literature.

On these literature, authors have proposed an automated system for brain tumor segmentation techniques that provides objective, reproducible segmentation that are close to the manual results. These automated brain tumor segmentation can help to alleviate the difficulties associated with manually analyzing brain tumors. This will speed-up the brain image analysis process, improve diagnosis outcome, and make easy the follow-up of the disease through evaluating tumor progression [34].

In this section, among the proposed brain tumor segmentation techniques in the literature; region growing, machine learning, and deep learning based techniques will be surveyed to identify the experimental dataset, pre-processing, feature extraction, segmentation algorithm, and the reported performance.

### 5.1. Region-Based and Shallow Unsupervised Machine Learning Approach

One of the most commonly used segmentation techniques in automated image processing applications is region-based segmentation. Regions in an image are a group of connected pixels that satisfy certain homogeneity criteria, such as pixel intensity values, shape, and texture [35]. In a region-based segmentation the image is partitioned into dissimilar regions so that the desired region is located precisely [36]. The region-based segmentation takes into account the pixel values, such as gray level difference and variance, and spatial proximity of pixels, such as Euclidean distance and region compactness in grouping pixels together. In brain tumor segmentation, region growing, and clustering algorithms are the most commonly used region based segmentation technique.

Clustering-based segmentation is one of the powerful region based segmentation techniques where an image is partitioned into a number of disjoint groups. In clustering based segmentation pixels with high similarity categorized in a given region whereas dissimilar pixels categorized into different regions [37]. Clustering techniques, which are an unsupervised learning method, have been widely investigated in medical image segmentation. However, in this survey work some of the most popular clustering methods, such as k-means and its varieties [38–44], fuzzy c-means [38,39,41,45], subtractive clustering (SC), and hybrid techniques [46–48].

K-means clustering is an unsupervised machine learning algorithm and it is commonly used to segment a region of interest from the remaining part of an image. K-means has been extensively tested in brain tumor segmentation and has shown acceptable accuracy [48]. The minimal computational requirement [37,48], simplicity to implement on large dataset [49], adaptation to new examples, and guaranteed convergence are some of the advantages that makes K-means popular segmentation algorithm. However, k-means suffers with incomplete delineation of the tumor region [49], selection of the initial centroid is not optimum [37,43], and it is sensitive to outliers [48,50]. Due to these limitations a number of solutions have been proposed, including, evenly spreading the initial cluster centers (k-means++), hybridizing k-means with other clustering techniques [49], adaptively initializing cluster centers, such as adaptive k-means [43], modified adaptive k-means (MAKM), and histogram based k-means.

Fuzzy c-means works by assigning membership values to each of the pixels in an image corresponding to the centers of the clusters depending on a certain similarity criteria [51]. In fuzzy c-means (FCM) clustering objects can belong to more than one cluster based on its degree of membership. Therefore, in such a type of soft clustering technique, image pixels can occupy multiple clusters. As a result, compared to hard-clustering techniques such as k-means, FCM performs better on relatively noise free images. However, in medical images such as brain MRI that can be easily affected by unknown noises, the FCM performance is severely affected [52]. A number of researches have been performed to improve the limitation of FCM [53–56].

In region growing brain tumor segmentation, tissues including tumorous regions are partitioned based on certain similarity criterion, such as homogeneity, texture, sharpness, and gray levels. The technique starts by selecting an initial seed based on predefined methods. Then, the neighboring pixels are added progressively to the seed pixel [57]. The region growing based segmentation can properly segment regions with similar properties and spatially separated regions. However, it is sensitive to noise and influenced by the similarity criterion [57]. Therefore, it may end up with disconnected regions and results in a hole in the segmented region. Furthermore, finding a good initial seed is not an easy task [57]. Region growing and conventional unsupervised machine learning based brain tumor segmentation techniques proposed in literature are summarized in Table 3. The table indicates the brain MRI dataset used in the experiment, the centroid initialization techniques, the objective function, and the segmentation performance.

### 5.2. Supervised Shallow Machine Learning Based Approach

Supervised machine learning-based brain tumor segmentation approaches transformed the image segmentation problem into a tumorous pixel classification problem. The input vector for these supervised learning models consisted of different extracted features, and the output is a vector of desired classes for segmentation. In brain tumor segmentation, where tumor regions are often scattered all over the image, pixel classification rather than classical segmentation methods are often preferable [65]. Therefore, the traditional supervised machine learning algorithms have been used in the segmentation of a brain tumor from a head MRI scan [66–76].

**Table 3.** Region growing and shallow unsupervised machine learning based brain tumor segmentation.

| Paper | Dataset | Segmentation Technique | Objective Function | Performance |
|---|---|---|---|---|
| [58] | BRATS 2015 BRATS-MICCAI | Multi-level thresholding with level-set segmentation | Euclidean distance | JI 81.94%, DSC 89.91% |
| [48] | https://radiopaedia.org/ (accessed on 3 May 2021) | K-means and FCM | Euclidean distance | ACC 56.4 % |
| [43] | BRATS | K-means with histogram peaks centroid initialization | Euclidean distance | - |
| [39] | BRATS | Patch based k-means with FCM | Euclidean distance | SI 91% |
| [42] | BRATS 2012 | Random | Sum of Squared Error | DSC 91% |
| [44] | MRI images collected by authors | Bi-secting (No initialization) | Sum of Squared Error | ACC 83.05% |
| [59] | BRATS | Force Clustering | Distance (in pixels) | - |
| [60] | BRATS 2017 | Random | Euclidean distance | DSC 62.5% |
| [61] | MRI images collected by authors | DPSO [1] | Euclidean distance | ACC 99.98%, SEN 95.02%, SPE 99.92% DSC 93.09% |

**Table 3.** *Cont.*

| Paper | Dataset | Segmentation Technique | Objective Function | Performance |
|---|---|---|---|---|
| [62] | MRI images collected by authors | FCM preceded by gross tumor volume segmentation with random centroid intialization | Inter-cluster variance | DSC 95.93 ± 4.23%, JI 92.81 ± 6.56%, SPE 95.31 ± 6.56%, SEN 98.09 ± 1.75% |
| [63] | MRI images collected by authors | DWT [2] based genetic algorithm (GA) | fitness function variance | ACC 97% |
| [64] | MRI images collected by authors | semi-automatic cellular automata seeded segmentation with morphological post-processing | pixel similarity function | DSC 90.88 ± 4.19%, JI 84.11 ± 6.74%, SPE 99.99 ± 0.01%, SEN 91.20 ± 7.00% |

[1] Darwinian Particle Swarm Optimization, [2] Discrete Wavelet Transform.

In this section, as shown in Table 4, most relevant literature on brain tumor segmentation using traditional machine learning algorithms, such as support vector machine (SVM), artificial neural network (ANN), random forest (RF) are surveyed to identify data used, the pre-processing, feature extraction techniques, the classifier model, and whether or not post-processing is implemented.

### 5.3. *Deep Learning-Based Approach*

Deep learning methodologies produce automatic features that avoid or minimize the need for handcrafted features. In the deep learning-based brain tumor segmentation approach, the general strategy is to pass an image through the pipeline of deep learning building blocks and input image segmentation is performed depending on the deep features. In literature, there are a variety of deep learning techniques proposed for segmenting brain tumors. Some of such blocks contain deep convolutional neural networks (DCNNs), convolutional neural network (CNN), recurrent neural networks (RNNs), long short-term memory (LSTM), deep neural networks (DNNs), deep autoencoders (AEs), and generative adversarial networks (GANs). In this section, literature in terms of these building blocks, the dataset used, and the reported performance are presented as shown in Table 5.

**Table 4.** Summary of a shallow machine learning based segmentation.

| Paper | Dataset | Preprocessing | Features | Model | Post-Processing | Performance |
|---|---|---|---|---|---|---|
| [66] | Clinically collected MRI | N4ITK | deep features from CNN | SVM | - | DSC 88%, SEN 89%, PR 83% |
| [67] | Clinically collected MRI | Registration | Intensity texture | Multi-kernel SVM | Region growing | TP 98.9%, FP 4.5%, FN 3.1% |
| [68] | BRATS 2013 | N4ITK, histogram matching, SLIC [1] | Gray statistical, GLCM | SVM | - | DSC 86.12%, SEN 79.69%, SPE 99.48% |
| [70] | BRATS 2015 | - | Intensity, texture | ANN, SVM | - | SVM: DSC 88.7%, IOU 79.7%, ANN: DSC 90.79%, IOU 83.1% |

**Table 4.** *Cont.*

| Paper | Dataset | Preprocessing | Features | Model | Post-Processing | Performance |
|---|---|---|---|---|---|---|
| [71] | BRATS 2015, [77–79] | - | Dual pathway tree based features | ccRF [2] | mpAC [3] | DSC 89%, SPE 90%, SEN 85% |
| [72] | BRATS 2012 | registration, normalization | intensity, similarity, blobness | RF | Independent connected component analysis | DSC 96.5% |
| [74] | [80] | N4ITK, normalization, histogram matching | intensity, gradient, context | RDF [4] | morphological filtering | DSC 86.41%, SEN 82%, PR 92.92% |
| [75] | BRATS 2015 | noise removal, enhancement | first higher order features, texture | RF | morphological other filtering | DSC 98.4%, SEN 97.9%, SPE 80.7%, ACC 97.7% |
| [76] | BRATS 2015 | histogram enhancement | Gabor wavelet, intensity | RF | morphological other filtering | DSC 85.5%, SEN 77.1%, SPE 99.3% |

[1] Simple Linear Iterative Clustering, [2] Concatenated and Connected Random Forest, [3] Multiscale Patch Driven Active Contour, [4] Random Decision Forest.

**Table 5.** Summary of deep learning based brain tumor segmentation techniques.

| Paper | Dataset | Preprocessing | Model Architecture | Performance |
|---|---|---|---|---|
| [81] | BRATS 2013 & 2015 | bias field correction, intensity and patch normalization, augmentation | Custom CNN | DSC 88%, SEN 89%, PR 87% |
| [82] | BRATS 2013 | intensity normalization, augmentation | HCNN + CRF-RRNN [1] | SEN 95%, SPE 95.5%, PR 96.5%, RE 97.8%, ACC 98.6% |
| [83] | BRATS 2015 | Z-score normalization on the image, | Residual Network+ Dilated convolution RDM-Net [2] | DSC 86% |
| [84] | BRATS 2015 | Z-score normalization | Stack Multi-connection Simple Reducing_Net (SMCSRNet) | DSC 83.42%, PR 78.96%, SEN 90.24% |
| [85] | BRATS 2019 | - | Ensemble of a 3D-CNN and U-net | DSC 90.6% |
| [86] | BRATS 2015 | Bias correction, intensity normalization | Two-PathGroup-CNN (2PG-CNN) | DSC 89.2%, PR 88.22%, SEN 88.32% |
| [87] | BRATS 2018 | - | Hybrid two track U-Net (HTTU-Net) | DSC 86.5%, SEN 88.3%, SPE 99.9% |

**Table 5.** *Cont.*

| Paper | Dataset | Preprocessing | Model Architecture | Performance |
|---|---|---|---|---|
| [88] | BRATS 2015 | - | P-Net with bounding box and image specific fine tunning (BIFSeg) | DSC 86.29% |
| [89] | ADNI | denoising, Skull stripping, sub-sampling | Multi-scale CNN (MSCNN) | ACC 90.1% |
| [90] | BRATS 2017 | Intensity normalization, resizing, Bias field correction | Cascaded 3D U-nets | DSC 89.4% |
| [91] | BRATS 2015 & 2017 | Down sampling | 3D Center-crop Dense Block | BRATS 2015: DSC 88.4%, SEN 83.8% BRATS 2017: DSC 88.7%, SEN 84.3% |
| [92] | BRATS 2018 & 2019 | Z-score normalization, cropping | 3D FCN [3] | BRATS 2018: DSC 90%, SEN 90.3, SPE 99.48%; BRATS 2019: DSC 89%, SEN 88.3%, SPE 99.51% |
| [93] | BRATS 2018 | intensity normalization, removing 1% of highest & lowest intensity | DCNN (Dense-MultiOCM [4]) | BRATS 2018: DSC 86.2%, SEN 84.8%, SPE 99.5% |
| [94] | TCIA | Image cropping, padding, resizing, intensity normalization | U-Net | DSC 84%, SEN 92%, SPE 92%, ACC 92% |
| [95] | BRATS 2013, 2015, 2018 | - | AFPNet [5] + 3D CRF | BRATS 2013 DSC 86%, BRATS 2015 DSC 82%, BRATS 2018 86.58% |
| [96] | BRATS 2015, 2017 | z-score normalization | Inception-based U-Net + up skip connection + cascaded training strategy | DSC 89%, PR 78.5%, SEN 89.5% |
| [97] | BRATS 2015, BrainWeb | cropping, z-score normalization, min-max normalization (BrainWeb) | Tripple intersecting UNets (TIU-Net) | BRATS 2015: DSC 85%, BrainWeb DSC 99.5% |
| [98] | BRATS 2015 | - | LSTM multi-modal UNet | DSC 73.09%, SEN 63.76%, PR 89.79% |

[1] Heterogeneous CNN + Conditional Random Fields-Recurrent Regression based Neural Network, [2] Deep Residual Dilate Network with Middle Supervision, [3] Fully Convolutional Neural Network, [4] OCcipito Module, [5] Atrous-Convolution Feature Pyramid.

## 6. Brain Tumor Classification Methods

Based on the WHO's classification of central nervous system (CNS) tumors, there are more than 150 types of CNS tumors that are mainly categorized into primary and metastatic (secondary) tumors [99]. The primary tumors originate from the brain or the immediate surrounding tissues. Whereas, metastatic tumors arise from other body parts and migrate to the brain through the bloodstream. Metastatic tumors are considered cancerous or malignant, while primary tumors can be benign or malignant.

A biopsy is the existing gold standard procedure in brain tumor classification. However, it usually requires definitive brain surgery to take a sample [100,101]. On the other hand, an automated brain tumor classification from an MRI is non-invasive so that it avoids tumor sample taking procedure and it is safer. In addition, the machine learning-based brain tumor classification from an MRI scan can improve the diagnosis and treatment planning [101]. As a result, an automatic brain tumor classification from MRI images using machine or deep learning techniques is an active research area, and promising results have been achieved [100,102–106].

### 6.1. Conventional Machine Learning Based Approach

Machine learning is a paradigm where a machine is given a task where its performance improves with experience. Machine learning techniques are commonly grouped into three major types: supervised, unsupervised, and reinforcement learning [107]. Supervised learning is based on training a data sample from the data source with correct classification already assigned by domain experts, whereas, in unsupervised learning, the algorithm finds hidden patterns from the unlabeled data. On the other hand, reinforcement learning is carried out by making a sequence of decisions using reward signals. Therefore, the algorithm learns through receiving either rewards or penalties for the actions it performs [107]. Machine learning has been used in the classification of brain tumors from MRI images, and promising classification performance has been reported [108–115].

The traditional machine learning-based brain tumor classification techniques often consist of preprocessing, segmentation, feature extraction, and classification stages.

#### 6.1.1. Pre-processing

Brain MRI scans are significantly affected by different types of noises, including salt and pepper, Gaussian, Rician, and speckle noise [116–118]. These noises impose challenges in machine learning-based applications [117,119]. Therefore, obtaining high-quality image denoising is one of the important tasks in the pre-processing stage. Each method used in MRI denoising has its advantages and disadvantages. Several methods have been developed for reducing noises based on statistical property and frequency spectrum distribution [119]. In addition to denoising, tasks such as removing tags, smoothing the foreground region, intensity inhomogeneity correction, maintaining relevant edges, resizing, cropping, and skull stripping are part of pre-processing [110–112].

#### 6.1.2. Region of Interest (ROI) Detection

In an MRI brain scan, the segmentation task labels each voxel in an MRI image to specify its tissue type and anatomical structure [119]. The objective of ROI detection in tumor classification is to locate the tumor region from an MRI scan, improve the visualization, and allow quantitative measurements of image structures in the feature extraction stage [108,112]. Brain tumor segmentation can be performed in three different ways, namely, manual segmentation, semi-automatic segmentation, and fully automatic segmentation [119]. The autonomous brain segmentation techniques have been briefly discussed in Section 5.

#### 6.1.3. Feature Extraction

The feature extraction techniques are mathematical models based on various image properties. The different types of features include texture, brightness, contrast, shape, Gabor transforms, gray-level co-occurrence matrix (GLCM), and wavelet-based features [115,120], histogram of local binary patterns (LBP) [121]. On the other hand, recently, deep features that are obtained from deep neural networks such as CNN have been used as input to SVM classifier to classify brain tumors [122]. In brain tumor classification, it is customary to fuse several features from different extraction models to improve the discrimination power of the machine learning model [123]. Furthermore, feature selection is applied for dimensionality reduction.

#### 6.1.4. Classification

Different classification techniques have been proposed by many authors for identifying tumor types from brain images. Different authors have classified tumor into a variety of ways, for instance meningioma, glioma, and pituitary [109,121,122,124,125]; astrocytoma, glioblastoma, and oligodendrogliamo [112]; glioma tumor grades (I–IV) [113]; benign and malignant stages(I–IV) [126–129]; diffuse midline glioma, medulloblastoma, pilocytic astrocytoma, and ependymoma [102]; multifocal, multicentric, and gliomatosis [130]; ependymoma and pilocytic astrocytoma [120].

In brain tumor classification, the most commonly used classifiers are neural network [108–111,131], support vector machines (SVM) [108,115,124,127–130,132,133], K-nearest neighbor (KNN) [112,121,130,134], Adaboost [126], and hybrid models [113,135,136]. The neural network was implemented using different architectures, such as feedforward neural network [110,125], multilayer perceptron neural network [109,137], and probabilistic neural network (PNN) [111,131]. Support vector machine (SVM) was commonly implemented using three kernels, linear, homogeneous polynomial, and Gaussian radial basis function (RBF) [108,115]. In the KNNclassifier, the testing feature vector is classified by finding the k-nearest training neighbor, that is, the classifier does not use any model to match and is only based on memory. However, KNN uses different measurements such as euclidean distance, city block, cosine, and correlation to find the nearest distance between the testing and training class feature vectors [134].

A summary of recent shallow machine learning-based brain tumor classification techniques is given on Table 6.

### 6.2. Deep Learning Approach

Even though promising progress has been made in classifying brain tumors into their respective types from an MRI brain scan using shallow supervised machine learning algorithms, there are still challenges in classifying brain tumors from an MRI scan. These challenges are mainly due to the ROI detection, and extracting descriptive information using traditionally handcrafted feature extraction techniques is not efficient [122]. This inefficiency mainly arises due to the complex structure of brain anatomy and the high-density nature of the brain.

Unlike shallow machine learning algorithms, deep learning is based on learning data representations and hierarchical feature learning. In deep learning-based brain tumor classification, the deep learning models discover the descriptive information that optimally represents different brain tumors. This nature of deep learning transforms the brain tumor classification from handcrafted feature-driven into data-driven problem [103]. Among the deep learning models, a convolutional neural network (CNN) is widely used in brain tumor classification tasks, and a substantial result has been achieved [100].

In the reviewed literature, there are differences in the techniques used for the classification of brain tumors. The difference encompasses: (i) the dataset used for classification including tumor types, (ii) the implemented pre-processing and data augmentation techniques, (iii) whether or not the ROI segmentation was used as a prior step in the classification, (iv) whether a pre-trained or custom-designed deep learning model is used.

**Table 6.** Summary of conventional ML based brain tumor classification techniques.

| Paper | Dataset | Preprocessing | ROI Detection | Feature Extraction | Classifier | Tumor Types | Performance |
|---|---|---|---|---|---|---|---|
| [108] | Local dataset | Median and weiner filter | k-means modified FCM | shape features, statistical features | ANN | Benign malignant stage (I-IV) | SPE 100%, SEN 98%, ACC 97.73%, BER 0.0294 |
| [109] | [138] | Median and weiner filter | manually | 2-D DWT 2-D Gabor feature | ANN | Glioma (GL), Meningioma (MG) Pituitary tumor (PT) | overall ACC 91.9%, SPE (GL) 96.29%, SPE (MG) 96%, SPE (PT) 96.2%, SEN (GL) 95.1%, SEN(MG) 86.97%, SEN(PT) 91.24% |
| [110] | Local dataset | resizing skull removing | Canny | Gabor filter, GLCM DWT | ANN | Benign and malignant stage (I-IV) | SPE 98.5%, SEN 99.1%, ACC 98.9% |
| [139] | Local dataset | resizing | - | PCA [1] | PNN | Benign malignant stage | SPE 100%, SEN 92.3%, ACC 97.4% |
| [112] | TCIA | resizing, cropping, median filtering | morphological, watersheed | shape features | KNN | Astrocytoma Glioblastoma Oligodendroglioma | ACC 89.5% |
| [115] | Local dataset | wavelets | thresholding | DWT coeficients statistical features | SVM | Benign malignant | ACC (linear) 92%, ACC (kernel) 99% |
| [134] | BRATS and Local dataset | enhancement median filter | Morphological | GLCM features | SVM | Benign malignant | BRATS:<br>SVM (linear):SPE 100%, SEN 72%, ACC 82.5%<br>SVM (Quadratic):SPE 73.3%, SEN 88%, ACC 82.5%<br>SVM (RBF): SPE 100%, SEN 76%, ACC 85%<br>Clinical:<br>SVM (linear):SPE 60%, SEN 76%, ACC 68%<br>SVM (Quadratic):SPE 88%, SEN 100%, ACC 94%<br>SVM (RBF): SPE 100%, SEN 92%, ACC 96% |
| [120] | Local dataset | | | Gabor transform texture wavelet | SVM | Ependymoma Pilocytic Astrocytoma | SPE 80%, SEN 93%, ACC 88%, AUC 0.86 |

**Table 6.** *Cont.*

| Paper | Dataset | Preprocessing | ROI Detection | Feature Extraction | Classifier | Tumor Types | Performance |
|---|---|---|---|---|---|---|---|
| [140] | BRATS-2015 | wavelet filters, inhomogeneity correction | edge detection, morphological operations | shape, texture, intensity | PSO [2]-SVM | Benign, malignant | SPE 94.8%, SEN 100% |
| [136] | - | median filtering skull removing | thresholding | GLCM | GA-SVM | Benign, malignant | - |
| [130] | REMBRANDT | - | - | texture features | SVM | Multifocal, Multicentric, Gliomatosis | PR 90%, SEN 90%, ACC 90%, F1-Score 90% |
| [133] | Local dataset | Image fusion with contourlet transform | Otsu's thresholding | curvlet transform GLCM features | SVM | Benign, Malignant | ACC 93% |
| [125] | [138] | min-max normalization, | - | NGIST features | RELM [3] | Meningioma, Glioma, Pituitary | ACC 94.23% |
| [126] | Local dataset | median filter | thresholding | GLCM texture features | Adaboost | Benign, Malignant | SPE 62.5%, SEN 88.25%, ACC 89.90% |
| [127] | Local dataset | resizing enhancement | morphological, thresholding | GLCM statistical texture features | SVM | Benign, Malignant | SPE 62.5%, SEN 88.25%, ACC 89.90% |
| [128] | Local dataset | noise removal, enhancement | Expectation maximization, levelset | GA, statistical features | SVM | Benign, Malignant | SPE 100%, SEN 98%, ACC 98.30% |

**Table 6.** *Cont.*

| Paper | Dataset | Preprocessing | ROI Detection | Feature Extraction | Classifier | Tumor Types | Performance |
|---|---|---|---|---|---|---|---|
| [124] | [138] | down sampling Gabor filter | - | statistical features | SVM | Meningioma, Glioma, Pituitary | Meningioma:<br>SVM (linear):RE 0.63, PR 0.66, ACC 82.38%<br>SVM (poly):RE 0.62,Pr. 0.73, ACC 84.33%<br>Glioma:<br>SVM (linear):RE 0.82, PR 0.82, ACC 83.01%<br>SVM (poly):RE 0.88, PR 0.79, ACC 84.01%<br>Pituitary:<br>SVM (linear):RE 0.94, PR 0.90, ACC 95.27%<br>SVM (poly):RE 0.91,PR 0.94, ACC 95.43% |
| [122] | Kaggle Brain Tumor Detection 2020 | cropping, resizing using bicubic interpolation | - | Deep features from pretrained CNN | SVM | Meningioma, Glioma, Pituitary | ACC 90.19% |

[1] Principal Component Analysis, [2] Particle Swarm Optimization, [3] Regularized Extreme Learning Machine.

For instance, Badža and Barjaktarović [100] used publicly available contrast-enhanced T1-weighted brain tumor MRI scans [138]. The dataset contains meningioma, glioma, and pituitary brain tumor types scanned along with the three anatomical views, i.e., axial, sagittal, and coronal. The images were preprocessed using techniques, such as normalization and resizing. In addition, images in the dataset are augmented with 90$^o$ rotation and vertical flipping to increase the training dataset. Furthermore, they used a custom-designed CNN model trained with Adam optimizer with a mini-batch size of 16 and tested with 10—fold cross-validation. The weights of the convolution layers are initialized using a Glorot initializer. The model performance was measure using sensitivity, specificity, accuracy, precision, recall, and F1-score. The sensitivity for meningioma, glioma, and pituitary is 89.8%, 96.2%, and 98.4%, respectively. The specificity of the model for meningioma, glioma, and pituitary is 90.2%, 95.5%, and 97.7%, respectively. Furthermore, the models' overall accuracy, average precision, average recall, and F1-score are 95.4%, 94.81%, 95.07%, and 94.94%, respectively. The summary of this and other literature is presented on Table 7.

**Table 7.** Summary of deep learning based brain tumor classification techniques.

| Paper | Dataset | Preprocessing | Classifier Model | Tumor Types | Performance |
|---|---|---|---|---|---|
| [100] | [138] | normalization, resizing, augmentation | Custom CNN model | Meningioma, Glioma, Pituitary | ACC 91.9%, precision 94.81%, RE 95.07%, F1-score 94.94%, SPE(GL) 96.2%, SPE(MG) 92%, SPE(PT) 97.7%, SEN(GL) 96.2%, SEN(MG) 89.8%, SEN(PT) 98.4% |
| [141] | [78,142] | Augmentation using GAN | Multi-stream 2D-CNN model | Glioma subtypes: Isocitrate dehydrogenase 1 mutation (IDH1), & IDH1 wild-type | mean ACC 88.82% mean SEN 81.81% mean SPE 92.17% |
| [143] | [138,144] | resizing augmentation | Custom CNN model | Meningioma, Glioma & Pituitary and Glioma (grade:II-IV) | MG: PR 95.8%, SEN 95.5%, SPE 98.7%, ACC 97.54%, GL: PR 97.2%, SEN 94.4%, SPE 95.1%, ACC 95.81%, PT: PR 95.2%, SEN 93.4%, SPE 97%, ACC 96.89% Grade II: PR 100%, SEN 100%, SPE 100%, ACC 100%, III: PR 100%, SEN 95%, SPE 100%, ACC 95%, IV:PR 96.3%, PR 100%, SEN 95%, SPE 100%, ACC 95%SEN 100%, SPE 98%, ACC 100% |
| [145] | [138] | - | CNNBCN [1] | Meningioma, Glioma& Pituitary | ACC 95.49% |
| [146] | [138] | - | BayesCap: captures prediction uncertainity | Meningioma, Glioma& Pituitary | mean ACC 73.9% CI [2]:(73.4%, 74.4%) |

**Table 7.** *Cont.*

| Paper | Dataset | Preprocessing | Classifier Model | Tumor Types | Performance |
| --- | --- | --- | --- | --- | --- |
| [147] | [138] | Image rotation, resizing | AutoML [3] | Meningioma, Glioma & Pituitary | MG: PR 94.51%, SEN 87.76%, SPE 98.7%, ACC 96.29%, F1-Score 91.01%, MCC [4] 88.77%, G-Mean 96.09% GL: PR 96.97%, SEN 95.32%, SPE 96.88%, ACC 96.08%, F1-Score 96.14%, MCC 92.17%, G-Mean 96.09% PT: PR 91.61%, SEN 99.24%, SPE 96.27%, ACC 97.14%, F1-Score 95.27%, MCC 93.38%, G-Mean 97.75% |
| [148] | [138] | - | Iception-V3 DensNet201 | Meningioma, Glioma& Pituitary | Iception-V3: ACC 99.34% DensNet201: ACC 99.51% |
| [149] | [138] | augmentation, contrast-stretching | AlexNet, GoogleNet & VGG16 [5] | Meningioma, Glioma& Pituitary | AlexNet: ACC 95.46% GoogleNet: ACC 98.04% VGG16 98.69% |
| [150] | [138] | - | ConvCaps | Meningioma, Glioma& Pituitary | ACC 93.5% |
| [151] | [138] | flipping, patching | CapsulNet | Meningioma, Glioma& Pituitary | MG: PR 85%, RE 94%, F1-Score 94, %GL: PR 85%, RE 94%, F1-Score 94%, PT: PR 85%, RE 94%, F1-Score 94% |
| [152] | [138] | - | G-ResNet | Meningioma, Glioma& Pituitary | ACC 95% |
| [153] | [138] | - | DDIRNet [6] | Meningioma, Glioma& Pituitary | ACC 99.69%, PR 99.6%, RE 99.4%, F1-score 99.4% |
| [103] | [138] | - | Multiscale CNN | Meningioma, Glioma& Pituitary | ACC 97.3% |
| [154] | [155] | DWT | DNN | Meningioma, Glioma& Pituitary | ACC 96.15%, PR 94.12%, AUC 98.75%,F1-score 96.97%, RE 100% |
| [156] | [138] | - | Custom CNN model | Meningioma, Glioma& Pituitary | ACC 84.19% |
| [157] | BraTS 2018 & 2019 | - | Pre-trained DenseNet201 | HGG [7] & LGG [8] | HGG: ACC 99.8%, LGG: ACC 99.3% |

**Table 7.** *Cont.*

| Paper | Dataset | Preprocessing | Classifier Model | Tumor Types | Performance |
| --- | --- | --- | --- | --- | --- |
| [158] | [138], [144,159] | - | Custom CNN model | Class 1: Normal, Metastatic, Meningioma, Glioma& Pituitary Class 2: Grade II, III & IV | Class 1: ACC 92.66% Class 2: ACC 98.14% |
| [160] | BraTS 2019 | - | Custom CNN model | Astrocytoma, Glioblastoma, Oligodendrogloma, | Class 1: ACC 92.66% Class 2: ACC 98.14% |
| [94] | TCIA | cropping, padding, resizing, normalization | VGG16 | Grade II & III | ACC 89%, SEN 87%, SPE 92% |

[1] Convolutional Neural Network based on Complex Networks, [2] Confidence Interval, [3] Automated Machine Learning, [4] Matthew's Correlation Coefficient, [5] Visual Geometry Group, [6] Deep Dense Inception Residual Network, [7] High Grade Glioma, [8] Low Grade Glioma.

## 7. Discussion

This paper presented a thorough survey of techniques used in brain tumor segmentation and classification. The survey encompasses several traditional machine learning and deep learning-based methods with their quantitative performance. The conventional image segmentation techniques, that is, region growing and unsupervised machine learning used in brain tumor segmentation are presented in Table 3. The region growing with all other conventional image processing segmentation techniques is the earliest approach applied in brain tumor segmentation [161]. It is mainly affected by noises, poor image quality, and initial seed point. To overcome these challenges, an automatic seed point selection by optimization techniques and artificial intelligence-based seed point selection has been proposed [162]. In addition, it has a limitation in segmenting tumors that appear scattered across the brain. In the second generation segmentation techniques which are based on shallow unsupervised machine learning, such as fuzzy c-means and k-means grouping of pixels into more than one class has been achieved. However, these methods are also highly sensitive to noise. Therefore, through incorporating additional information and adaptively selecting the centroid, the segmentation performance of medical images can be improved [6]. In addition, the inherent ambiguous boundaries between normal tissues and brain tumors pose a significant challenge for conventional and clustering segmentation techniques. Therefore, to address this challenge, pixel-level classification-based segmentation techniques using traditional supervised machine learning have been proposed [70]. These methods are often accompanied by feature engineering, where the tumor descriptive pieces of information are extracted to train the model. Furthermore, the supervised machine learning segmentation output is further improved through post-processing [71,76].

Nowadays, conventional image processing and shallow machine learning-based brain tumor segmentation techniques are becoming obsolete due to the advent of deep learning-based techniques. The deep learning-based approach performs an end-to-end tumor segmentation by passing an MRI image through the pipeline of its building blocks. These models often extract tumor descriptive information automatically and avoid the need for handcrafted features. However, the need for a large dataset to train the models and the difficulty in interpreting the models hinders their usage in medical fields [163]. In terms of segmentation performance, it is evident from Tables 4 and 5 that the deep learning-based and supervised shallow machine learning-based with post-processing has comparable performances. Asummary of the number of brain tumor segmentation techniques surveyed in this is given on Figure 1.

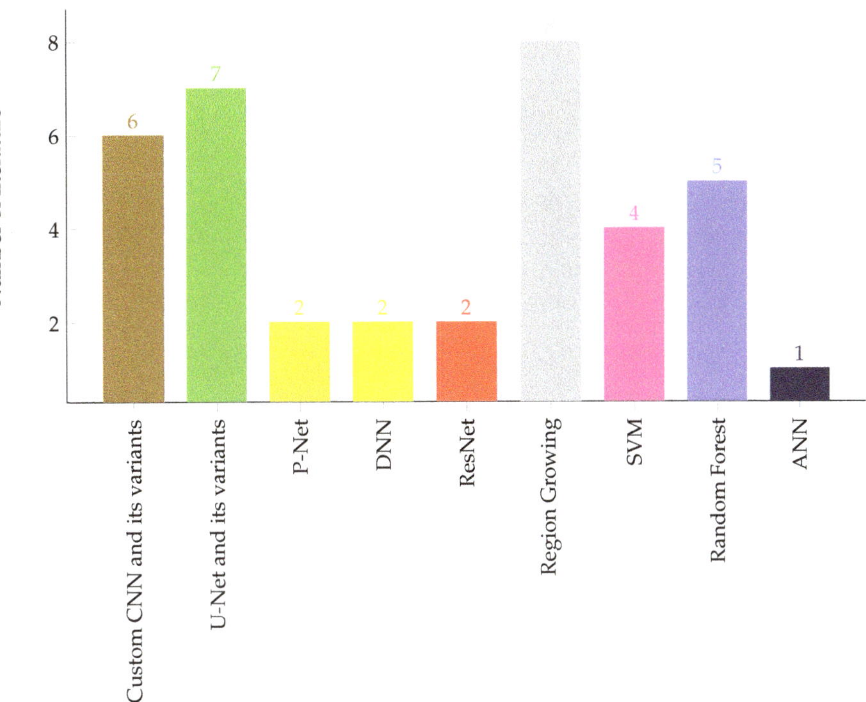


**Figure 1.** Number of brain tumor segmentation methods.

Aside from segmentation of brain tumor region from head MRI scan, classification of tumor into their respective histological type has great importance in diagnosis and treatment planning which actually requires biopsy procedure in today's medical practice [158]. Several methods which encompass shallow machine learning and deep learning have been proposed for brain tumor classification. The conventional shallow machine learning algorithms often consist of preprocessing, ROI detection, and feature extraction. However, due to the inherent noise sensitivity of MRI image acquisition, variations in the shape, size, location, and contrast of tumor tissue cells, extracting descriptive information is a challenging task. Therefore, nowadays, deep learning techniques are becoming the state-of-the-art approach to classify different types of brain tumors, such as astrocytoma, glioma, meningioma, and pituitary. Several brain tumor classifications have been discussed in this survey, and a summary of the number of brain tumor classification techniques surveyed in this paper are given on Figure 2.

Several brain tumor datasets that are collected by researchers datasets and those that are available on repositories were used in the training and testing of brain tumor classification models. The publicly available dataset provided by J. Cheng et al. [138], which contains meningioma, glioma, and pituitary tumor in T1-WC MRI-images is one of the most commonly used datasets in the training and testing classifier models. Using this dataset, Gumaei, A. et al. [125] has achieved a classification accuracy of 94.23% using a regularized extreme learning machine, while the Kokkalla, S. et al. [153] have reported a classification accuracy of 99.69% using custom modified deep-dense inception residual network (DDIRNet). These results indicate that the deep learning-based model outweighs the shallow machine learning-based techniques for this particular dataset.

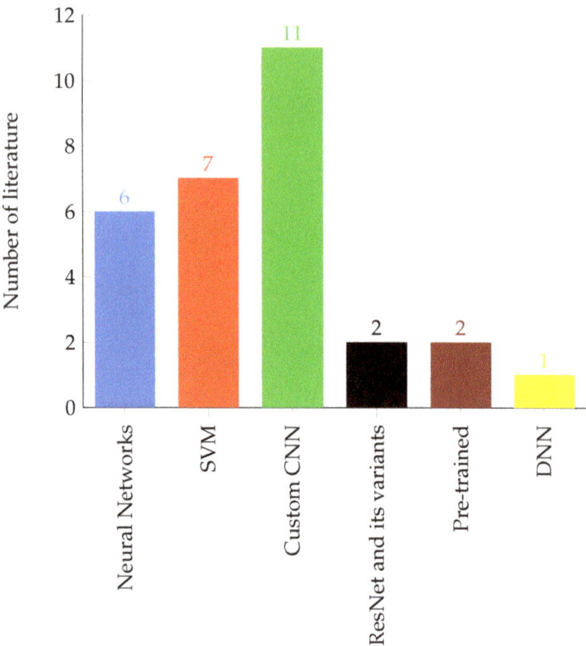


**Figure 2.** Number of brain tumor classification methods.

*Challenges in Automatic Brain Tumor Segmentation and Classification*

The development of autonomous brain tumor segmentation and classification models using MRI images is still a challenging task. The challenges are due to several constraints including the effect of different types of noises embedded in the brain MRI images [116–118], motion and metal artifacts during image acquisition [164], low-resolution MRI images [165], and lack of deep learning models interpretability and transparency [166,167].

One of the most common challenges in machine learning-based brain tumor segmentation and classification is the noisiness of an MRI image. Therefore, noise estimation and denoising MRI images is a crucial pre-processing task for improving the accuracy of brain tumor segmentation and classification models. Therefore, several techniques have been proposed for denoising MRI images, such as modified iterative grouping median filter [118], Wiener filter and wavelet transform [168], non-local means [169], and deep learning-based approaches [170,171]. However, a robust denoising technique for MRI images is still challenging and the pursuit to obtain an efficient denoising technique has been an active research area [170]. Similarly, motion, metal, and other artifacts are also a source of challenge to the robustness of machine learning-based brain tumor segmentation and classification. Recently, deep learning-based solutions for minimizing the effects of these artifacts have been proposed [164,172]. MRI provides a high fidelity brain scan image compared to other imaging techniques. However, post-acquisition image processing techniques, including deep learning-based methods have been used to increase the resolution of MR images so that the efficiency of autonomous brain tumor segmentation and classification models improved[165,173]. The other major challenge is the lack of deep models' interpretability, and often they are perceived as black-box. As a result, attaining any evidence regarding the process they perform is difficult. However, the transparency and interpretability of deep learning techniques are crucial for the complete integration into medical diagnosis [166].

## 8. Conclusions

Automating the brain tumor segmentation and classification task has tremendous benefits in improving the diagnosis, treatment planning, and follow-up of patients. Through applying various techniques, including conventional image processing, shallow machine learning, and deep learning techniques, undeniable progress have been achieved in automating brain tumor segmentation and classification tasks. However, building a fully autonomous system that can be used on clinical floors is still a challenging task.

Compared to region-growing and shallow machine learning algorithms, automating the brain tumor segmentation and classification using deep learning techniques have huge benefits. This is mainly due to the powerful feature learning ability of deep learning techniques. In addition, as can be shown in Figures 1 and 2, deep learning-based brain tumor segmentation and classification techniques are becoming the most active research area. In this paper, a comprehensive survey on region growing, shallow machine learning, and deep learning-based brain tumor segmentation and classification methods are presented. These methods are structurally categorized and summarized to give an insight to the reader of the dataset used, pre-processing, feature extraction, segmentation, classification, post-processing, and the reported model performances in the literature. Furthermore, the pros and cons of the methods and the model evaluation metrics have been discussed.

**Author Contributions:** Conceptualization, E.S.B.; Methodology, E.S.B.; Validation, Y.M.A., F.S., T.G.D.; Writing—original draft preparation, E.S.B.; Writing—review and editing, E.S.B., Y.M.A., F.S., T.G.D. All authors have read and agreed to the published version of the manuscript

**Funding:** This research received no external funding.

**Institutional Review Board Statement:** Not applicable.

**Informed Consent Statement:** Not applicable.

**Data Availability Statement:** Not applicable.

**Conflicts of Interest:** The authors declare no conflict of interest.

## References

1. Afework, Y.K.; Debelee, T.G. Detection of Bacterial Wilt on Enset Crop Using Deep Learning Approach. *Int. J. Eng. Res. Afr.* **2020**, *51*, 131–146. [CrossRef]
2. Debelee, T.G.; Schwenker, F.; Ibenthal, A.; Yohannes, D. Survey of deep learning in breast cancer image analysis. *Evol. Syst.* **2019**, *11*, 143–163. [CrossRef]
3. Debelee, T.G.; Kebede, S.R.; Schwenker, F.; Shewarega, Z.M. Deep Learning in Selected Cancers' Image Analysis—A Survey. *J. Imaging* **2020**, *6*, 121. [CrossRef] [PubMed]
4. Debelee, T.G.; Amirian, M.; Ibenthal, A.; Palm, G.; Schwenker, F. Classification of Mammograms Using Convolutional Neural Network Based Feature Extraction. In *Lecture Notes of the Institute for Computer Sciences, Social Informatics and Telecommunications Engineering*; Springer International Publishing: Berlin/Heidelberg, Germany,2018; pp. 89–98. [CrossRef]
5. Debelee, T.G.; Gebreselasie, A.; Schwenker, F.; Amirian, M.; Yohannes, D. Classification of Mammograms Using Texture and CNN Based Extracted Features. *J. Biomimetics Biomater. Biomed. Eng.* **2019**, *42*, 79–97. [CrossRef]
6. Debelee, T.G.; Schwenker, F.; Rahimeto, S.; Yohannes, D. Evaluation of modified adaptive k-means segmentation algorithm. *Comput. Vis. Media* **2019**, *5*, 347–361. [CrossRef]
7. Kebede, S.R.; Debelee, T.G.; Schwenker, F.; Yohannes, D. Classifier Based Breast Cancer Segmentation. *J. Biomimetics Biomater. Biomed. Eng.* **2020**, *47*, 41–61. [CrossRef]
8. Megersa, Y.; Alemu, G. Brain tumor detection and segmentation using hybrid intelligent algorithms. In Proceedings of the AFRICON 2015, Addis Ababa, Ethiopia, 14–17 September 2015. [CrossRef]
9. Roberts, T.A.; Hyare, H.; Agliardi, G.; Hipwell, B.; d'Esposito, A.; Ianus, A.; Breen-Norris, J.O.; Ramasawmy, R.; Taylor, V.; Atkinson, D.; et al. Noninvasive diffusion magnetic resonance imaging of brain tumour cell size for the early detection of therapeutic response. *Sci. Rep.* **2020**, *10*. [CrossRef]
10. Villanueva-Meyer, J.E.; Mabray, M.C.; Cha, S. Current Clinical Brain Tumor Imaging. *Neurosurgery* **2017**, *81*, 397–415. [CrossRef]
11. Rosenbloom, M.J.; Pfefferbaum, A. Magnetic resonance imaging of the living brain: evidence for brain degeneration among alcoholics and recovery with abstinence. *Alcohol Res. Health J. Natl. Inst. Alcohol Abus. Alcohol.* **2008**, *31*, 362–37.
12. Charles R. Noback, Norman L. Strominger, R.J.; A. Ruggiero, D. *The Human Nervous System: Structure and Function*; Humana Press: Totowa, NJ, USA,2005.

13. Louis D.N., Ohgaki H., W.O. *WHO Classification of Tumors of the Central Nervous System*; International Agency for Research on Cancer (IARC): Lyon, France, 2007.
14. Kayode, A.A.; Shahzadi, A.; Akram, M.; Anwar, H.; Kayode, O.T.; Akinnawo, O.O.; Okoh, S.O. Brain Tumor: An overview of the basic clinical manifestations and treatment. *Glob. J. Cancer Ther.* **2020**, *2020*, 38–41. [CrossRef]
15. Johnson, D.R.; Guerin, J.B.; Giannini, C.; Morris, J.M.; Eckel, L.J.; Kaufmann, T.J. 2016 Updates to the WHO Brain Tumor Classification System: What the Radiologist Needs to Know. *RadioGraphics* **2017**, *37*, 2164–2180. [CrossRef]
16. Roth, P.; Pace, A.; Rhun, E.L.; Weller, M.; Ay, C.; Moyal, E.C.J.; Coomans, M.; Giusti, R.; Jordan, K.; Nishikawa, R.; et al. Neurological and vascular complications of primary and secondary brain tumours: EANO-ESMO Clinical Practice Guidelines for prophylaxis, diagnosis, treatment and follow-up. *Ann. Oncol.* **2021**, *32*, 171–182. [CrossRef] [PubMed]
17. Jan C. Buckner, e.a. Central Nervous System Tumors. In *Mayo Clinic Proceedings*; Elsevier: Amsterdam, The Netherlands, 2007; Volume 82, pp. 1271–1286.
18. Smithuis, R. Neuroradiology: Brain Index. Available online: https://radiologyassistant.nl/neuroradiology/brain (accessed on 3 March 2021).
19. Alves, A.F.F.; de Arruda Miranda, J.R.; Reis, F.; de Souza, S.A.S.; Alves, L.L.R.; de Moura Feitoza, L.; de Souza de Castro, J.T.; de Pina, D.R. Inflammatory lesions and brain tumors: is it possible to differentiate them based on texture features in magnetic resonance imaging? *J. Venom. Anim. Toxins Incl. Trop. Dis.* **2020**, *26*. [CrossRef] [PubMed]
20. Kasban, H. A Comparative Study of Medical Imaging Techniques. *Int. J. Inf. Sci. Intell. Syst.* **2015**, *4*, 37–58.
21. Ammari, S.; Pitre-Champagnat, S.; Dercle, L.; Chouzenoux, E.; Moalla, S.; Reuze, S.; Talbot, H.; Mokoyoko, T.; Hadchiti, J.; Diffetocq, S.; et al. Influence of Magnetic Field Strength on Magnetic Resonance Imaging Radiomics Features in Brain Imaging, an In Vitro and In Vivo Study. *Front. Oncol.* **2021**, *10*. [CrossRef] [PubMed]
22. Rajasekaran, K.A.; Gounder, C.C. Advanced Brain Tumour Segmentation from MRI Images. In *High-Resolution Neuroimaging—Basic Physical Principles and Clinical Applications*; InTech: Vienna, Austria, 2018. [CrossRef]
23. Foltz, W.D.; Jaffray, D.A. Principles of Magnetic Resonance Imaging. *Radiat. Res.* **2012**, *177*, 331–348. [CrossRef] [PubMed]
24. Hornark, J.P. The Basics of MRI. Available online: http://www.cis.rit.edu/htbooks/mri (accessed on 20 March 2021).
25. Mustafa, W.F.; Abbas, M.; Elsorougy, L. Role of diffusion-weighted imaging in differentiation between posterior fossa brain tumors. *Egypt. J. Neurol. Psychiatry Neurosurg.* **2020**, *56*. [CrossRef]
26. Salama, G.R.; Heier, L.A.; Patel, P.; Ramakrishna, R.; Magge, R.; Tsiouris, A.J. Diffusion Weighted/Tensor Imaging, Functional MRI and Perfusion Weighted Imaging in Glioblastoma—Foundations and Future. *Front. Neurol.* **2018**, *8*. [CrossRef]
27. Fink, J.R.; Muzi, M.; Peck, M.; Krohn, K.A. Multimodality Brain Tumor Imaging: MR Imaging, PET, and PET/MR Imaging. *J. Nucl. Med.* **2015**, *56*, 1554–1561. [CrossRef]
28. Luo, Q.; Li, Y.; Luo, L.; Diao, W. Comparisons of the accuracy of radiation diagnostic modalities in brain tumor. *Medicine* **2018**, *97*, e11256. [CrossRef]
29. Sharma, P.; Shukla, A.P. A Review on Brain Tumor Segmentation and Classification for MRI Images. In Proceedings of the 2021 International Conference on Advance Computing and Innovative Technologies in Engineering (ICACITE), Greater Noida, India, 30–31 December 2021. [CrossRef]
30. Rao, C.S.; Karunakara, K. A comprehensive review on brain tumor segmentation and classification of MRI images. *Multimed. Tools Appl.* **2021**, *80*, 17611–17643. [CrossRef]
31. Magadza, T.; Viriri, S. Deep Learning for Brain Tumor Segmentation: A Survey of State-of-the-Art. *J. Imaging* **2021**, *7*, 19. [CrossRef]
32. Tiwari, A.; Srivastava, S.; Pant, M. Brain tumor segmentation and classification from magnetic resonance images: Review of selected methods from 2014 to 2019. *Pattern Recognit. Lett.* **2020**, *131*, 244–260. [CrossRef]
33. Kumari, N.; Saxena, S. Review of Brain Tumor Segmentation and Classification. In Proceedings of the 2018 International Conference on Current Trends towards Converging Technologies (ICCTCT), Coimbatore, India, 1–3 March 2018. [CrossRef]
34. Meier, R.; Knecht, U.; Loosli, T.; Bauer, S.; Slotboom, J.; Wiest, R.; Reyes, M. Clinical Evaluation of a Fully-automatic Segmentation Method for Longitudinal Brain Tumor Volumetry. *Sci. Rep.* **2016**, *6*. [CrossRef]
35. Pohle, R.; Toennies, K.D. Segmentation of medical images using adaptive region growing. In *Medical Imaging 2001: Image Processing*; Sonka, M.; Hanson, K.M., Eds.; SPIE: Bellingham, WA, USA,2001. [CrossRef]
36. Dey, N.; Ashour, A.S. Computing in Medical Image Analysis. In *Soft Computing Based Medical Image Analysis*; Elsevier: Amsterdam, The Netherlands,2018; pp. 3–11. [CrossRef]
37. Dhanachandra, N.; Manglem, K.; Chanu, Y.J. Image Segmentation Using K -means Clustering Algorithm and Subtractive Clustering Algorithm. *Procedia Comput. Sci.* **2015**, *54*, 764–771. [CrossRef]
38. Hooda, H.; Verma, O.P.; Singhal, T. Brain tumor segmentation: A performance analysis using K-Means, Fuzzy C-Means and Region growing algorithm. In Proceedings of the 2014 IEEE International Conference on Advanced Communications, Control and Computing Technologies, Ramanathapuram, India, 8–10 May 2014. [CrossRef]
39. Bal, A.; Banerjee, M.; Sharma, P.; Maitra, M. Brain Tumor Segmentation on MR Image Using K-Means and Fuzzy-Possibilistic Clustering. In Proceedings of the 2018 2nd International Conference on Electronics, Materials Engineering & Nano-Technology (IEMENTech), Kolkata, India, 4–5 April 2018. [CrossRef]

40. Kumar, D.V.; Krishniah, V.J.R. Segmentation of Brain Tumor Using K-Means Clustering Algorithm. *J. Eng. Appl. Sci.* **2018**, *13*, 3942–3945.
41. Selvakumar, J.; Lakshmi, A.; Arivoli, T. Brain tumor segmentation and its area calculation in brain MR images using K-mean clustering and Fuzzy C-mean algorithm. In Proceedings of the IEEE-International Conference on Advances in Engineering, Science And Management (ICAESM-2012), Nagapattinam, India, 30–31 March 2012; pp. 186–190.
42. Shanker, R.; Singh, R.; Bhattacharya, M. Segmentation of tumor and edema based on K-mean clustering and hierarchical centroid shape descriptor. In Proceedings of the 2017 IEEE International Conference on Bioinformatics and Biomedicine (BIBM), Kansas City, MO, USA, 13–16 November 2017, pp. 1105–1109. [CrossRef]
43. Kaur, N.; Sharma, M. Brain tumor detection using self-adaptive K-means clustering. In Proceedings of the 2017 International Conference on Energy, Communication, Data Analytics and Soft Computing (ICECDS),Chennai, India, 1–2 August 2017, pp. 1861–1865. [CrossRef]
44. Mahmud, M.R.; Mamun, M.A.; Hossain, M.A.; Uddin, M.P. Comparative Analysis of K-Means and Bisecting K-Means Algorithms for Brain Tumor Detection. In Proceedings of the 2018 International Conference on Computer, Communication, Chemical, Material and Electronic Engineering (IC4ME2), Rajshahi, Bangladesh, 8–9 February 2018, pp. 1–4. [CrossRef]
45. Shasidhar, M.; Raja, V.S.; Kumar, B.V. MRI Brain Image Segmentation Using Modified Fuzzy C-Means Clustering Algorithm. In Proceedings of the 2011 International Conference on Communication Systems and Network Technologies, Katra, India, 3–5 June 2011, pp. 473–478. [CrossRef]
46. Agrawal, R.; Sharma, M.; Singh, B.K. Segmentation of Brain Tumour Based on Clustering Technique: Performance Analysis. *J. Intell. Syst.* **2019**, *28*, 291–306. [CrossRef]
47. Pitchai, R.; Supraja, P.; Victoria, A.H.; Madhavi, M. Brain Tumor Segmentation Using Deep Learning and Fuzzy K-Means Clustering for Magnetic Resonance Images. *Neural Process. Lett.* **2020**. [CrossRef]
48. Almahfud, M.A.; Setyawan, R.; Sari, C.A.; Setiadi, D.R.I.M.; Rachmawanto, E.H. An Effective MRI Brain Image Segmentation using Joint Clustering (K-Means and Fuzzy C-Means). In Proceedings of the 2018 International Seminar on Research of Information Technology and Intelligent Systems (ISRITI), Yogyakarta, Indonesia, 21-22 Nov. 2018, pp. 11-16, [CrossRef]
49. Abdel-Maksoud, E.; Elmogy, M.; Al-Awadi, R. Brain tumor segmentation based on a hybrid clustering technique. *Egypt. Informatics J.* **2015**, *16*, 71–81. [CrossRef]
50. Mannor, S.; Jin, X.; Han, J.; Jin, X.; Han, J.; Jin, X.; Han, J.; Zhang, X. K-Medoids Clustering. In *Encyclopedia of Machine Learning*; Springer: New York, NY, USA, 2011; pp. 564–565. [CrossRef]
51. Bezdek, J.C.; Hall, L.O.; Clarke, L.P. Review of MR image segmentation techniques using pattern recognition. *Med. Phys.* **1993**, *20*, 1033–1048. [CrossRef] [PubMed]
52. Blessy, S.A.P.S.; Sulochana, C.H. Performance analysis of unsupervised optimal fuzzy clustering algorithm for MRI brain tumor segmentation. *Technol. Health Care* **2014**, *23*, 23–35. [CrossRef]
53. Arakeri, M.P.; Reddy, G.R.M. Efficient Fuzzy Clustering Based Approach to Brain Tumor Segmentation on MR Images. In *Communications in Computer and Information Science*; Springer: Berlin/Heidelberg, Germany,2011; pp. 790–795. [CrossRef]
54. Dubey, Y.K.; Mushrif, M.M. FCM Clustering Algorithms for Segmentation of Brain MR Images. *Adv. Fuzzy Syst.* **2016**, *2016*, 1–14. [CrossRef]
55. Badmera, M.S.; Nilawar, A.P.; Karwankar, A.R. Modified FCM approach for MR brain image segmentation. In Proceedings of the 2013 International Conference on Circuits, Power and Computing Technologies (ICCPCT), Nagercoil, India, 20–21 March 2013, pp. 891–896. [CrossRef]
56. Sheela, C.J.J.; Suganthi, G. Automatic Brain Tumor Segmentation from MRI using Greedy Snake Model and Fuzzy C-Means Optimization. *J. King Saud Univ. Comput. Inf. Sci.* **2019**. [CrossRef]
57. Wang, Y. *Tutorial: Image Segmentation*; Graduate Institute of Communication Engineering National Taiwan University: Taipei, Taiwan, 2010.
58. Rajinikanth, V.; Fernandes, S.L.; Bhushan, B.; Harisha.; Sunder, N.R. Segmentation and Analysis of Brain Tumor Using Tsallis Entropy and Regularised Level Set. In Proceedings of 2nd International Conference on Micro-Electronics, Electromagnetics and Telecommunications; Springer Singapore, 7 September 2017; pp. 313–321. [CrossRef]
59. Cabria, I.; Gondra, I. Automated Localization of Brain Tumors in MRI Using Potential-K-Means Clustering Algorithm. In Proceedings of the 2015 12th Conference on Computer and Robot Vision, Halifax, NS, Canada, 3–5 June 2015, pp.125–132. [CrossRef]
60. Suraj, N.S.S.K.; Muppalla, V.; Sanghani, P.; Ren, H. Comparative Study of Unsupervised Segmentation Algorithms for Delineating Glioblastoma Multiforme Tumour. In Proceedings of the 2018 3rd International Conference on Advanced Robotics and Mechatronics (ICARM), Singapore, 18–20 July 2018, pp. 468–473. [CrossRef]
61. Mehidi, I.; Belkhiat, D.E.C.; Jabri, D. An Improved Clustering Method Based on K-Means Algorithm for MRI Brain Tumor Segmentation. In Proceedings of the 2019 6th International Conference on Image and Signal Processing and their Applications (ISPA), Mostaganem, Algeria, 24–25 November 2019, pp. 1–4. [CrossRef]
62. Rundo, L.; Militello, C.; Tangherloni, A.; Russo, G.; Vitabile, S.; Gilardi, M.C.; Mauri, G. NeXt for neuro-radiosurgery: A fully automatic approach for necrosis extraction in brain tumor MRI using an unsupervised machine learning technique. *Int. J. Imaging Syst. Technol.* **2017**, *28*, 21–37. [CrossRef]
63. Chandra, G.R.; Rao, K.R.H. Tumor Detection In Brain Using Genetic Algorithm. *Procedia Comput. Sci.* **2016**, *79*, 449–457. [CrossRef]

64. Rundo, L.; Militello, C.; Russo, G.; Vitabile, S.; Gilardi, M.C.; Mauri, G. GTVcut for neuro-radiosurgery treatment planning: an MRI brain cancer seeded image segmentation method based on a cellular automata model. *Nat. Comput.* **2017**, *17*, 521–536. [CrossRef]
65. Ayachi, R.; Ben Amor, N. Brain Tumor Segmentation Using Support Vector Machines. In *Symbolic and Quantitative Approaches to Reasoning with Uncertainty*; Sossai, C.; Chemello, G., Eds.; Springer: Berlin/Heidelberg, Germany, 2009; pp. 736–747.
66. Cui, B.; Xie, M.; Wang, C. A Deep Convolutional Neural Network Learning Transfer to SVM-Based Segmentation Method for Brain Tumor. In Proceedings of the 2019 IEEE 11th International Conference on Advanced Infocomm Technology (ICAIT), Jinan, China,18–20 October 2019, pp. 1–5. [CrossRef]
67. Zhang, N.; Ruan, S.; Lebonvallet, S.; Liao, Q.; Zhu, Y. Multi-kernel SVM based classification for brain tumor segmentation of MRI multi-sequence. In Proceedings of the 2009 16th IEEE International Conference on Image Processing (ICIP), Cairo, Egypt, 7–10 November 2009, pp. 3373–3376. [CrossRef]
68. Chen, W.; Qiao, X.; Liu, B.; Qi, X.; Wang, R.; Wang, X. Automatic brain tumor segmentation based on features of separated local square. In Proceedings of the 2017 Chinese Automation Congress (CAC), 20–22 October 2017, Jinan, China. [CrossRef]
69. Chithambaram, T.; Perumal, K. Brain tumor segmentation using genetic algorithm and ANN techniques. In Proceedings of the 2017 IEEE International Conference on Power, Control, Signals and Instrumentation Engineering (ICPCSI), 21–22 September 2017, Chennai, India. [CrossRef]
70. Bougacha, A.; Boughariou, J.; Slima, M.B.; Hamida, A.B.; Mahfoudh, K.B.; Kammoun, O.; Mhiri, C. Comparative study of supervised and unsupervised classification methods: Application to automatic MRI glioma brain tumors segmentation. In Proceedings of the 2018 4th International Conference on Advanced Technologies for Signal and Image Processing (ATSIP), 21–24 March 2018, Tunisia. [CrossRef]
71. Ma, C.; Luo, G.; Wang, K. Concatenated and Connected Random Forests With Multiscale Patch Driven Active Contour Model for Automated Brain Tumor Segmentation of MR Images. *IEEE Trans. Med. Imaging* **2018**, *37*, 1943–1954. [CrossRef]
72. Tang, H.; Lu, H.; Liu, W.; Tao, X. Tumor segmentation from single contrast MR images of human brain. In Proceedings of the 2015 IEEE 12th International Symposium on Biomedical Imaging (ISBI), 16–19 April 2015, New York Marriott at Brooklyn Bridge, NY, USA. [CrossRef]
73. Csaholczi, S.; Kovacs, L.; Szilagyi, L. Automatic Segmentation of Brain Tumor Parts from MRI Data Using a Random Forest Classifier. In Proceedings of the 2021 IEEE 19th World Symposium on Applied Machine Intelligence and Informatics (SAMI), 21–23 January 2021, Herl'any, Slovakia. [CrossRef]
74. Pinto, A.; Pereira, S.; Dinis, H.; Silva, C.A.; Rasteiro, D.M.L.D. Random decision forests for automatic brain tumor segmentation on multi-modal MRI images. In Proceedings of the 2015 IEEE 4th Portuguese Meeting on Bioengineering (ENBENG), 26–28 February 2015, Porto, Portugal. [CrossRef]
75. Hatami, T.; Hamghalam, M.; Reyhani-Galangashi, O.; Mirzakuchaki, S. A Machine Learning Approach to Brain Tumors Segmentation Using Adaptive Random Forest Algorithm. In Proceedings of the 2019 5th Conference on Knowledge Based Engineering and Innovation (KBEI), 28 February–1 March 2019, Tehran, Iran. [CrossRef]
76. Fulop, T.; Gyorfi, A.; Csaholczi, S.; Kovacs, L.; Szilagyi, L. Brain Tumor Segmentation from Multi-Spectral MRI Data Using Cascaded Ensemble Learning. In Proceedings of the 2020 IEEE 15th International Conference of System of Systems Engineering (SoSE), 2–4 June 2020 Budapest, Hungary. [CrossRef]
77. Bakas, S.; Akbari, H.; Sotiras, A.; Bilello, M.; Rozycki, M.; Kirby, J.S.; Freymann, J.B.; Farahani, K.; Davatzikos, C. Advancing The Cancer Genome Atlas glioma MRI collections with expert segmentation labels and radiomic features. *Sci. Data* **2017**, *4*. [CrossRef]
78. Bakas, S.; Akbari, H.; Sotiras, A.; Bilello, M.; Rozycki, M.; Kirby, J.; Freymann, J.; Farahani, K.; Davatzikos, C. Segmentation Labels for the Pre-operative Scans of the TCGA-GBM collection. *Cancer Imaging Arch.* **2017**. [CrossRef]
79. Tobon-Gomez, C.; Geers, A.J.; Peters, J.; Weese, J.; Pinto, K.; Karim, R.; Ammar, M.; Daoudi, A.; Margeta, J.; Sandoval, Z.; et al. Benchmark for Algorithms Segmenting the Left Atrium From 3D CT and MRI Datasets. *IEEE Trans. Med Imaging* **2015**, *34*, 1460–1473. [CrossRef] [PubMed]
80. Menze, B.H.; Jakab, A.; Bauer, S.; Kalpathy-Cramer, J.; Farahani, K.; Kirby, J.; Burren, Y.; Porz, N.; Slotboom, J.; Wiest, R.; et al. The Multimodal Brain Tumor Image Segmentation Benchmark (BRATS). *IEEE Trans. Med Imaging* **2015**, *34*, 1993–2024. [CrossRef] [PubMed]
81. Pereira, S.; Pinto, A.; Alves, V.; Silva, C.A. Brain Tumor Segmentation Using Convolutional Neural Networks in MRI Images. *IEEE Trans. Med. Imaging* **2016**, *35*, 1240–1251. [CrossRef] [PubMed]
82. Deng, W.; Shi, Q.; Wang, M.; Zheng, B.; Ning, N. Deep Learning-Based HCNN and CRF-RRNN Model for Brain Tumor Segmentation. *IEEE Access* **2020**, *8*, 26665–26675. [CrossRef]
83. Ding, Y.; Li, C.; Yang, Q.; Qin, Z.; Qin, Z. How to Improve the Deep Residual Network to Segment Multi-Modal Brain Tumor Images. *IEEE Access* **2019**, *7*, 152821–152831. [CrossRef]
84. Ding, Y.; Chen, F.; Zhao, Y.; Wu, Z.; Zhang, C.; Wu, D. A Stacked Multi-Connection Simple Reducing Net for Brain Tumor Segmentation. *IEEE Access* **2019**, *7*, 104011–104024. [CrossRef]
85. Ali, M.; Gilani, S.O.; Waris, A.; Zafar, K.; Jamil, M. Brain Tumour Image Segmentation Using Deep Networks. *IEEE Access* **2020**, *8*, 153589–153598. [CrossRef]

86. Razzak, M.I.; Imran, M.; Xu, G. Efficient Brain Tumor Segmentation With Multiscale Two-Pathway-Group Conventional Neural Networks. *IEEE J. Biomed. Health Inform.* **2019**, *23*, 1911–1919. [CrossRef]
87. Aboelenein, N.M.; Songhao, P.; Koubaa, A.; Noor, A.; Afifi, A. HTTU-Net: Hybrid Two Track U-Net for Automatic Brain Tumor Segmentation. *IEEE Access* **2020**, *8*, 101406–101415. [CrossRef]
88. Wang, G.; Li, W.; Zuluaga, M.A.; Pratt, R.; Patel, P.A.; Aertsen, M.; Doel, T.; David, A.L.; Deprest, J.; Ourselin, S.; Vercauteren, T. Interactive Medical Image Segmentation Using Deep Learning With Image-Specific Fine Tuning. *IEEE Trans. Med. Imaging* **2018**, *37*, 1562–1573. [CrossRef]
89. Hao, J.; Li, X.; Hou, Y. Magnetic Resonance Image Segmentation Based on Multi-Scale Convolutional Neural Network. *IEEE Access* **2020**, *8*, 65758–65768. [CrossRef]
90. Zhou, T.; Canu, S.; Ruan, S. Fusion based on attention mechanism and context constraint for multi-modal brain tumor segmentation. *Comput. Med Imaging Graph.* **2020**, *86*, 101811. [CrossRef] [PubMed]
91. Ye, F.; Zheng, Y.; Ye, H.; Han, X.; Li, Y.; Wang, J.; Pu, J. Parallel pathway dense neural network with weighted fusion structure for brain tumor segmentation. *Neurocomputing* **2021**, *425*, 1–11. [CrossRef]
92. Sun, J.; Peng, Y.; Guo, Y.; Li, D. Segmentation of the multimodal brain tumor image used the multi-pathway architecture method based on 3D FCN. *Neurocomputing* **2021**, *423*, 34–45. [CrossRef]
93. Ben naceur, M.; Akil, M.; Saouli, R.; Kachouri, R. Fully automatic brain tumor segmentation with deep learning-based selective attention using overlapping patches and multi-class weighted cross-entropy. *Med. Image Anal.* **2020**, *63*, 101692. [CrossRef] [PubMed]
94. Naser, M.A.; Deen, M.J. Brain tumor segmentation and grading of lower-grade glioma using deep learning in MRI images. *Comput. Biol. Med.* **2020**, *121*, 103758. [CrossRef]
95. Zhou, Z.; He, Z.; Jia, Y. AFPNet: A 3D fully convolutional neural network with atrous-convolution feature pyramid for brain tumor segmentation via MRI images. *Neurocomputing* **2020**, *402*, 235–244. [CrossRef]
96. Li, H.; Li, A.; Wang, M. A novel end-to-end brain tumor segmentation method using improved fully convolutional networks. *Comput. Biol. Med.* **2019**, *108*, 150–160. [CrossRef]
97. Zhang, J.; Zeng, J.; Qin, P.; Zhao, L. Brain tumor segmentation of multi-modality MR images via triple intersecting U-Nets. *Neurocomputing* **2021**, *421*, 195–209. [CrossRef]
98. Xu, F.; Ma, H.; Sun, J.; Wu, R.; Liu, X.; Kong, Y. LSTM Multi-modal UNet for Brain Tumor Segmentation. In Proceedings of the 2019 IEEE 4th International Conference on Image, Vision and Computing (ICIVC), Xiamen, China, 5–7 July 2019, pp. 236–240. [CrossRef]
99. Kleihues, P.; Louis, D.N.; Scheithauer, B.W.; Rorke, L.B.; Reifenberger, G.; Burger, P.C.; Cavenee, W.K. The WHO Classification of Tumors of the Nervous System. *J. Neuropathol. Exp. Neurol.* **2002**, *61*, 215–225. [CrossRef]
100. Badža, M.M.; Barjaktarović, M.Č. Classification of Brain Tumors from MRI Images Using a Convolutional Neural Network. *Appl. Sci.* **2020**, *10*, 1999. [CrossRef]
101. Tandel, G.S.; Biswas, M.; Kakde, O.G.; Tiwari, A.; Suri, H.S.; Turk, M.; Laird, J.; Asare, C.; Ankrah, A.A.; Khanna, N.N.; et al. A Review on a Deep Learning Perspective in Brain Cancer Classification. *Cancers* **2019**, *11*, 111. [CrossRef] [PubMed]
102. Quon, J.; Bala, W.; Chen, L.; Wright, J.; Kim, L.; Han, M.; Shpanskaya, K.; Lee, E.; Tong, E.; Iv, M.; et al. Deep Learning for Pediatric Posterior Fossa Tumor Detection and Classification: A Multi-Institutional Study. *Am. J. Neuroradiol.* **2020**. [CrossRef]
103. Díaz-Pernas, F.J.; Martínez-Zarzuela, M.; Antón-Rodríguez, M.; González-Ortega, D. A Deep Learning Approach for Brain Tumor Classification and Segmentation Using a Multiscale Convolutional Neural Network.*Healthcare* **2021**, *9*, 153. [CrossRef] [PubMed]
104. Deepak, S.; Ameer, P. Brain tumor classification using deep CNN features via transfer learning. *Comput. Biol. Med.* **2019**, *111*, 103345. [CrossRef]
105. Paul, J.S.; Plassard, A.J.; Landman, B.A.; Fabbri, D. Deep learning for brain tumor classification. In *Medical Imaging 2017: Biomedical Applications in Molecular, Structural, and Functional Imaging*; Krol, A.; Gimi, B., Eds.; SPIE: Bellingham, WA, USA, 2017. [CrossRef]
106. Khan, H.A.; Jue, W.; Mushtaq, M.; Mushtaq, M.U. Brain tumor classification in MRI image using convolutional neural network. *Math. Biosci. Eng.* **2020**, *17*, 6203–6216. [CrossRef] [PubMed]
107. Dangeti, P. *Statistics for Machine Learning*; Packt Publishing: Birmingham, UK,2017.
108. Ahmmed, R.; Swakshar, A.S.; Hossain, M.F.; Rafiq, M.A. Classification of tumors and it stages in brain MRI using support vector machine and artificial neural network. In Proceedings of the 2017 International Conference on Electrical, Computer and Communication Engineering (ECCE), Cox's Bazar, Bangladesh,16–18 Feb. 2017, pp. 229–234. [CrossRef]
109. Ismael, M.R.; Abdel-Qader, I. Brain Tumor Classification via Statistical Features and Back-Propagation Neural Network. In Proceedings of the 2018 IEEE International Conference on Electro/Information Technology (EIT), Rochester, MI, USA, 3–5 May 2018, pp. 0252–0257. [CrossRef]
110. Sathi, K.A.; Islam, M.S. Hybrid Feature Extraction Based Brain Tumor Classification using an Artificial Neural Network. In Proceedings of the 2020 IEEE 5th International Conference on Computing Communication and Automation (ICCCA), Greater Noida, India,30–31 October 2020, pp. 155–160. [CrossRef]
111. Shree, N.V.; Kumar, T.N.R. Identification and classification of brain tumor MRI images with feature extraction using DWT and probabilistic neural network. *Brain Inform.* **2018**, *5*, 23–30. [CrossRef]

112. Ramdlon, R.H.; Kusumaningtyas, E.M.; Karlita, T. Brain Tumor Classification Using MRI Images with K-Nearest Neighbor Method. In Proceedings of the 2019 International Electronics Symposium (IES), Surabaya, Indonesia, 27-28 Sept. 2019, pp. 660-667, [CrossRef]
113. Garg, G.; Garg, R. Brain Tumor Detection and Classification based on Hybrid Ensemble Classifier. *arXiv* **2021**, arXiv:2101.00216.
114. N., E.; M., N.; Al-Atabany, W. Evaluating the Efficiency of different Feature Sets on Brain Tumor Classification in MR Images. *Int. J. Comput. Appl.* **2018**, *180*, 1–7. [CrossRef]
115. Gurbina, M.; Lascu, M.; Lascu, D. Tumor Detection and Classification of MRI Brain Image using Different Wavelet Transforms and Support Vector Machines. In Proceedings of the 2019 42nd International Conference on Telecommunications and Signal Processing (TSP), Budapest, Hungary,1–3 July 2019, pp. 505–508. [CrossRef]
116. Ali, H.M. MRI Medical Image Denoising by Fundamental Filters. In *High-Resolution Neuroimaging—Basic Physical Principles and Clinical Applications*; InTech: Vienna, Austria,2018. [CrossRef]
117. Liu, L.; Yang, H.; Fan, J.; Liu, R.W.; Duan, Y. Rician noise and intensity nonuniformity correction (NNC) model for MRI data. *Biomed. Signal Process. Control* **2019**, *49*, 506–519. [CrossRef]
118. Ramesh, S.; Sasikala, S.; Paramanandham, N. Segmentation and classification of brain tumors using modified median noise filter and deep learning approaches. *Multimed. Tools Appl.* **2021**, *80*, 11789–11813. [CrossRef]
119. Ravikumar Gurusamy, D.V.S. A Machine Learning Approach for MRI Brain Tumor Classification. *Comput. Mater. Contin.* **2017**, *53*, 91–108. [CrossRef]
120. Li, M.; Wang, H.; Shang, Z.; Yang, Z.; Zhang, Y.; Wan, H. Ependymoma and pilocytic astrocytoma: Differentiation using radiomics approach based on machine learning. *J. Clin. Neurosci.* **2020**, *78*, 175–180. [CrossRef] [PubMed]
121. Kaplan, K.; Kaya, Y.; Kuncan, M.; Ertunç, H.M. Brain tumor classification using modified local binary patterns (LBP) feature extraction methods. *Med. Hypotheses* **2020**, *139*, 109696. [CrossRef] [PubMed]
122. Kang, J.; Ullah, Z.; Gwak, J. MRI-Based Brain Tumor Classification Using Ensemble of Deep Features and Machine Learning Classifiers. *Sensors* **2021**, *21*, 2222. [CrossRef] [PubMed]
123. Amin, J.; Sharif, M.; Raza, M.; Saba, T.; Rehman, A. Brain Tumor Classification: Feature Fusion. In Proceedings of the 2019 International Conference on Computer and Information Sciences (ICCIS), Sakaka, Saudi Arabia, 3–4 April 2019, pp. 1–6.[CrossRef]
124. Baranwal, S.K.; Jaiswal, K.; Vaibhav, K.; Kumar, A.; Srikantaswamy, R. Performance analysis of Brain Tumour Image Classification using CNN and SVM. In Proceedings of the 2020 Second International Conference on Inventive Research in Computing Applications (ICIRCA), Coimbatore, India, 15– 17 July 2020, pp. 537-542, [CrossRef]
125. Gumaei, A.; Hassan, M.M.; Hassan, M.R.; Alelaiwi, A.; Fortino, G. A Hybrid Feature Extraction Method With Regularized Extreme Learning Machine for Brain Tumor Classification. *IEEE Access* **2019**, *7*, 36266–36273. [CrossRef]
126. Minz, A.; Mahobiya, C. MR Image Classification Using Adaboost for Brain Tumor Type. In Proceedings of the 2017 IEEE 7th International Advance Computing Conference (IACC), Hyderabad, India, 5–7 January 2017, pp. 701–705. [CrossRef]
127. Gayathri, S.; Wise, D.J.W.; Janani, V.; Eleaswari, M.; Hema, S. Analyzing, Detecting and Automatic Classification of Different Stages of Brain Tumor Using Region Segmentation and Support Vector Machine. In Proceedings of the 2020 International Conference on Electronics and Sustainable Communication Systems (ICESC),Coimbatore, India, 2–4 July 2020, pp. 404–408. [CrossRef]
128. Sarkar, A.; Maniruzzaman, M.; Ahsan, M.S.; Ahmad, M.; Kadir, M.I.; Islam, S.M.T. Identification and Classification of Brain Tumor from MRI with Feature Extraction by Support Vector Machine. In Proceedings of the 2020 International Conference for Emerging Technology (INCET), Belgaum, India, 5–7 June 2020, pp. 1–4. [CrossRef]
129. Mathew, A.R.; Anto, P.B. Tumor detection and classification of MRI brain image using wavelet transform and SVM. In Proceedings of the 2017 International Conference on Signal Processing and Communication (ICSPC), Coimbatore, India, 28–29 July 2017, pp. 75–78. [CrossRef]
130. Cinarer, G.; Emiroglu, B.G. Classificatin of Brain Tumors by Machine Learning Algorithms. In Proceedings of the 2019 3rd International Symposium on Multidisciplinary Studies and Innovative Technologies (ISMSIT), Ankara, Turkey,11–13 October 2019, pp. 1–4 [CrossRef]
131. Lavanyadevi, R.; Machakowsalya, M.; Nivethitha, J.; Kumar, A.N. Brain tumor classification and segmentation in MRI images using PNN. In Proceedings of the 2017 IEEE International Conference on Electrical, Instrumentation and Communication Engineering (ICEICE), Karur, India, 27–28 April 2017, pp. 1–6. [CrossRef]
132. Amin, J.; Sharif, M.; Yasmin, M.; Fernandes, S.L. A distinctive approach in brain tumor detection and classification using MRI. *Pattern Recognit. Lett.* **2020**, *139*, 118–127. [CrossRef]
133. Prabha, S.; Raghav, R.; Moulya, C.; Preethi, K.G.; Sankaran, K. Fusion based Brain Tumor Classification using Multiscale Transform Methods. In Proceedings of the 2020 International Conference on Communication and Signal Processing (ICCSP), Chennai, India, 28–30 July 2020, pp. 1390–1393. [CrossRef]
134. Wasule, V.; Sonar, P. Classification of brain MRI using SVM and KNN classifier. In Proceedings of the 2017 Third International Conference on Sensing, Signal Processing and Security (ICSSS), Chennai, India, 4–5 May 2017, pp. 218–223. [CrossRef]
135. Sachdeva, J.; Kumar, V.; Gupta, I.; Khandelwal, N.; Ahuja, C.K. A package-SFERCB-"Segmentation, feature extraction, reduction and classification analysis by both SVM and ANN for brain tumors". *Appl. Soft Comput.* **2016**, *47*, 151–167. [CrossRef]

136. Keerthana, K.; Xavier, S. An Intelligent System for Early Assessment and Classification of Brain Tumor. In Proceedings of the 2018 Second International Conference on Inventive Communication and Computational Technologies (ICICCT), Coimbatore, India, 20–21 April 2018; pp. 1265–1268. [CrossRef]
137. Yin, B.; Wang, C.; Abza, F. New brain tumor classification method based on an improved version of whale optimization algorithm. *Biomed. Signal Process. Control* **2020**, *56*, 101728. [CrossRef]
138. Cheng, J. Brain Tumor Dataset. 2017. Available online: https://figshare.com/articles/dataset/brain_tumor_dataset/1512427 (accessed on 2 June 2021).
139. B.Gaikwad, S.; Joshi, M.S. Brain Tumor Classification using Principal Component Analysis and Probabilistic Neural Network. *Int. J. Comput. Appl.* **2015**, *120*, 5–9. [CrossRef]
140. Kumar, A.; Ashok, A.; Ansari, M.A. Brain Tumor Classification Using Hybrid Model Of PSO And SVM Classifier. In Proceedings of the 2018 International Conference on Advances in Computing, Communication Control and Networking (ICACCCN), Greater Noida, India,12–13 October 2018; pp. 1022–1026. [CrossRef]
141. Ge, C.; Gu, I.Y.H.; Jakola, A.S.; Yang, J. Enlarged Training Dataset by Pairwise GANs for Molecular-Based Brain Tumor Classification. *IEEE Access* **2020**, *8*, 22560–22570. [CrossRef]
142. Bakas, S.; Akbari, H.; Sotiras, A.; Bilello, M.; Rozycki, M.; Kirby, J.; Freymann, J.; Farahani, K.; Davatzikos, C. Segmentation Labels for the Pre-operative Scans of the TCGA-LGG collection. *Cancer Imaging Arch.* **2017**. [CrossRef]
143. Sultan, H.H.; Salem, N.M.; Al-Atabany, W. Multi-Classification of Brain Tumor Images Using Deep Neural Network. *IEEE Access* **2019**, *7*, 69215–69225. [CrossRef]
144. Scarpace, L.; Flanders, A.E.; Jain, R.; Mikkelsen, T.; Andrews, D.W. Data from Rembrandt. 2019. Available online: https://wiki.cancerimagingarchive.net/display/Public/REMBRANDT (accessed on 3 May 2021).
145. Huang, Z.; Du, X.; Chen, L.; Li, Y.; Liu, M.; Chou, Y.; Jin, L. Convolutional Neural Network Based on Complex Networks for Brain Tumor Image Classification With a Modified Activation Function. *IEEE Access* **2020**, *8*, 89281–89290. [CrossRef]
146. Afshar, P.; Mohammadi, A.; Plataniotis, K.N. BayesCap: A Bayesian Approach to Brain Tumor Classification Using Capsule Networks. *IEEE Signal Process. Lett.* **2020**, *27*, 2024–2028. [CrossRef]
147. Ucuzal, H.; YASAR, S.; Colak, C. Classification of brain tumor types by deep learning with convolutional neural network on magnetic resonance images using a developed web-based interface. In Proceedings of the 2019 3rd International Symposium on Multidisciplinary Studies and Innovative Technologies (ISMSIT), Ankara, Turkey, 2019; pp. 1–5[CrossRef]
148. Noreen, N.; Palaniappan, S.; Qayyum, A.; Ahmad, I.; Imran, M.; Shoaib, M. A Deep Learning Model Based on Concatenation Approach for the Diagnosis of Brain Tumor. *IEEE Access* **2020**, *8*, 55135–55144. [CrossRef]
149. Rehman, A.; Naz, S.; Razzak, M.I.; Akram, F.; Imran, M. A Deep Learning-Based Framework for Automatic Brain Tumors Classification Using Transfer Learning. *Circuits Syst. Signal Process.* **2019**, *39*, 757–775. [CrossRef]
150. Cheng, Y.; Qin, G.; Zhao, R.; Liang, Y.; Sun, M. ConvCaps: Multi-input Capsule Network for Brain Tumor Classification. In *Neural Information Processing*; Springer International Publishing: Berlin/Heidelberg, Germany,2019; pp. 524–534. [CrossRef]
151. Kurup, R.V.; Sowmya, V.; Soman, K.P. Effect of Data Pre-processing on Brain Tumor Classification Using Capsulenet. In *ICICCT 2019 – System Reliability, Quality Control, Safety, Maintenance and Management*; Springer: Singapore, 2019; pp. 110–119. [CrossRef]
152. Liu, D.; Liu, Y.; Dong, L. G-ResNet: Improved ResNet for Brain Tumor Classification. In *Neural Information Processing*; Springer International Publishing: Berlin/Heidelberg, Germany, 2019; pp. 535–545. [CrossRef]
153. Kokkalla, S.; Kakarla, J.; Venkateswarlu, I.B.; Singh, M. Three-class brain tumor classification using deep dense inception residual network. *Soft Comput.* **2021**. [CrossRef]
154. Çinarer, G.; Emiroğlu, B.G.; Yurttakal, A.H. Prediction of Glioma Grades Using Deep Learning with Wavelet Radiomic Features. *Appl. Sci.* **2020**, *10*, 6296. [CrossRef]
155. Erickson, B.; Akkus, Z.; Sedlar, J.; Korfiatis, P. Data from LGG-1p19qDeletion. 2017. Available online: https://wiki.cancerimagingarchive.net/display/Public/LGG-1p19qDeletion (accessed on 3 May 2021). [CrossRef]
156. Abiwinanda, N.; Hanif, M.; Hesaputra, S.T.; Handayani, A.; Mengko, T.R. Brain Tumor Classification Using Convolutional Neural Network. In *IFMBE Proceedings*; Springer: Singapore, 2018; pp. 183–189. [CrossRef]
157. Sharif, M.I.; Khan, M.A.; Alhussein, M.; Aurangzeb, K.; Raza, M. A decision support system for multimodal brain tumor classification using deep learning. *Complex Intell. Syst.* **2021**. [CrossRef]
158. Irmak, E. Multi-Classification of Brain Tumor MRI Images Using Deep Convolutional Neural Network with Fully Optimized Framework. *Iran. J. Sci. Technol. Trans. Electr. Eng.* **2021**. [CrossRef]
159. Clark, K.; Vendt, B.; Smith, K.; Freymann, J.; Kirby, J.; Koppel, P.; Moore, S.; Phillips, S.; Maffitt, D.; Pringle, M.; Tarbox, L.; Prior, F. The Cancer Imaging Archive (TCIA): Maintaining and Operating a Public Information Repository. *J. Digit. Imaging* **2013**, *26*, 1045–1057. [CrossRef]
160. Pei, L.; Vidyaratne, L.; Rahman, M.M.; Iftekharuddin, K.M. Context aware deep learning for brain tumor segmentation, subtype classification, and survival prediction using radiology images. *Sci. Rep.* **2020**, *10*. [CrossRef]
161. Kumar, S.N.; Fred, A.L.; Varghese, P.S. An Overview of Segmentation Algorithms for the Analysis of Anomalies on Medical Images. *J. Intell. Syst.* **2018**, *29*, 612–625. [CrossRef]
162. Biratu, E.S.; Schwenker, F.; Debelee, T.G.; Kebede, S.R.; Negera, W.G.; Molla, H.T. Enhanced Region Growing for Brain Tumor MR Image Segmentation. *J. Imaging* **2021**, *7*, 22. [CrossRef]

163. Miotto, R.; Wang, F.; Wang, S.; Jiang, X.; Dudley, J.T. Deep learning for healthcare: review, opportunities and challenges. *Briefings Bioinform.* **2017**, *19*, 1236–1246. [CrossRef]
164. Zhu, G.; Jiang, B.; Tong, L.; Xie, Y.; Zaharchuk, G.; Wintermark, M. Applications of Deep Learning to Neuro-Imaging Techniques. *Front. Neurol.* **2019**, *10*. [CrossRef]
165. Sert, E.; Özyurt, F.; Doğantekin, A. A new approach for brain tumor diagnosis system: Single image super resolution based maximum fuzzy entropy segmentation and convolutional neural network.*Med. Hypotheses* **2019**, *133*, 109413. [CrossRef]
166. Natekar, P.; Kori, A.; Krishnamurthi, G. Demystifying Brain Tumor Segmentation Networks: Interpretability and Uncertainty Analysis. *Front. Comput. Neurosci.* **2020**, *14*. [CrossRef]
167. Saleem, H.; Shahid, A.R.; Raza, B. Visual interpretability in 3D brain tumor segmentation network. *Comput. Biol. Med.* **2021**, *133*, 104410. [CrossRef] [PubMed]
168. Zeng, Y.; Zhang, B.; Zhao, W.; Xiao, S.; Zhang, G.; Ren, H.; Zhao, W.; Peng, Y.; Xiao, Y.; Lu, Y.; Zong, Y.; Ding, Y. Magnetic Resonance Image Denoising Algorithm Based on Cartoon, Texture, and Residual Parts. *Comput. Math. Methods Med.* **2020**, *2020*, 1–10. [CrossRef]
169. Heo, Y.C.; Kim, K.; Lee, Y. Image Denoising Using Non-Local Means (NLM) Approach in Magnetic Resonance (MR) Imaging: A Systematic Review. *Appl. Sci.* **2020**, *10*, 7028. [CrossRef]
170. López, M.M.; Frederick, J.M.; Ventura, J. Evaluation of MRI Denoising Methods Using Unsupervised Learning. *Front. Artif. Intell.* **2021**, *4*. [CrossRef]
171. Kidoh, M.; Shinoda, K.; Kitajima, M.; Isogawa, K.; Nambu, M.; Uetani, H.; Morita, K.; Nakaura, T.; Tateishi, M.; Yamashita, Y.; Yamashita, Y. Deep Learning Based Noise Reduction for Brain MR Imaging: Tests on Phantoms and Healthy Volunteers. *Magn. Reson. Med Sci.* **2020**, *19*, 195–206. [CrossRef]
172. Higaki, T.; Nakamura, Y.; Tatsugami, F.; Nakaura, T.; Awai, K. Improvement of image quality at CT and MRI using deep learning. *Jpn. J. Radiol.* **2018**, *37*, 73–80. [CrossRef]
173. Kim, K.H.; Do, W.J.; Park, S.H. Improving resolution of MR images with an adversarial network incorporating images with different contrast. *Med. Phys.* **2018**, *45*, 3120–3131. [CrossRef]

Journal of *Imaging*

MDPI

*Article*

# A Computational Study on Temperature Variations in MRgFUS Treatments Using PRF Thermometry Techniques and Optical Probes

**Carmelo Militello [1,*], Leonardo Rundo [2,3], Fabrizio Vicari [4], Luca Agnello [5], Giovanni Borasi [4], Salvatore Vitabile [5] and Giorgio Russo [1]**

1. Institute of Molecular Bioimaging and Physiology, Italian National Research Council (IBFM-CNR), Cefalu, 90015 Palermo, Italy; giorgio.russo@ibfm.cnr.it
2. Department of Radiology, University of Cambridge, Cambridge CB2 0QQ, UK; lr495@cam.ac.uk
3. Cancer Research UK Cambridge Centre, Cambridge CB2 0RE, UK
4. LAboratorio di Tecnologie Oncologiche (LATO), Cefalu, 90015 Palermo, Italy; fabrizio.vicari.plus@gmail.com (F.V.); giovanni.borasi@gmail.com (G.B.)
5. Department of Biomedicine, Neuroscience and Advanced Diagnostics (BiND), University of Palermo, 90127 Palermo, Italy; luca.agnello@gmail.com (L.A.); salvatore.vitabile@unipa.it (S.V.)

* Correspondence: carmelo.militello@ibfm.cnr.it

**Abstract:** Structural and metabolic imaging are fundamental for diagnosis, treatment and follow-up in oncology. Beyond the well-established diagnostic imaging applications, ultrasounds are currently emerging in the clinical practice as a noninvasive technology for therapy. Indeed, the sound waves can be used to increase the temperature inside the target solid tumors, leading to apoptosis or necrosis of neoplastic tissues. The Magnetic resonance-guided focused ultrasound surgery (MRgFUS) technology represents a valid application of this ultrasound property, mainly used in oncology and neurology. In this paper; patient safety during MRgFUS treatments was investigated by a series of experiments in a tissue-mimicking phantom and performing ex vivo skin samples, to promptly identify unwanted temperature rises. The acquired MR images, used to evaluate the temperature in the treated areas, were analyzed to compare classical proton resonance frequency (PRF) shift techniques and referenceless thermometry methods to accurately assess the temperature variations. We exploited radial basis function (RBF) neural networks for referenceless thermometry and compared the results against interferometric optical fiber measurements. The experimental measurements were obtained using a set of interferometric optical fibers aimed at quantifying temperature variations directly in the sonication areas. The temperature increases during the treatment were not accurately detected by MRI-based referenceless thermometry methods, and more sensitive measurement systems, such as optical fibers, would be required. In-depth studies about these aspects are needed to monitor temperature and improve safety during MRgFUS treatments.

**Keywords:** MRgFUS; proton resonance frequency shift; temperature variations; referenceless thermometry; RBF neural networks; interferometric optical fibers



**Citation:** Militello, C.; Rundo, L.; Vicari, F.; Agnello, L.; Borasi, G.; Vitabile, S.; Russo, G. A Computational Study on Temperature Variations in MRgFUS Treatments Using PRF Thermometry Techniques and Optical Probes. *J. Imaging* **2021**, 7, 63. https://doi.org/10.3390/jimaging7040063

Academic Editor: Reyer Zwiggelaar

Received: 28 January 2021
Accepted: 23 March 2021
Published: 25 March 2021



## 1. Introduction

Image-guided thermal ablations are increasingly employed in minimally invasive treatments in patients with cancer [1–4]. In the last decades, a large number of high-intensity focused ultrasound (HIFU) [5,6] devices have been used in oncology to cover a wide range of cancer types, such as prostate [7], bone metastases [8], liver [9], breast [10], thyroid [11], uterine fibroids [12,13], liver and pancreas [14], and brain [15]; as well as psychiatric disorders [16] and essential tremor [17,18].

Considering the imaging modalities that currently guide HIFU treatments, two possible methodologies are available: (i) ultrasound-guided therapeutic focused ultrasound

(USgFUS) [19,20], which uses the shift of the echo timing related to the temperature variation of the treated tissues [21]; and (ii) magnetic resonance-guided focused ultrasound surgery (MRgFUS) [22], which leverages the intrinsic dependence of the temperature with respect to some fundamental parameters, such as the apparent diffusion coefficient (ADC) of water molecules, the spin-lattice relaxation time (T1), and the water proton resonance frequency (PRF) [23].

In order to evaluate the incidence and severity of adverse reactions to the USgFUS ablation of uterine fibroids, Chen et al. [24] performed a multicenter, large-scale retrospective study involving 9988 patients with uterine fibroids or adenomyosis. Even though all the required procedures were applied, including skin preparation, 26 of the patients had blisters or tangerine pericarp-like burns in their abdominal skin, and two of them required surgical removal of the necrotic tissue. In [25], a preliminary report on bone metastasis pain-palliation therapy with MRgFUS, an unusual second-degree skin burn occurred on the body side opposite to the transducer position. The authors argued that this accident occurred due to a series of energetically intense sonications that may not have been totally included inside the patient's body, causing a far-field energy accumulation at the air–skin interface [26,27]. In the case of MRgFUS capsulotomy, safety and clinical efficacy need to be carefully assessed by considering issues related to skull heating [16].

With particular interest in MRgFUS, automated techniques for uterine fibroid MR image segmentation have been recently devised to improve treatment planning [28] and evaluation [29,30], thus increasing the result repeatability and reliability [31]. Importantly, the attention of manufacturers to MRgFUS treatment safety has increased in recent years; therefore, multicenter studies have been performed to propose effective solutions. For instance, a modified clinical MRgFUS fibroid therapy system, called Sonalleve (Philips Healthcare, Vantaa, Finland), was integrated with a 1.5 T magnetic resonance imaging (MRI) scanner (Achieva, Philips Healthcare, Best, The Netherlands). This system directly relied upon a skin-cooling device for the treatment of symptomatic uterine fibroids [32]. In the experiments conducted, involving eight patients, no adverse effects were reported when this cooling device was integrated with the patient table to keep the transducer–patient interface at a fixed temperature of 20 °C.

The aim of this work is to explore the sensitivity of MRI guidance to monitor the temperature increase for patient safety [26,27]. In particular, we simulated the temperature variations in a fibroid treatment on a tissue-mimicking phantom, acquiring temperature measurements using thermal imaging provided by the operating console of the MRgFUS ExAblate 2100 (Insightec Ltd., Carmel, Israel), as well as interferometric optical probes. The temperature maps were obtained using classic PRF and referenceless thermometry methods and compared against the measurements.

## 2. Materials and Methods

In our experiments, an Insightec ExAblate 2100 HIFU transducer integrated with a Signa HTxt MRI scanner (General Electric Medical Systems, Milwaukee, WI, USA) was used. The same clinical device is employed at the Foundation Institute "G. Giglio", Cefalù (PA), Italy, for uterine fibroid treatment and bone metastasis pain-palliative therapy. This system exploits MRI to acquire temperature maps of treated tissues by quantifying the phase variation resulting from the temperature-dependent changes in the resonance frequency. The phase differences are proportional to temperature-dependent PRF shifts, thus enabling the assessment of temperature rises [33]. Temperature maps derived from MRI can be obtained using gradient recalled echo (GRE) imaging sequences. The console operator monitors the temperature rise taking into consideration: (i) the thermal map of a chosen slice (Figure 1a); and (ii) the temperature plots concerning the selected point (by means of a crosshair cursor) and a small neighboring region (Figure 1b). These methods were successfully used to model the thermal dose delivery [34] strictly related to tissue thermo-ablation [35,36]. Any unwanted temperature increase outside the "target" is due to an energy accumulation, caused by acoustic impedance discontinuity in the ultrasound wave-propagation path [37–39].

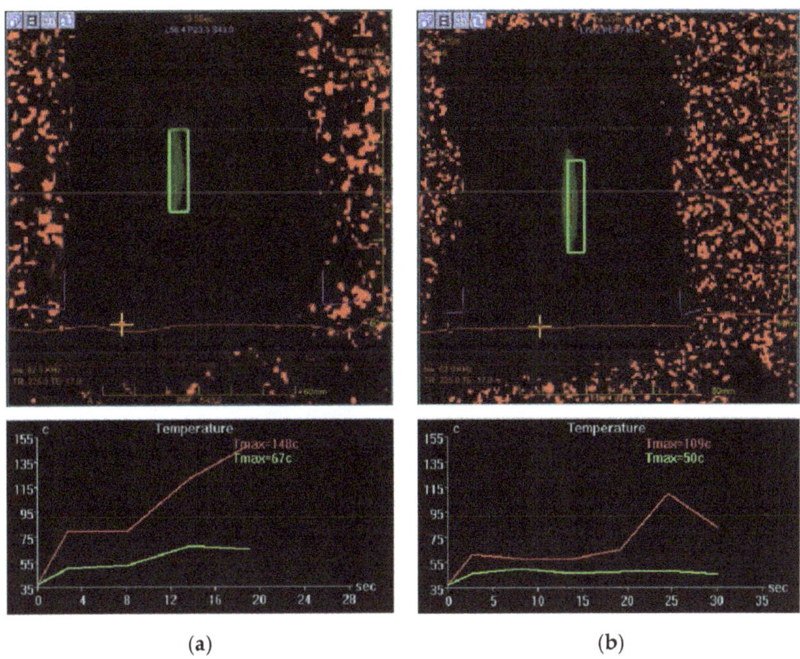


(a) (b)

**Figure 1.** (**a**) Thermal map of a sonication during a treatment. The crosshair cursor, selected by the operator, represents the point of interest for temperature trend control. (**b**) Temperature plot of a single pixel (red line) and a small neighboring region around the crosshair cursor (green line).

### 2.1. MRgFUS Treatments

The experimental measurements were carried out using the ExAblate uterine-fibroids protocol, considering a real fibroid treatment as reference.

Prior to MRgFUS treatments for uterine fibroid ablation, the patient was sedated to minimize her movements, but nevertheless she could constantly provide feedback on the perception of pain and heat during the treatment. The MR images were acquired to localize the fibroid position and to plan the treatment with the most suitable ultrasound beam path, and sonication size and number. The treatment was planned by software that analyzed the region of treatment (ROT)—i.e., the region that will undergo the ultrasound beams—and the limited energy density regions (LEDRs)—i.e., the regions containing the organs at risk (OARs). Treatment planning aims to deliver the sonications in the entire ROT, making sure that the ultrasound beam does not cross the LEDRs.

To verify the focus-position accuracy, a preliminary sonication at sublethal energy was delivered. Some MR images were acquired to detect the temperature distribution in the neighborhood of the focus point. Using an iterative procedure, the operator can modify the wave characteristics to improve the target accuracy and the temperature increase. As a result, the treatment was performed by delivering sonications with lethal energy. Each sonication typically lasted 20–40 s, with a cooling time of 80–90 s between two successive sonications.

At the end of the treatment, the patient, in the position she had during treatment, underwent a diagnostic MR examination with gadolinium-based contrast medium, aimed to evaluate the nonperfused volume (NPV), which was the uterine fibroid area covered by sonications. Moreover, the skin was examined to evaluate any side effects due to the temperature increase during the treatment.

In this work, to quantify temperature increases in the interface area suffering from acoustic impedance discontinuity in the ultrasound wave-propagation path, we used a configuration composed of: (i) a standard phantom tissue mimicking the daily quality assurance (DQA) routine, as previously proposed by Zucconi et al. [40]; (ii) an ex vivo porcine skin sample placed under the phantom to simulate the patient's skin; and (iii) a gel pad (between the porcine skin and ExAblate bed). A set of interferometric probes was also used to monitor the skin temperature, over the probes and gel pad (Figure 2). We assumed that the porcine skin would respond to the temperature increases like the human skin.

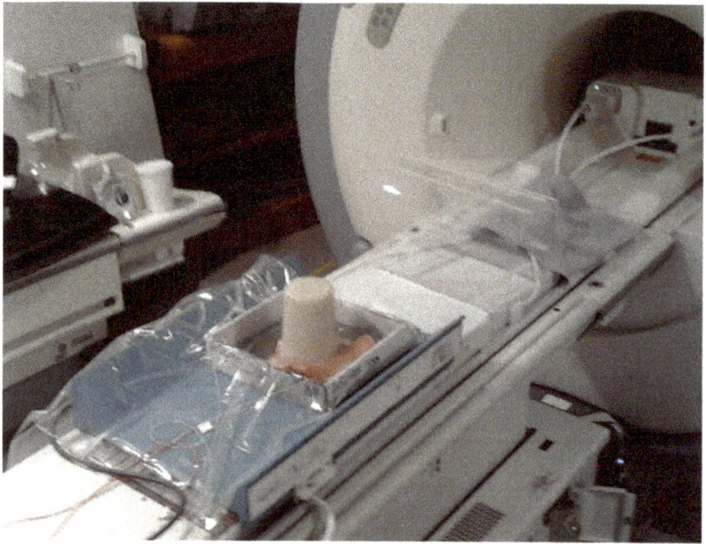

**Figure 2.** The realized configuration: the daily quality assurance (DQA) phantom over a skin portion. Although barely noticeable, the gel pad was placed under the skin to ensure acoustic coupling between the ExAblate bed and skin. In the left area of the image, the two interferometric probes are visible.

A ROT of 78.7 $cm^3$ was defined inside the phantom and automatically covered by the system with 56 sonications. Neglecting absorption and attenuation in the propagation path [41], an average energy of 2353 $\pm$ 611 J can be attributed to the sonications emitted by the 208 elements of the phased-array HIFU transducer [42], for an average duration of 20.0 $\pm$ 2.9 s (with an elongated beam geometry). The time cooling was set at 85 s and the ultrasound frequency at 1.1 MHz.

The software distributed the sonications over the ROT, forming s-shaped paths, in order to prevent local overheating.

### 2.2. Optical Thermometry

For continuous temperature monitoring during the MRgFUS sonications, an MR-compatible instrumentation was required. The AccuSens interferometric signal conditioner (Opsens Inc., Québec, QC, Canada) equipped with an OTP-M birefringent crystal sensor was chosen. The main characteristics are reported in Table 1.

**Table 1.** Characteristics of the AccuSens interferometric signal conditioner.

| Characteristic | Value |
|---|---|
| Temperature operating range | 0 °C to 85 °C |
| Specific calibrated range | 20 °C to 45 °C standard (other ranges available) |
| Resolution | 0.01 °C |
| Accuracy (specific calibrated range) | ±0.15 °C @ ±3.3 σ limit (99.9% confidence level) |
| Response time | <1 s |
| Operating humidity range | 0–100% |

The bottom surface of the phantom was divided into two portions: a circular crown, which was never crossed by the ultrasound, and an inner area covered by the HIFU. One of the OTP-M probes was inserted into the middle of the circular region, and the tip of another one on the boundary between these two regions (Figure 3). Using this configuration, a mask for the relative positioning of sensors and phantom on the gel pad was designed. Then, this mask was reproduced on a plastic drape included in the "patient accessory set" necessary for the treatment, since this material did not introduce any acoustical impedance discontinuity.

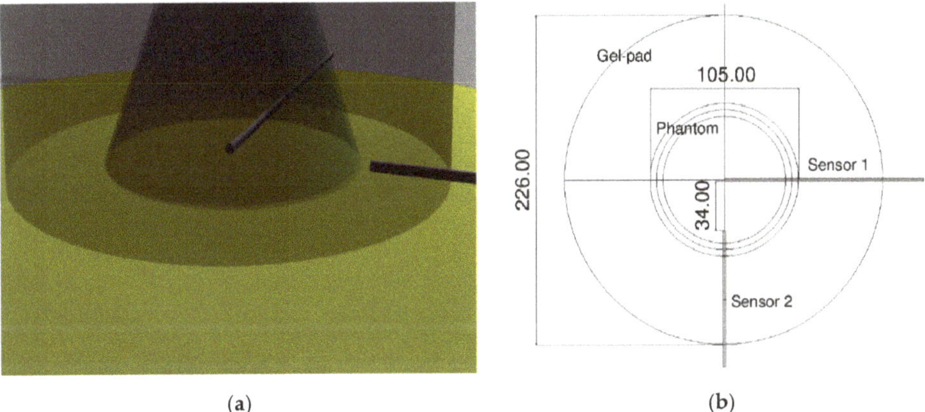


**Figure 3.** Probe positions relative to the ultrasound field. (**a**) 3D model with the two probes positioned; (**b**) schematics of the positioning/coupling apparatus.

*2.3. Signal-to-Noise-Ratio Estimation*

In order to evaluate if there was an adequate signal within the interface region necessary to quantify temperatures, the signal-to-noise ratio (SNR) was calculated according to Gorny et al. [43]. The investigated areas were the phantom, the skin interface, and the gel pad.

Some sample MR images were evaluated; in particular, the images of the phantom relative to sonication 4 and 5 were examined. Each acquired region was characterized by an overall thickness of 16 mm, and was acquired in different locations with respect to the phantom size (circular base with a diameter of 105 mm, as shown in Figure 3).

As shown in Figure 4, the acquired region of sonication #4 (red area) ranged from +35 mm to +51 mm, while the region of sonication #5 (orange area) ranged from −14.4 mm to +1.6 mm.

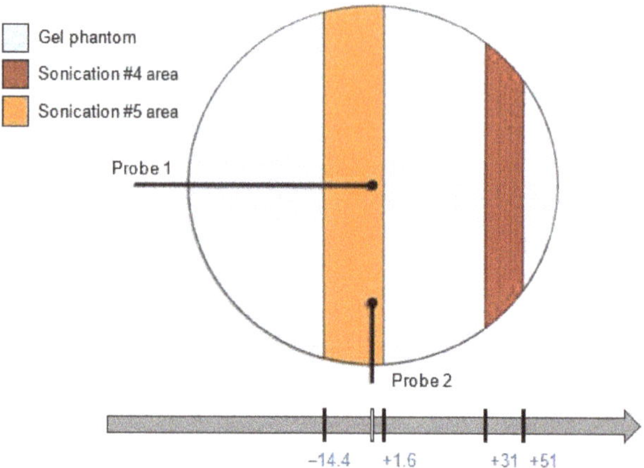


**Figure 4.** The MRI acquisition locations of sonications #4 and #5.

The images related to each region were acquired in subsequent temporal frames of 3 s, which allowed us to reconstruct the temporal trend of the temperature rise for each acquired area. The SNR value was calculated according to Equation (1):

$$\mathrm{SNR} = \frac{0.655 \cdot \mu\left(\mathrm{Signal}_{\mathrm{object}}\right)}{\sigma\left(\mathrm{Signal}_{\mathrm{background}}\right)}, \quad (1)$$

where the ratio between the mean signal value ($\mu$) of the object (i.e., phantom, skin, and gel pad) region of interest (ROI) and the standard deviation ($\sigma$) of an area that contains only background noise (e.g., air) were considered. The 0.655 factor was due to the Rician distribution of the background noise in a magnitude image, which tended to a Rayleigh distribution as the SNR tended to zero [44].

The three ROIs investigated for the SNR estimation are represented in Figure 5. The signal intensity of the phantom, the skin interface, and gel pad areas were compared to a region where the signal was ideally zero (i.e., the background ROI).

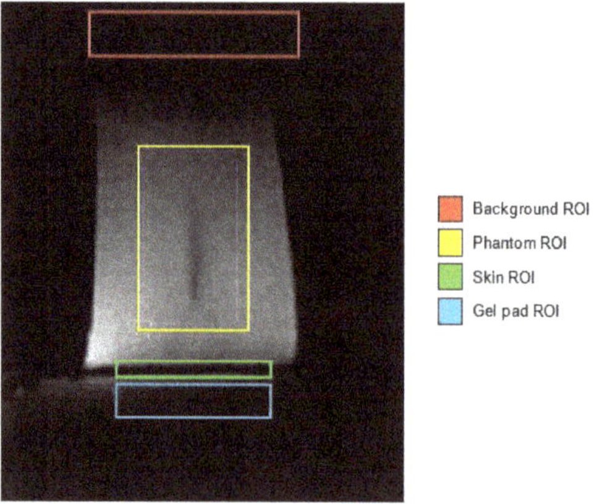


**Figure 5.** The ROIs investigated for the SNR estimation. The different ROIs that were drawn were the tissue-mimicking phantom (yellow), skin interface (green), gel pad (cyan), and background area (red).

### 2.4. Referenceless Thermometry

Classical PRF shift thermometry—in which one or more baseline images are acquired before the thermal therapy and then are subtracted pixel-by-pixel from the images acquired during heating—is affected by artifacts, which could lead to unrealistic temperature increases [13,45,46]. These temperature-independent artifacts are mainly due to movements of the anatomical region undergoing MRgFUS treatments, or to magnetic field inhomogeneities. With the goal of reducing these issues, referenceless thermometry could be used, thus allowing us to estimate the heating caused by an MRgFUS treatment without using a baseline image as temperature reference.

With the goal of accurately estimating the temperature variations, referenceless thermometry methods were developed; in particular, we devised an interpolation method based on artificial neural networks (ANNs) to reconstruct the original baseline phase image and reliably evaluate temperature variations in the sonication area [47,48]. In fact, assuming that the phase image surrounding the treated region has a smooth trend (even under the heated area), referenceless (or self-referenced) thermometry techniques estimate the temperature variations by means of a set of smooth low-order polynomial functions to the surrounding phase, or to a complex magnitude image with the same phase using a weighted least-squares fit [49]. The extrapolation of the polynomial inside the heated region is used as background phase estimation, which is subtracted from the actual phase to evaluate the phase difference before and after heating caused by ultrasound sonications and, successively, quantify the temperature increase.

In the referenceless phase estimation, an ROI has to be delineated around the area to be heated. First of all, two regions (namely, outer and inner) must be selected in the phase image to perform the interpolation. Figure 6 shows the phase map and the outer baseline region around the sonicated area (after the removal of the inner ROI containing the heated region). It is essential to choose the outer ROI outside the heated region because the temperature changes within the ROI affect the reconstruction of the background phase.

The most straightforward computational approach to solve this problem is to fit the data with a polynomial function [50]. However, an invertible system that uniquely defines the interpolant is not guaranteed for all positions of the interpolation points, and often it could show spurious bumps. The background phase in the frame ROI is reconstructed by means of an ANN exploiting radial basis functions (RBFs) as kernel [51,52].

In particular, a 3-layer feed-forward ANN was designed (with 1 input layer, 1 output layer and 1 hidden layer) in which each hidden node implemented an RBF. ANNs are well-suited for interpolation purposes, especially if there are large areas of missing data, and the RBF approximation method allows several advantages with respect to polynomial interpolants: (i) the network training finds the optimal weights from the input to the hidden layer, and then the weights from the hidden to the output layer are calculated; and (ii) the geometry of the input points is not restricted to a regular grid.

#### Radial Basis Function Theory

Let $f : \mathbb{R}^d \to \mathbb{R}$ be a real valued function of $d$ variables that has to be approximated by $s : \mathbb{R}^d \to \mathbb{R}$, given the values $\{f(X_i) : i = 1, 2, \ldots, n\}$, where $\{X_i : i = 1, 2, \ldots, n\}$ is a set of $n$ distinct points in $\mathbb{R}^d$ called the interpolation nodes. We will consider an approximation of the form:

$$s(X) = p_m(X) + \sum_{i=1}^{n} \lambda_i \varphi(\|X - X_i\|_2), \; X \in \mathbb{R}^d, \lambda_i \in \mathbb{R}, \tag{2}$$

where: $p_m$ is a low-degree polynomial that can be also omitted, $\|\cdot\|_2$ denotes the Euclidean norm, and $\varphi$ is a fixed function from $\mathbb{R}$ to $\mathbb{R}$. Thus, the radial basis function $s(\cdot)$ is a linear combination of translations of the single radially symmetric function $\varphi(\|\cdot\|_2)$, plus a low-degree polynomial. We will denote with $\pi_m^d$ the space of all polynomials of degree $m$ at most in $d$ variables. The coefficients $\lambda_i$, which represent the weights of the approximation

$s$, are determined by requiring that $s$ satisfies the interpolation conditions expressed in the following Equation (3):

$$s(X_j) \equiv f(X_j),\ j = 1, 2, \ldots, n, \tag{3}$$

together with the side conditions:

$$\sum_{i=1}^{n} \lambda_i q(X_j) = 0,\ \forall q \in \pi_m^d. \tag{4}$$

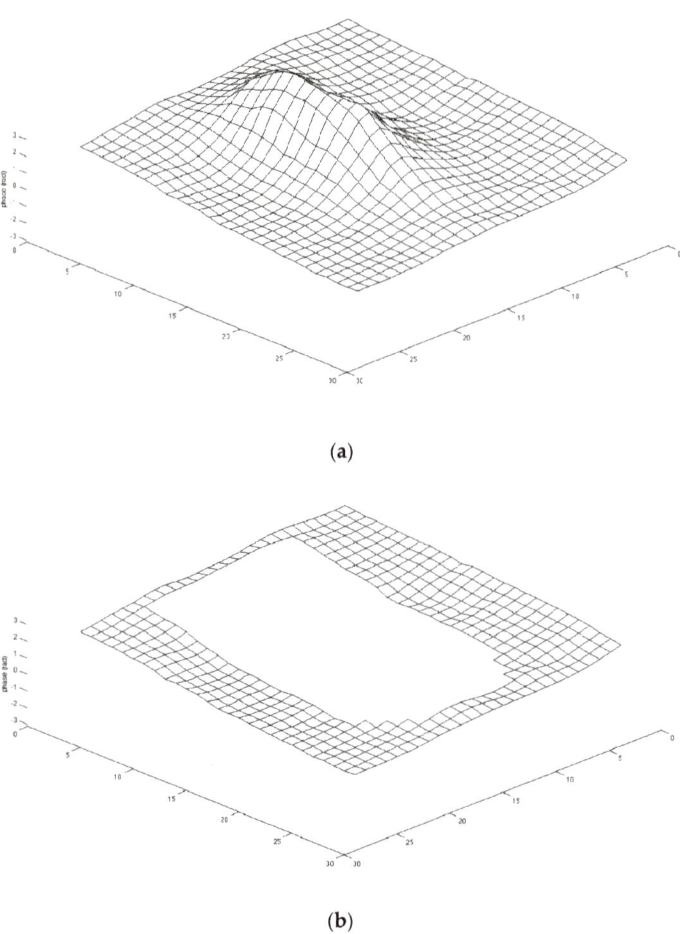

**Figure 6.** (**a**) 3D plot of a phase map with sonicated area; (**b**) 3D plot of the outer region of the phase map in (**a**) after removing the sonicated area.

Some typical conditions on the nodes under which the interpolation conditions (3) and (4) uniquely specify the radial basis function (2) are given in Table 2. In this context "not coplanar" means that the nodes do not all lie in a single hyperplane, or equivalently that no linear polynomial in $d$-variables vanishes at all the nodes. The surveys presented in [53] and [54] are excellent references to these and other properties of radial basis functions.

**Table 2.** Conditions imposed on nodes for various radial basis interpolants.

| Function Type | Spatial Dimension $d$ | Polynomial Degree $m$ | Restriction on Nodes |
|---|---|---|---|
| linear RBF | any | 1 | not coplanar |
| thin-plate spline | 2 | 1 | not coplanar |
| Gaussian | any | absent | none |
| multiquadratic RBF | any | absent | none |

## 3. Results

The selected ROIs were propagated for all the temporal sequences and in all the depths, so the SNR value was calculated on every acquired 3D volume. As depicted in Figure 7, the MR images of the sonications #4 and #5 showed the impulsive noise in the area surrounding the phantom, especially in the skin interface and in the gel pad.

The signal acquired using the thermometric MRI protocol can be acceptable for aqueous tissues (such as the regions treated with MRgFUS), but unsatisfactory for fatty tissues. In fact, as widely stated in [55], the tissue-type temperature independence of the PRF shift is almost true for aqueous tissues, while the dependence in adipose tissues is affected by susceptibility effects. Consequently, the temperature sensitivity of fat is extremely low [56], indicating that MRI-based thermometry inside fatty tissues (such as the skin interface taken into account here) is difficult.

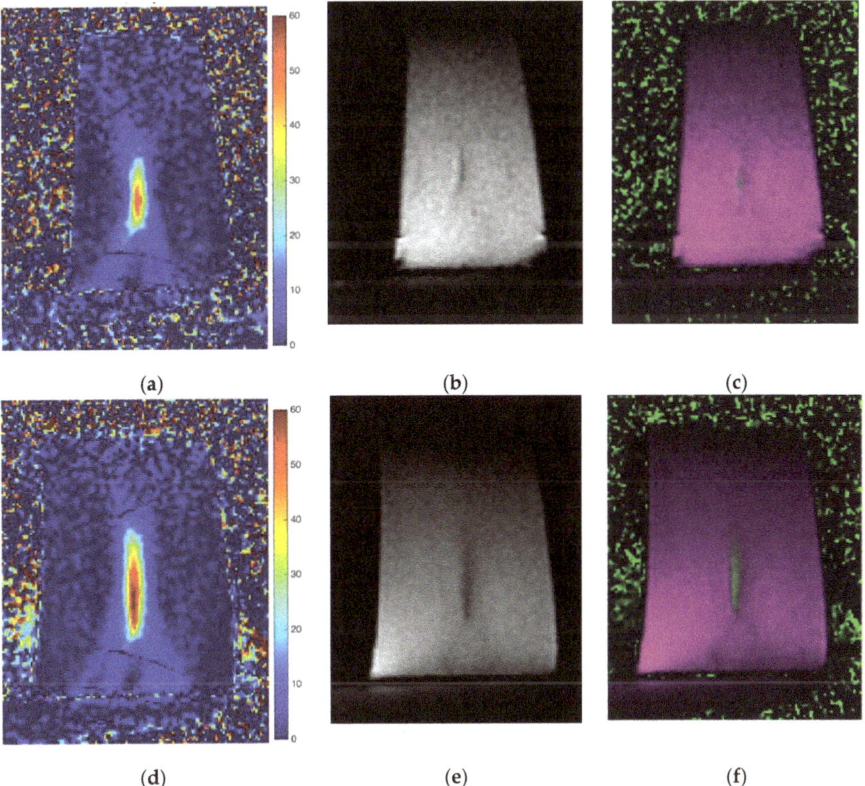


**Figure 7.** Sonications #4 (first row) and #5 (second row) morphological and thermal map examples: (**a**) and (**d**) temperature reconstruction; (**b**) and (**e**) morphological image; (**c**) and (**f**) temperature image overlapped on the morphological image. It is possible to estimate the noise in the gel pad and in the skin interface by observing the low SNR in those areas.

These insights also were confirmed by our experimental findings, which showed that SNRs inside the area near the gel pad and the porcine skin were relatively low when compared to the SNR inside the phantom. Figure 8 shows that the signal was globally low in all three acquired MRI volumes. The phantom area showed a higher signal compared to the skin layer and the gel pad, where the signal appeared very poor.

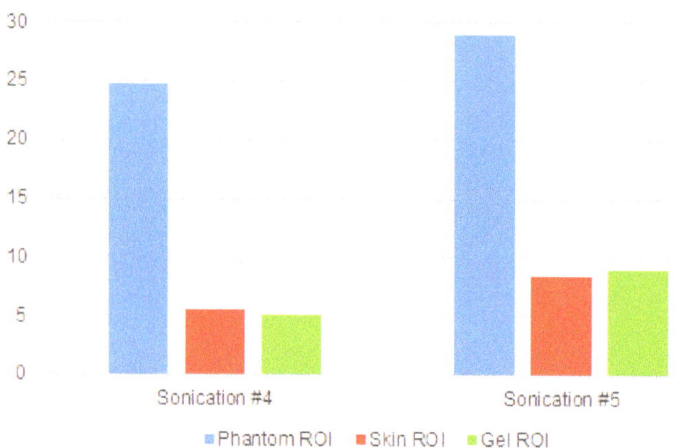


**Figure 8.** SNR values calculated for the phantom, skin, and gel ROIs. The phantom signal had the highest SNR values, while the gel area and the skin-interface area had the lowest SNR values.

The treatment was performed in about 2 h. The interferometric probes under the porcine skin, positioned according to the scheme on Figure 3, measured a large amount of temperature data. Figure 9 shows the maximum temperature rise recorded by the probes in all the sonications. This is a clear confirmation that the probes were actually placed as planned: the first probe was in the middle of the phantom and received more heat than the second one, which was in a more decentralized position than the ROT.

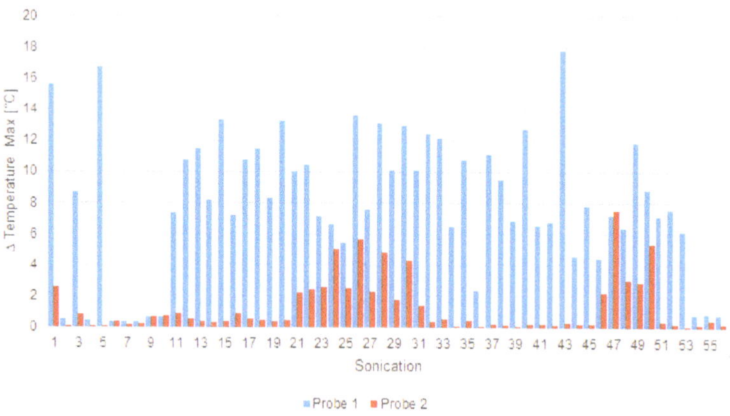


**Figure 9.** Maximum rise of temperature in each sonication for probe 1 (blue) and probe 2 (red).

In some sonications, temperature-rising measurements were weakly perceived ($\Delta T < 1$ °C) for the relative position along the hypersonic field; this was the case in the fourth sonication (Figure 10a). In other cases, like the fifth sonication, the temperature rose about 16 °C (Figure 10b).

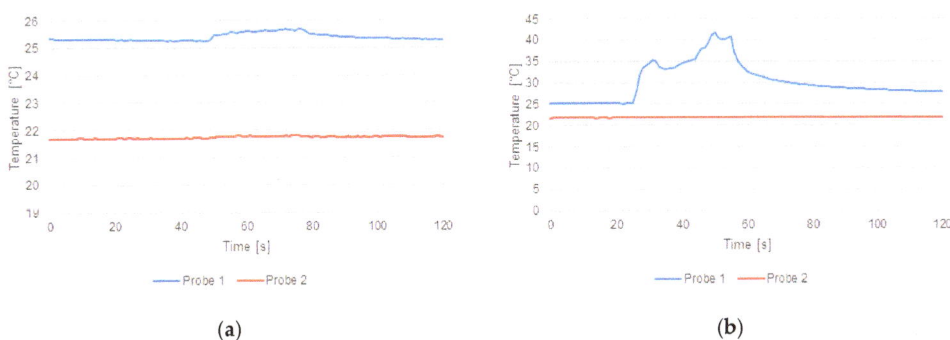


**Figure 10.** Temperature measured by the optical probe 1 (blue) and probe 2 (red) during sonications #4 (**a**) and #5 (**b**).

Our analysis, coupled with the PRF-based temperature quantification provided by the ExAblate control console, was employed by considering referenceless thermometry on 2D phase map data, by means of ANNs using different interpolants RBF kernels (i.e., linear, thin-plate spline, and multiquadratic) [47]. In these cases, it also was not possible to detect meaningful temperature increases.

RBF and polynomial interpolations were applied on the data set; the former showed a "bump-like" tendency and the latter overestimated the temperature, because the analyzed area was characterized by a low signal intensity where the noise was a significant component (Figure 11).

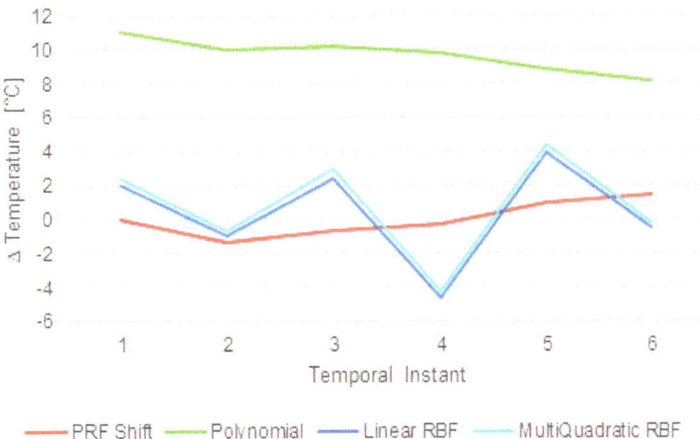


**Figure 11.** The interpolated temperature errors compared to PRF-based temperature measurements (which does not show any significant temperature rise). The polynomial (green) line overestimated the data, while the linear RBF (blue) and multiquadratic RBF (cyan) lines had a "bump-like" trend caused by the presence of noisy data.

To show the differences measured by the two probes, a two-sided Wilcoxon signed rank test on paired data [57] was performed with the null hypothesis that the samples came from continuous distributions with equal medians. In all the tests, a significance level of 0.05 was considered. More details are provided in what follows: (i) the distributions of the temperature increases measured by the probes (Figure 9) were statistically significant considering all the sonications ($p = 1.719 \times 10^{-10}$); (ii) the distributions of the temperature measured over time by the probes (Figure 10) were statistically significant for sonications for both sonications #4 ($p = 2.095 \times 10^{-24}$) and #5 ($p = 6.601 \times 10^{-44}$); and (iii) the polynomial interpolation (Figure 11) significantly overestimated the data ($p = 0.031$), while the linear RBF and multiquadratic RBF interpolations were not statistically different from the PRF shift data ($p = 0.687$ in both cases).

## 4. Discussion

Starting from the current issues concerning patient safety related to undesired temperature variations that can cause skin burns, an MRgFUS fibroid treatment was simulated using an ex vivo porcine skin and a DQA tissue-mimicking phantom. The treatment consisted of 56 ultrasound sonications and a maximum temperature increment ($\Delta T = 17.78$ °C, given in the 43th sonication), as shown in Figure 9. Even if the temperature increase was obtained intentionally through bad acoustic coupling and by considering the interference of the probes, the obtained results showed how it is quite difficult for a clinical operator to detect a possible (and naturally unwanted) temperature increase by relying only on the operating console that displays MR thermographic images. According to the study of Moritz and Henriques [58], the relationship between time and temperature for this sonication is not intense enough to cause a skin burn, but the authors showed how a repetition of five times could lead to complete and irreversible epidermal necrosis. The same results can be obtained using more recent model-based classification approaches [59]. PRF-based temperature monitoring is not useful with this kind of tissue, which was also confirmed by using referenceless thermometry with polynomial and RBF interpolation models. This can be attributed to the small thickness of the skin in the axial and sagittal planes compared to: (i) the spatial resolution of the acquired MR images, (ii) the difficulty of catching the skin on a coronal slice in the low-quality (to guarantee the appropriate acquisition speed for real-time temperature monitoring) MR images acquired during the treatment, and (iii) the thermometry system developed for clinical applications that is not optimized for such a purpose. Moreover, the bump-like tendency of the RBF interpolation errors (see Figure 11) could be due to a low SNR in the analyzed area, where the noise represented a significant component while the signal was practically negligible, as shown in Figure 8.

While attempts have been made to reduce temperature increases on patients' skin through the quantification of the near-field (between the ultrasound transducer and target) heating [60], a real-time temperature monitoring could give a better control during the treatment. It might be necessary to develop novel image-processing algorithms and methods to enhance phase-map acquisition in PRF-based thermometry techniques, as well as MRI sequences with a higher pixel resolution, to improve the temperature monitoring and limit any unwanted hot spots.

## 5. Conclusions and Future Work

In this work, the potential side effects regarding patient safety due to temperature increases that rarely affect MRgFUS treatments were assessed. Along with the classical PRF shift thermometry, a novel approach that exploited a referenceless technique based on the RBF interpolation was used to evaluate the skin temperature during sonications. Moreover, in this study, we also used two interferometric probes to measure the reached temperatures. In a simulation of a real uterine fibroid treatment, only the probes were able to detect temperature increases, while no important temperature changes were revealed by the used interpolation methods. The achieved results showed that these methods, based on the PRF shift thermometry, could be unsuitable to detect temperature increases on the skin.

One of the issues to consider in our analysis is the low SNR value in the investigated region. New hardware and software solutions need to be studied to increase the temperature-detector sensitivity by rising the SNR in order to also enhance MRgFUS treatment safety and effectiveness.

In the future, more temporal instants should be considered for temperature measurements and increases. Multiple repetitions of the experiments will increase the statistical robustness of the experimental findings.

Moreover, the planned experiments could be designed to reliably simulate a configuration for clinical environments. To address the issues related to the acoustic interference generated by the optical fibers across the ultrasound propagation, other techniques that are able to accurately measure the skin temperature in real time and with a good time resolution could be employed. For instance, thermoscanners have a high temperature accuracy (±0.3 °C), a very high recognition speed (<300 ms), and a temperature range (25–45 °C) that are sufficient to evaluate skin temperature increases in real time. Some systems could be also optically coupled to monitor the skin's irradiated area for all tests. After extensive ex vivo tests, the developed systems could be employed during clinical treatments.

**Author Contributions:** Conceptualization, C.M., L.R., F.V., L.A., G.B. and G.R.; methodology, C.M., F.V., L.A. and G.R.; software, C.M., F.V. and L.A.; validation, C.M., L.R., F.V. and G.R.; formal analysis, C.M., L.R., F.V., L.A. and G.R.; investigation, C.M., L.R., F.V., L.A., G.B. and G.R.; resources, C.M. and G.R.; data curation, C.M., L.R., F.V. and L.A.; writing—original draft preparation, C.M., L.R. and F.V.; writing—review and editing, G.B., S.V. and G.R.; visualization, C.M., L.R., F.V. and L.A.; supervision, G.B., S.V. and G.R. All authors have read and agreed to the published version of the manuscript.

**Funding:** This work was supported by the "Sviluppo di una piattaforma tecnologica per il trattamento non invasivo di patologie oncologiche e infettive basate sull'uso di ultrasuoni focalizzati" MIUR project (PON01_01059), approved by MIUR D.D. n. 655/Ric. This work also was supported by the "Smart Health 2.0" MIUR project (PON 04a2_C), approved by MIUR D.D. n. 626/Ric and n. 703/Ric.

**Institutional Review Board Statement:** Not applicable.

**Informed Consent Statement:** Not applicable.

**Data Availability Statement:** Data sharing is not applicable to this article.

**Conflicts of Interest:** The authors declare no conflict of interest.

## References

1. Mauri, G.; Nicosia, L.; Varano, G.M.; Bonomo, G.; Della Vigna, P.; Monfardini, L.; Orsi, F. Tips and tricks for a safe and effective image-guided percutaneous renal tumour ablation. *Insights Imaging* **2017**, *8*, 357–363. [CrossRef]
2. Mainini, A.P.; Monaco, C.; Pescatori, L.C.; De Angelis, C.; Sardanelli, F.; Sconfienza, L.M.; Mauri, G. Image-guided thermal ablation of benign thyroid nodules. *J. Ultrasound* **2017**, *20*, 11–22. [CrossRef]
3. Seifabadi, R.; Li, M.; Xu, S.; Chen, Y.; Squires, A.; Negussie, A.H.; Bakhutashvili, I.; Choyke, P.; Turkbey, I.B.; Tse, Z.T.H.; et al. MRI Robot for Prostate Focal Laser Ablation: An Ex Vivo Study in Human Prostate. *J. Imaging* **2018**, *4*, 140. [CrossRef]
4. Ning, G.; Zhang, X.; Zhang, Q.; Wang, Z.; Liao, H. Real-time and multimodality image-guided intelligent HIFU therapy for uterine fibroid. *Theranostics* **2020**, *10*, 4676–4693. [CrossRef] [PubMed]
5. Lynn, J.G.; Zwemer, R.L.; Chick, A.J.; Miller, A.E. A new method for the generation and use of focused ultrasound in experimental biology. *J. Gen. Physiol.* **1942**, *26*, 179–193. [CrossRef] [PubMed]
6. Doan, V.H.M.; Nguyen, V.T.; Choi, J.; Park, S.; Oh, J. Fuzzy Logic Control-Based HIFU System Integrated with Photoacoustic Imaging Module for Ex Vivo Artificial Tumor Treatment. *Appl. Sci.* **2020**, *10*, 7888. [CrossRef]
7. Wilt, T.J.; MacDonald, R.; Rutks, I.; Shamliyan, T.A.; Taylor, B.C.; Kane, R.L. Systematic review: Comparative effectiveness and harms of treatments for clinically localized prostate cancer. *Ann. Intern. Med.* **2008**, *148*, 435–448. [CrossRef]
8. Napoli, A.; Mastantuono, M.; Marincola, B.C.; Anzidei, M.; Zaccagna, F.; Moreschini, O.; Passariello, R.; Catalano, C. Osteoid Osteoma: MR-guided Focused Ultrasound for Entirely Noninvasive Treatment. *Radiology* **2013**, *267*, 514–521. [CrossRef]
9. Li, C.-X.; Xu, G.-L.; Jiang, Z.-Y.; Li, J.-J.; Luo, G.-Y.; Shan, H.-B.; Zhang, R.; Li, Y. Analysis of clinical effect of high-intensity focused ultrasound on liver cancer. *World J. Gastroenterol.* **2004**, *10*, 2201–2204. [CrossRef]
10. Merckel, L.G.; Bartels, L.W.; Köhler, M.O.; van den Bongard, H.J.G.D.; Deckers, R.; Mali, W.P.T.M.; Binkert, C.A.; Moonen, C.T.; Gilhuijs, K.G.A.; van den Bosch, M.A.A.J. MR-Guided High-Intensity Focused Ultrasound Ablation of Breast Cancer with a Dedicated Breast Platform. *Cardiovasc. Interv. Radiol.* **2013**, *36*, 292–301. [CrossRef]
11. Gharib, H.; Hegedüs, L.; Pacella, C.M.; Baek, J.H.; Papini, E. Nonsurgical, Image-Guided, Minimally Invasive Therapy for Thyroid Nodules. *J. Clin. Endocrinol. Metab.* **2013**, *98*, 3949–3957. [CrossRef]
12. Hindley, J.; Gedroyc, W.M.; Regan, L.; Stewart, E.; Tempany, C.; Hynnen, K.; Macdanold, N.; Inbar, Y.; Itzchak, Y.; Rabinovici, J.; et al. MRI Guidance of Focused Ultrasound Therapy of Uterine Fibroids:Early Results. *Am. J. Roentgenol.* **2004**, *183*, 1713–1719. [CrossRef] [PubMed]
13. Militello, C.; Rundo, L.; Gilardi, M.C. Applications of imaging processing to MRgFUS treatment for fibroids: A review. *Transl. Cancer Res.* **2014**, *3*, 472–482. [CrossRef]

14. Zaccagna, F.; Anzidei, M.; Sandolo, F.; Marincola, B.C.; Palla, C.; Leonardi, A.; Caliolo, G.; Andreani, F.; De Soccio, V.; Catalano, C.; et al. MRgFUS for liver and pancreas cancer treatments: The Umberto I hospital experience. *Transl. Cancer Res.* **2014**, *3*. [CrossRef]
15. Coluccia, D.; Figueiredo, C.A.; Wu, M.Y.; Riemenschneider, A.N.; Diaz, R.; Luck, A.; Smith, C.; Das, S.; Ackerley, C.; O'Reilly, M.; et al. Enhancing glioblastoma treatment using cisplatin-gold-nanoparticle conjugates and targeted delivery with magnetic resonance-guided focused ultrasound. *Nanomed. Nanotechnol. Biol. Med.* **2018**, *14*, 1137–1148. [CrossRef] [PubMed]
16. Davidson, B.; Hamani, C.; Huang, Y.; Jones, R.M.; Meng, Y.; Giacobbe, P.; Lipsman, N. Magnetic Resonance-Guided Focused Ultrasound Capsulotomy for Treatment-Resistant Psychiatric Disorders. *Oper. Neurosurg.* **2020**. [CrossRef] [PubMed]
17. Kapadia, A.N.; Elias, G.J.B.; Boutet, A.; Germann, J.; Pancholi, A.; Chu, P.; Zhong, J.; Fasano, A.; Munhoz, R.; Chow, C.; et al. Multimodal MRI for MRgFUS in essential tremor: Post-treatment radiological markers of clinical outcome. *J. Neurol. Neurosurg. Psychiatry* **2020**, *91*, 921–927. [CrossRef]
18. Bruno, F.; Catalucci, A.; Arrigoni, F.; Sucapane, P.; Cerone, D.; Cerrone, P.; Ricci, A.; Marini, C.; Masciocchi, C. An experience-based review of HIFU in functional interventional neuroradiology: Transcranial MRgFUS thalamotomy for treatment of tremor. *Radiol. Med.* **2020**, *125*, 877–886. [CrossRef]
19. Abel, M.; Ahmed, H.; Leen, E.; Park, E.; Chen, M.; Wasan, H.; Price, P.; Monzon, L.; Gedroyc, W.; Abel, P. Ultrasound-guided trans-rectal high-intensity focused ultrasound (HIFU) for advanced cervical cancer ablation is feasible: A case report. *J. Ther. Ultrasound* **2015**, *3*, 1–4. [CrossRef]
20. Gross, D.; Coutier, C.; Legros, M.; Bouakaz, A.; Certon, D. A CMUT Probe for Ultrasound-Guided Focused Ultrasound Targeted Therapy. *IEEE Trans. Ultrason. Ferroelectr. Freq. Control* **2015**, *62*, 1145–1160. [CrossRef]
21. Ye, G.; Smith, P.P.; Noble, A.; Mayia, F. A Model Based Approach to Monitor Temperature During HIFU Thermal Therapy. In *AIP Conference Proceedings*; American Institute of Physics: College Park, MD, USA, 2007.
22. Napoli, A.; Anzidei, M.; Ciolina, F.; Marotta, E.; Marincola, B.C.; Brachetti, G.; Di Mare, L.; Cartocci, G.; Boni, F.; Noce, V.; et al. MR-Guided High-Intensity Focused Ultrasound: Current Status of an Emerging Technology. *Cardiovasc. Interv. Radiol.* **2013**, *36*, 1190–1203. [CrossRef] [PubMed]
23. Jolesz, F.A.; Hynynen, K.H. *MRI-Guided Focused Ultrasound Surgery*; CRC Press: Boca Raton, FL, USA, 2007; ISBN 9781420019933.
24. Chen, J.; Chen, W.; Zhang, L.; Li, K.; Peng, S.; He, M.; Hu, L. Safety of ultrasound-guided ultrasound ablation for uterine fibroids and adenomyosis: A review of 9988 cases. *Ultrason. Sonochem.* **2015**, *27*, 671–676. [CrossRef] [PubMed]
25. Joo, B.; Park, M.-S.; Lee, S.H.; Choi, H.J.; Lim, S.T.; Rha, S.Y.; Rachmilevitch, I.; Lee, Y.H.; Suh, J.-S. Pain Palliation in Patients with Bone Metastases Using Magnetic Resonance-Guided Focused Ultrasound with Conformal Bone System: A Preliminary Report. *Yonsei Med. J.* **2015**, *56*, 503–509. [CrossRef]
26. Kim, S.J.; Kim, K.A. Safety issues and updates under MR environments. *Eur. J. Radiol.* **2017**, *89*, 7–13. [CrossRef]
27. Epistatou, A.C.; Tsalafoutas, I.A.; Delibasis, K.K. An Automated Method for Quality Control in MRI Systems: Methods and Considerations. *J. Imaging* **2020**, *6*, 111. [CrossRef]
28. Antila, K.; Nieminen, H.J.; Sequeiros, R.B.; Ehnholm, G. Automatic segmentation for detecting uterine fibroid regions treated with MR-guided high intensity focused ultrasound (MR-HIFU). *Med. Phys.* **2014**, *41*, 73502. [CrossRef]
29. Rundo, L.; Militello, C.; Vitabile, S.; Casarino, C.; Russo, G.L.; Midiri, M.; Gilardi, M.C. Combining split-and-merge and multi-seed region growing algorithms for uterine fibroid segmentation in MRgFUS treatments. *Med. Biol. Eng. Comput.* **2016**, *54*, 1071–1084. [CrossRef]
30. Militello, C.; Vitabile, S.; Rundo, L.; Russo, G.; Midiri, M.; Gilardi, M.C. A fully automatic 2D segmentation method for uterine fibroid in MRgFUS treatment evaluation. *Comput. Biol. Med.* **2015**, *62*, 277–292. [CrossRef]
31. Rundo, L.; Tangherloni, A.; Cazzaniga, P.; Nobile, M.S.; Russo, G.; Gilardi, M.C.; Vitabile, S.; Mauri, G.; Besozzi, D.; Militello, C. A novel framework for MR image segmentation and quantification by using MedGA. *Comput. Methods Programs Biomed.* **2019**, *176*, 159–172. [CrossRef]
32. Ikink, M.E.; Van Breugel, J.M.M.; Schubert, G.; Nijenhuis, R.J.; Bartels, L.W.; Moonen, C.T.W.; van den Bosch, M.A.A.J. Volumetric MR-Guided High-Intensity Focused Ultrasound with Direct Skin Cooling for the Treatment of Symptomatic Uterine Fibroids: Proof-of-Concept Study. *BioMed Res. Int.* **2015**, *2015*, 1–10. [CrossRef]
33. Ishihara, Y.; Calderon, A.; Watanabe, H.; Okamoto, K.; Suzuki, Y.; Kuroda, K.; Suzuki, Y. A precise and fast temperature mapping using water proton chemical shift. *Magn. Reson. Med.* **1995**, *34*, 814–823. [CrossRef]
34. Sapareto, S.A.; Dewey, W.C. Thermal dose determination in cancer therapy. *Int. J. Radiat. Oncol. Biol. Phys.* **1984**, *10*, 787–800. [CrossRef]
35. O'Neill, D.P.; Peng, T.; Stiegler, P.; Mayrhauser, U.; Koestenbauer, S.; Tscheliessnigg, K.; Payne, S.J. A Three-State Mathematical Model of Hyperthermic Cell Death. *Ann. Biomed. Eng.* **2010**, *39*, 570–579. [CrossRef]
36. Yung, J.P.; Shetty, A.; Elliott, A.; Weinberg, J.S.; McNichols, R.J.; Gowda, A.; Hazle, J.D.; Stafford, R.J. Quantitative comparison of thermal dose models in normal canine brain. *Med. Phys.* **2010**, *37*, 5313–5321. [CrossRef]
37. Wu, F.; Wang, Z.-B.; Chen, W.-Z.; Zou, J.-Z.; Bai, J.; Zhu, H.; Li, K.-Q.; Jin, C.-B.; Xie, F.-L.; Su, H.-B. Advanced Hepatocellular Carcinoma: Treatment with High-Intensity Focused Ultrasound Ablation Combined with Transcatheter Arterial Embolization. *Radiology* **2005**, *235*, 659–667. [CrossRef] [PubMed]
38. Li, J.-J.; Xu, G.-L.; Gu, M.-F.; Luo, G.-Y.; Rong, Z.; Wu, P.-H.; Xia, J.-C. Complications of high intensity focused ultrasound in patients with recurrent and metastatic abdominal tumors. *World J. Gastroenterol.* **2007**, *13*, 2747–2751. [CrossRef] [PubMed]

39. Jung, S.E.; Cho, S.H.; Jang, J.H.; Han, J.-Y. High-intensity focused ultrasound ablation in hepatic and pancreatic cancer: Complications. *Abdom. Imaging* **2011**, *36*, 185–195. [CrossRef]
40. Zucconi, F.; Colombo, P.E.; Pasetto, S.; Lascialfari, A.; Ticca, C.; Torresin, A. Analysis and reduction of thermal dose errors in MRgFUS treatment. *Phys. Med.* **2014**, *30*, 111–116. [CrossRef] [PubMed]
41. Kinsler, L.E.; Frey, A.R.; Coppens, A.B.; Sanders, J.V. *Fundamentals of Acoustics*, 4th ed.; Wiley India Private Ltd.: New Delhi, India, 2009; ISBN 9788126521999.
42. National Council on Radiation. *Protection and Measurements Biological Effects of Ultrasound: Mechanisms and Clinical Implications*; National Council on Radiation: Bethesda, MD, USA, 1983.
43. Gorny, K.R.; Hangiandreou, N.J.; Ward, H.A.; Hesley, G.K.; Brown, D.L.; Felmlee, J.P. The utility of pelvic coil SNR testing in the quality assurance of a clinical MRgFUS system. *Phys. Med. Biol.* **2009**, *54*, N83–N91. [CrossRef]
44. Firbank, M.J.; Harrison, R.M.; Williams, E.D.; Coulthard, A. Quality assurance for MRI: Practical experience. *Br. J. Radiol.* **2000**, *73*, 376–383. [CrossRef]
45. Ross, J.C.; Tranquebar, R.; Shanbhag, D. Real-Time Liver Motion Compensation for MRgFUS. *Comput. Vis.* **2008**, *11*, 806–813. [CrossRef]
46. Jenne, J.W.; Tretbar, S.H.; Hewener, H.J.; Speicher, D.; Barthscherer, T.; Sarti, C.; Bongers, A.; Schwaab, J.; Günther, M. Ultrasonography-based motion tracking for MRgFUS. In *AIP Conference Proceedings*; AIP Publishing LLC: Melville, NY, USA, 2017.
47. Agnello, L.; Militello, C.; Gagliardo, C.; Vitabile, S. Referenceless thermometry using radial basis function interpolation. In Proceedings of the 2014 World Symposium on Computer Applications & Research (WSCAR), Sousse, Tunisia, 18–20 January 2014.
48. Agnello, L.; Militello, C.; Gagliardo, C.; Vitabile, S. Radial Basis Function Interpolation for Referenceless Thermometry Enhancement. In *Advances in Neural Networks: Computational and Theoretical Issues*; Springer: Cham, Switzerland, 2015; pp. 195–206.
49. Kuroda, K.; Kokuryo, D.; Kumamoto, E.; Suzuki, K.; Matsuoka, Y.; Keserci, B. Optimization of self-reference thermometry using complex field estimation. *Magn. Reson. Med.* **2006**, *56*, 835–843. [CrossRef] [PubMed]
50. Rieke, V.; Vigen, K.K.; Sommer, G.; Daniel, B.L.; Pauly, J.M.; Butts, K. Referenceless PRF shift thermometry. *Magn. Reson. Med.* **2004**, *51*, 1223–1231. [CrossRef]
51. Beatson, R.; Newsam, G. Fast evaluation of radial basis functions: I. *Comput. Math. Appl.* **1992**, *24*, 7–19. [CrossRef]
52. Carr, J.C.; Fright, W.R.; Beatson, R.K. Surface interpolation with radial basis functions for medical imaging. *IEEE Trans. Med Imaging* **1997**, *16*, 96–107. [CrossRef]
53. Powell, M.J.D.; Light, W.A. *Advances in Numerical Analysis III Wavelets Subdivision Algorithms and Radial Basis Functions. The Theory of Radial Basis Function*; Oxford University Press on Demand: Oxford, UK, 1992.
54. Light, W.A. Some Aspects of Radial Basis Function Approximation. *Approx. Theory Spline Funct. Appl.* **1992**, 163–190. [CrossRef]
55. Rieke, V.; Pauly, K.B. MR thermometry. *J. Magn. Reson. Imaging* **2008**, *27*, 376–390. [CrossRef] [PubMed]
56. Kuroda, K.; Oshio, K.; Mulkern, R.V.; Jolesz, F.A. Optimization of chemical shift selective suppression of fat. *Magn. Reson. Med.* **1998**, *40*, 505–510. [CrossRef]
57. Wilcoxon, F. Individual Comparisons by Ranking Methods. *Biom. Bull.* **1945**, *1*, 80. [CrossRef]
58. Moritz, A.R.; Henriques, F.C. Studies of Thermal Injury: II. The Relative Importance of Time and Surface Temperature in the Causation of Cutaneous Burns. *Am. J. Pathol.* **1947**, *23*, 695–720.
59. Viglianti, B.L.; Dewhirst, M.W.; Abraham, J.P.; Gorman, J.M.; Sparrow, E.M. Rationalization of thermal injury quantification methods: Application to skin burns. *Burns* **2014**, *40*, 896–902. [CrossRef] [PubMed]
60. Mougenot, C.; Köhler, M.O.; Enholm, J.; Quesson, B.; Moonen, C. Quantification of near-field heating during volumetric MR-HIFU ablation. *Med. Phys.* **2010**, *38*, 272–282. [CrossRef] [PubMed]

*Article*

# Bucket of Deep Transfer Learning Features and Classification Models for Melanoma Detection

**Mario Manzo [1,*] and Simone Pellino [2]**

[1] Information Technology Services, University of Naples "L'Orientale", 80121 Naples, Italy
[2] Department of Applied Science, I.S. Mattei Aversa M.I.U.R., 81031 Rome, Italy; simonepellino@gmail.com
* Correspondence: mmanzo@unior.it

Received: 9 October 2020; Accepted: 23 November 2020; Published: 26 November 2020 

**Abstract:** Malignant melanoma is the deadliest form of skin cancer and, in recent years, is rapidly growing in terms of the incidence worldwide rate. The most effective approach to targeted treatment is early diagnosis. Deep learning algorithms, specifically convolutional neural networks, represent a methodology for the image analysis and representation. They optimize the features design task, essential for an automatic approach on different types of images, including medical. In this paper, we adopted pretrained deep convolutional neural networks architectures for the image representation with purpose to predict skin lesion melanoma. Firstly, we applied a transfer learning approach to extract image features. Secondly, we adopted the transferred learning features inside an ensemble classification context. Specifically, the framework trains individual classifiers on balanced subspaces and combines the provided predictions through statistical measures. Experimental phase on datasets of skin lesion images is performed and results obtained show the effectiveness of the proposed approach with respect to state-of-the-art competitors.

**Keywords:** melanoma detection; deep learning; transfer learning; ensemble classification

---

## 1. Introduction

Among the types of malignant cancer, melanoma is the deadliest form of skin cancer and its incidence rate is growing rapidly around the world. Early diagnosis is particularly important since melanoma can be cured with a simple excision. In the majority, due to the similarity of the various skin lesions (melanoma and not-melanoma) [1], the visual analysis could be unsuitable and would lead to a wrong diagnosis. In this regard, image processing and artificial intelligence tools can provide a fundamental aid to a step of automatic classification [2]. Further improvement in diagnosis is provided by dermoscopy technique [3]. Dermoscopy technique can be applied to the skin, in order to capture illuminated and magnified images of the skin lesion in a non invasive way to highlight areas containing spots. Furthermore, the visual effect of the deeper skin layer can be improved if the skin surface reflection is removed. Anyhow, classification of melanoma dermoscopy images is a difficult task for different issues. First, the degree of similarity between melanoma and not-melanoma lesions. Second, the segmentation, and, therefore, the identification of the affected area is very complicated because of the variations in terms of texture, size, color, shape and location. The last issue and not the least, is the additional skin conditions such as hair, veins or variations due to image capturing. To this end, many solutions have been provided to improve the task. For example, low-level hand-crafted features [4] are adopted to discriminate non-melanoma and melanoma lesions. In some cases, these types of features are unable to discriminate clearly, leading to results that are sometimes not very relevant [5]. Differently, segmentation is adopted to isolate the foreground elements from the background ones [6]. Consequently, the segmentation includes low-level features with a low representational power that provides unsatisfactory results [7]. In recent years, deep learning has become an effective solution for the extraction of significant features on large data. In particular, the diffusion of deep neural networks, applied to the image classification task, is connected to various factors such as the availability of software in terms of open source license, the constant growth of hardware power and the availability of large datasets [8]. Deep learning has proven effective for the management, analysis, representation and classification of medical images [9]. Specifically, for the treatment of melanoma, deep neural networks were adopted both in segmentation and classification phases [10]. However, the high variation of the types of melanoma and the imbalance of the data have a decisive impact on performance [11], hindering the generalization of the model and leading to over-fitting [12]. In order to overcome the aforementioned issues, in this paper, we introduce a novel framework based on transfer deep learning and ensemble classification for melanoma detection. It works based on three integrated stages. A first, which performs image preprocessing operations. A second, which extracts features using transfer deep learning. A third, including a layer of ensemble learning, in which different classification algorithms and features extracted are combined with the aim of making the best decision (melanoma/not-melanoma). Our approach provides the following main contributions:

- A deep and ensemble learning-based framework, to simultaneously address inter-class variation and class imbalance for the task of melanoma classification.
- A framework that, in the classification phase, at the same time, creates multiple image representation models, based on features extracted with deep transfer learning.
- The demonstration of how the choice of multiple features can enrich image representation by leading a lesion assessment like a skilled dermatologist.
- Some experimental greater improvements over existing methods on different state of art datasets about melanoma detection task.

The paper is structured as follows. Section 2 provides an overview of state-of-the-art about melanoma classification approaches. Section 3 describes in detail proposed framework. Section 4 provides a wide experimental phase, while Section 5 concludes the paper.

## 2. Related Work

In this section, we briefly analyze the most important approaches of skin lesions recognition literature. In this field are included numerous works that address the issue according to different aspects. Some works offer an important contribution about image representation, by implementing segmentation algorithms or new descriptors. Instead, others implement complex mechanisms of learning and classification.

In Reference [13], a novel boundary descriptor based on the color variation of the skin lesion input images, achieved with standard cameras, is introduced. Furthermore, in order to reach higher performance, a set of textural and morphological features is added. Multilayer perceptron neural network as classifier is adopted.

In Reference [14], authors propose a complex framework that implements an illumination correction and features extraction on skin image lesions acquired using normal consumer-grade cameras. Applying a multi-stage illumination improvement algorithm and defining a set of high-level intuitive features (HLIF), that quantifies the level of asymmetry and border irregularity about a lesion, the proposed model can be used to classify accurate skin lesion diagnoses.

While in Reference [15], authors, to properly evaluate contents of the concave contours, introduce a novel border descriptor named boundary intersection-based signature (BIBS). Shape signature is a one-dimensional illustration of shape border and cannot contribute to a proper description for concave borders that have more than one intersection points. For this reason, BIBS analyzes boundary contents of shape especially shapes with concave contours. Support vector machine (SVM) for classification process is adopted.

Another descriptor for the individualization of skin lesions is named high-level intuitive features (HLIFs) [16]. HLIFs are created to simulate a model of human-observable characteristics. It captures specific characteristics that are significant to the given application: color asymmetry—analyzing and clustering pixels colors, structural asymmetry—applying the Fourier descriptors of the shape, border irregularity—using morphological opening and closing, color characteristics—transforming the image to a perceptually uniform color space, building color-spatial representations that model the color information for a patch of pixels, clustering the patch representations into $k$ color clusters, quantifying the variance found using the original lesion and the $k$ representative colors.

A texture analysis method of Local Binary Patterns (LBP) and Block Difference of Inverse Probabilities is proposed in Reference [17]. A comparison is provided with classification results obtained by taking the raw pixel intensity values as input. Classification stage is achieved generating an automated model obtained by both convolutional neural networks (CNN) and SVM.

In Reference [18], authors propose a system that automatically extracts the lesion regions, using non-dermoscopic digital images, and then computes color and texture descriptors. Extracted features are adopted for automatic prediction step. The classification is managed using a majority vote of all predictions.

In Reference [19], non-dermoscopic clinical images to assist a dermatologist in early diagnosis of melanoma skin cancer are adopted. Images are preprocessed in order to reduce artifacts like noise effects. Subsequently, images are analyzed through a pretrained CNN which is a member of deep learning models. CNNs are trained by a large number of training samples in order to distinguish between melanoma and benign cases.

In Reference [20], Predict-Evaluate-Correct K-fold (PECK) algorithm is presented. The algorithm works by merging deep CNNs with SVM and random forest classifiers to achieve an introspective learning method. In addition, authors provides a novel segmentation algorithm, named Synthesis and Convergence of Intermediate Decaying Omnigradients (SCIDOG), to accurately detect lesion contours in non-dermoscopic images, even in the presence of significant noise, hair and fuzzy lesion boundaries.

In Reference [21], authors propose a novel solution to improve melanoma classification by defining a new feature that exploits the border-line characteristics of the lesion segmentation mask combining gradients with LBP. These border-line features are used together with the conventional ones and lead to higher accuracy in classification stage.

In Reference [22], an objective features extraction function for CNN is proposed. The goal is to acquire the variation separability as opposed to the categorical cross entropy which maximizes according to the target labels. The deep representative features increase the variance between the images making it more discriminative. In addition, the idea is to build a CNN and perform principal component analysis (PCA) during the training phase.

In Reference [23], a deep learning computer aided diagnosis system for automatic segmentation and classification of melanoma lesions is proposed. The system extracts CNN and statistical and contrast location features on the results of raw image segmentation. The combined features are utilized to obtain the final classification of melanoma, malignant or benign.

In Reference [24], authors propose an efficient algorithm for prescreening of pigmented skin lesions for malignancy using general-purpose digital cameras. The proposed method enhances borders and extracts a broad set of dermatologically important features. These discriminative features allow classification of lesions into two groups of melanoma and benign.

In Reference [25], a skin lesion detection system optimized to run entirely on the resource constrained smartphone is described. The system combines a lightweight method for skin detection with a hierarchical segmentation approach including two fast segmentation algorithms and proposes novel features to characterize a skin lesion. Furthermore, the system implements an improved features selection algorithm to determine a small set of discriminative features adopted by the final lightweight system.

Multiple-instance learning (MIL)-based approaches are of great interest in recent years. MIL is a type of supervised learning and works by receiving a set of instances, named bags, individually labeled. In Reference [26], authors present an MIL approach with application to melanoma detection. The goal was to discriminate between positive and negative sets of items. The main rule concerns a bag that is positive if at least one of its instances is positive and it is negative if all its instances are negative. Differently in Reference [27], MIL approaches are described with purpose to discriminate melanoma from dysplastic nevi. Specifically, authors introduce an MIL approach that adopts spherical separation surfaces. Finally, in Reference [28], a preliminary comparison between two different approaches, SVM and MIL, is proposed, focusing on the key role played by the feature selection (color and texture). In particular, the authors are inspired by the good results obtained applying MIL techniques for classifying some medical dermoscopic images.

## 3. Materials and Methods

In this section, we describe the proposed framework which includes two well known methodologies: deep neural network and ensemble learning. The main idea is to combine algorithms of features extraction and classification. The result is a set of competitive models providing a range of confidential decisions useful for making choices during classification. The framework is composed of three levels. A first, which performs preprocessing operations such as image resize and data balancing. A second, of transfer learning, which extracts features using deep neural networks. A third level, of ensemble learning, in which different classification algorithms (SVM [29], Logistic Label Propagation (LLP) [30], KNN [31]) and features extracted are combined with the aim of making the best decision. Adopted classifiers are trained and tested through a bootstrapping policy. Finally, the framework iterates through a predetermined number of times in a supervised learning context. Figure 1 shows a graphic overview of the proposed framework.

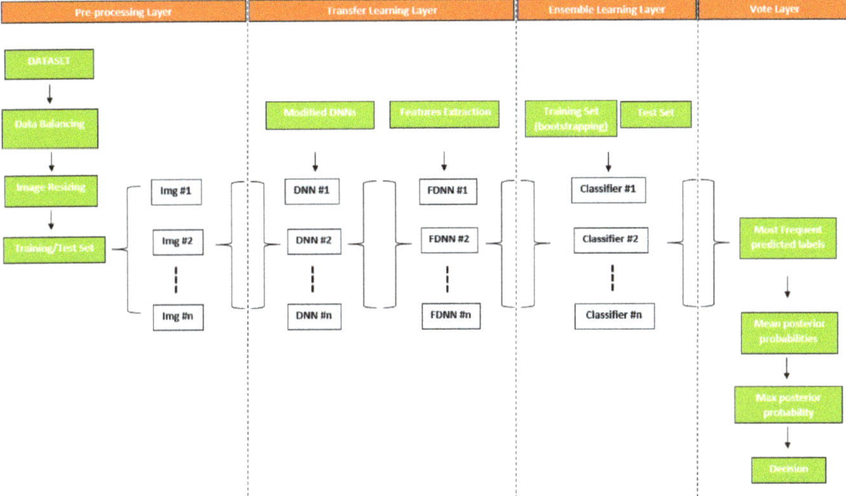


**Figure 1.** Overview of the proposed framework.

### 3.1. Data Balancing

Melanoma lesion analysis and classification is connected with accurate segmentation with purpose to isolate areas of the image containing information of interest. Moreover, the wide variety of skin lesions and the unpredictable obstructions on the skin make traditional segmentation an ineffective tool, especially for non-dermoscopic images. Furthermore, the problem of imbalance, present in many datasets, makes the classification difficult to address, especially when the samples of the minority class are very underrepresented. In the case under consideration, to compensate the strong imbalance between the two classes, a balancing phase was performed. The goal was to isolate segments of the image that could contain melanoma. In particular, the resampling of the minority class is performed by adding images altered through the application of K-Means color segmentation algorithm [32]. The application of segmentation algorithms for image augmentation [33], and consequently to provide a balancing between classes, represented a good compromise for this stage of the pipeline.

### 3.2. Image Resize

Images to be processed have been resized based on the dimension, related to the input layer, claimed by the deep neural networks (details can be found in Table 1 column 5). Many of the networks require this type of step but it does not alter the image information content in any way. This normalization step is essential because images of different or large dimensions cannot be processed for the features extraction stage.

**Table 1.** Description of the adopted pretrained network.

| Network | Depth | Size (MB) | Parameters (Millions) | Input Size | Features Layer |
|---|---|---|---|---|---|
| Alexnet | 8 | 227 | 61 | 227 × 227 | fc7 |
| Googlenet | 8 | 27 | 7 | 224 × 224 | pool5-7x7_s1 |
| Resnet18 | 18 | 44 | 11.7 | 224 × 224 | pool5 |
| Resnet50 | 50 | 96 | 25.6 | 224 × 224 | avg_pool |

### 3.3. Transfer Learning and Features Extraction

The transfer learning approach has been chosen for features extraction purpose. Commonly, a pretrained network is adopted as starting point to learn a new task. It is the easiest and fastest way to exploit the representational power of pretrained deep networks. It is usually much faster and easier to tune a network with transfer learning than training a new network from scratch with randomly initialized weights. We have selected deep learning architectures for image classification based on their structure and performance skills. The goal was to extract features from images through neural networks by redesigning their structures in the final layer according to the needs of the addressed task (two outgoing classes: melanoma and not-melanoma). The features extraction is performed through a chosen layer (different for each network and specified in the Table 1), placed in the final part of the structure. The image will be encoded through a vector of real numbers produced by consecutive convolution steps, from the input layer to the layer chosen for the representation. Below, a description of the adopted networks is reported.

Alexnet [8] consists of 5 convolutional layers and 3 fully connected layers. It includes the non-saturating ReLU activation function, better then tanh and sigmoid during training phase. For features extraction, we have chosen a fully connected 7 (fc7) layer composed of 4096 neurons.

Googlenet [34] is composed of 22 layers deep. The network is inspired by LeNet [35] but implemented a novel element which is dubbed an inception module. This module is based on several very small convolutions in order to drastically reduce the number of parameters. Their architecture reduced the number of parameters from 60 million (AlexNet) to 4 million. Furthermore, it includes batch normalization, image distortions and Root Mean Square Propagation algorithm. For features extraction, we have chosen global average pooling (pool5-7x7_s1) layer composed of 1024 neurons.

Resnet18 and Resnet50 [36] are inspired by pyramidal cells contained in the cerebral cortex. They use particular skip connections or shortcuts to jump over some layers. They are composed of 18 and 50 layers deep, which with the help of a technique known as skip connection has paved the way for residual networks. For feature extraction, we have chosen two global average pooling (pool5 and avg-pool) layers composed of 512 and 2048 neurons, respectively.

### 3.4. Network Design

The adopted networks have been adapted to the melanoma classification problem. Originally, they have been trained on the Imagenet dataset [37], composed of a million images and classified into 1000 classes. The result is a rich features representation for a wide range of images. The network processes an image and provides a label along with probabilities for each of the classes. Commonly, the first layer of the network is the image input layer. This requires input images with 3 color channels. Just after, convolutional layers work to extract image features in which the last learnable layer and the final classification layer adopt to classify the input image. In order to make suitable the pretrained network to classify new images, the two last layers with new layers are replaced. In many cases, the last layer, including learnable weights, is a fully connected layer. This is replaced with a new fully connected layer related to the number of outputs equal to the number of classes of new data. Moreover, to speedup the learning in the new layer with respect to transferred layers, it is recommended to increase the learning rate factors. As an optional choice, the weights of earlier layers can be frozen by setting the related learning rate to zero. This setting produces a failure of update of the weights during the training, and a consequent lowering of the execution time as the gradients of the related layers must not be calculated. This aspect is very interesting to avoid overfitting in the case of small datasets.

### 3.5. Ensemble Learning

The contribution of different transfer learning features and classifiers can be mixed in an ensemble context. Considering the set of images, with cardinality $k$, belonging to $x$ classes, to be classified

$$Imgs = \{i_1, i_2, \ldots, i_k\} \tag{1}$$

each element of the set will be treated with the procedure below. Let us consider the set $C$ composed of $n$ classifiers

$$C = \{\beta_1, \beta_2, \ldots, \beta_n\} \tag{2}$$

and set $F$ composed of $m$ vectors of transferred learning features

$$F = \{\Theta_1, \Theta_2, \ldots. \Theta_m\} \tag{3}$$

the goal is the combination each element of the set $C$ with the elements of the set $F$. The set of combinations can be defined as $CF$

$$CF = \begin{bmatrix} \beta_1\Theta_1 & \ldots & \beta_1\Theta_m \\ \vdots & \ddots & \\ \beta_n\Theta_1 & & \beta_n\Theta_m \end{bmatrix} \tag{4}$$

each combination provides a decision $i \in I\{-1, 1\}$, where 1 stands for melanoma and $-1$ for not-melanoma, related to image of the set $Imgs$. The set of decisions $D$ can be defined as follows

$$D = \begin{bmatrix} d_{\beta_1\Theta_1} & \ldots & d_{\beta_1\Theta_m} \\ \vdots & \ddots & \\ d_{\beta_n\Theta_1} & & d_{\beta_n\Theta_m} \end{bmatrix} \tag{5}$$

Each $d_{\beta_i\Theta_j}$ value represents a decision based on the combination of sets $C$ and $F$. In addition, the set of scores S can be defined as follows

$$S = \begin{bmatrix} P(i|x)_{d_{\beta_1\Theta_1}} & \ldots & P(i|x)_{d_{\beta_1\Theta_m}} \\ \vdots & \ddots & \\ P(i|x)_{d_{\beta_n\Theta_1}} & & P(i|x)_{d_{\beta_n\Theta_m}} \end{bmatrix} \tag{6}$$

a score value, $s \in S\{0, \ldots, 1\}$, is associated with each decision $d$ and represents the posterior probability $P(i|x)$ that an image $i$ belongs to class $x$. At this point, let us introduce the concept of mode, defined as the value which is repeatedly occurred in a given set

$$mode = l + \left(\frac{f_1 - f_0}{2f_1 - f_0 - f_2}\right) \times h \tag{7}$$

where $l$ is the lower limit of the modal class, $h$ is the size of the class interval, $f_1$ is the frequency of the modal class, $f_0$ is the frequency of the class which precedes the modal class and $f_2$ is the frequency of the class which successes the modal class. The columns of matrix $D$ are analyzed with the mode, in order to obtain the values of the most frequent decisions. This step is carried out in order to verify the best response of the different classifiers, contained in the set $C$, which adopt the same type of features. Moreover, the *mode* provides two indications. The most frequent value and its occurrences (indices). For each most frequent occurrence, modal value, the corresponding score of the matrix $S$ is extracted. In this regard, a new vector is generated

$$DS = \{ds_{P(i|x)_{d_{\beta_{1,\dots,n}\Theta_1}}}, \dots, ds_{P(i|x)_{d_{\beta_{1,\dots,n}\Theta_m}}}\}, \quad (8)$$

where each element $ds$ contains the average of the scores that have a higher frequency, extracted through the *mode*, in the related column of the matrix $D$. In addition, the modal value of each column of the matrix $D$ is stored in the vector $DM$

$$DM = \{dm_{d_{\beta_{1,\dots,n}\Theta_1}}, \dots, dm_{d_{\beta_{1,\dots,n}\Theta_m}}\}, \quad (9)$$

the final decision will consist in the selection of the element of the vector $DM$ with the same position of the maximum score value of the vector $DS$. This last step verifies the best prediction based on the different features adopted, essentially the best features suitable for the classification of the image.

### 3.6. Train and Test Strategy: Bootstrapping

Bootstrapping is a statistical technique which consists of creating samples of size $B$, named bootstrap samples, from a dataset of size $N$. The bootstrap samples are randomly inserted with replacement on the dataset. This strategy has important statistical properties. First, subsets can be considered as directly extracted from the original distribution, independently of each others, containing representative and independent samples, almost independent and identically distributed (idd). Two considerations must be made in order to validate the hypotheses. First, the $N$ dimension of the original dataset should be large enough to detect the underlying distribution. Sampling the original data is a good approximation of real distribution (representativeness). Second, the $N$ dimension of the dataset should be better than the $B$ dimension of the bootstrap samples so that the samples are not too correlated (independence). Commonly, considering the samples to be truly independent means requiring too much data compared to the amount actually available. This strategy can be adopted to generate several bootstrap samples that can be considered nearly representative and almost independent (almost iid samples). In the proposed framework, bootstrapping is applied to set $F$ (Equation (3)) in order to perform the training and testing stages of classifiers. This strategy seemed suitable for the problem faced in order to create a competitive environment capable of providing the best performance.

## 4. Experimental Results

This section describes the experiments performed on public datasets. In order to produce compliant performance, the settings included in well-known melanoma classification methods, in which the main critical issue concerns the features extraction for image representation, are adopted.

### 4.1. Datasets

The first adopted dataset is MED-NODE (http://www.cs.rug.nl/~imaging/databases/melanoma_naevi/). It was created by the Department of Dermatology of the University Medical Center Groningen (UMCG). The dataset was initially used to train the MED-NODE computer assisted melanoma detection system [18]. It is composed of 170 non-dermoscopic images, where 70 are melanoma and 100 are nevi. The image dimensions vary greatly, ranging from 201 × 257 to 3177 × 1333 pixels.

The second adopted dataset, Skin-lesion (from now), is described in Reference [16]. It is composed of 206 images of skin lesion, which were obtained using standard consumer-grade cameras in varying and unconstrained environmental conditions. These images were extracted from the online public databases Dermatology Information System (http://www.dermis.net) and DermQuest (http://www.dermquest.com). Of these images, 119 are melanomas, and 87 are not-melanoma. Each image contains a single lesion of interest.

### 4.2. Settings

The framework consists of different modules written in Matlab language. Moreover, we applied pretrained networks available which are included in the ImageNet Large-Scale Visual Recognition Challenge (ILSVRC) [38]. Among all the computational stages, the features extraction process, described in Section 3.3, was certainly the most expensive. As is certainly known, the networks are composed of fully connected layers that make the structure extremely dense and complex. This aspect certainly increases the computational load. Alexnet, Googlenet, Resnet50 are adopted to extract features on MED-NODE dataset. Differently, Resnet50 and Resnet18 are adopted for Skin-lesion dataset. The choice is not random but was made based on two criteria. Primarily, a study about the network specifications and characteristics most suitable for the problem faced in literature and, secondly, the performance obtained. A different combination did not provide expected feedback. In the Table 1, some important details related to the layers chosen for feature extraction are shown. Networks were trained by setting the mini batch size to 5, the maximum epochs to 10, the initial learning rate to $3 \times 10^{-4}$ and the optimizer is stochastic gradient descent with momentum (SGDM) algorithm. For both experimental procedures, in order to train the classifiers, 80% and 20% of images are included in train and test sets, respectively, for a number of iterations equal to 10. Table 2 enumerates classification algorithms included in the framework and related settings (some algorithms appear more times with different configurations). For completeness and clarity, and in order to demonstrate the best solution, both results of combinations adopted, even if they did not provide the best performance, are indicated in Tables 4 and 5.

**Table 2.** Classification algorithms and related settings.

| Algorithms | Setting |
|---|---|
| SVM [29] | KernelFunction:polynomial, KernelScale: auto |
| SVM [29] | KernelFunction: Gaussian, KernelScale: auto |
| LLP [30] | KernelFunction: rbf, Regularization parameter: 1, init: 0, maxiter: 1000 |
| KNN [31] | NumNeighbors: 3, Distance: spearman |
| KNN [31] | NumNeighbors: 4, Distance: correlation |

### 4.3. Discussion

Table 3 shows the metrics adopted for the performance evaluation, in order to provide a uniform comparison with algorithms working on the same task.

**Table 3.** Evaluation metrics adopted during the relevance feedback stage.

| Metric | Equation |
|---|---|
| True Positive Rate | $TPR = \frac{TP}{TP+FN}$ |
| True Negative Rate | $TNR = \frac{TN}{TN+FP}$ |
| Positive Predictive Value | $PPV = \frac{TP}{TP+FP}$ |
| Negative Predictive Value | $NPV = \frac{TN}{TN+FN}$ |
| Accuracy | $ACC = \frac{TP+FN}{TP+FP+TN+FN}$ |
| $F_1$-Score(Positive) | $F_1^P = \frac{2 \cdot PPV \cdot TPR}{PPV+TPR}$ |
| $F_1$-Score(Negative) | $F_1^N = \frac{2 \cdot NPV \cdot TNR}{NPV+TNR}$ |
| Matthew's Correlation Coefficient | $MCC = \frac{TP \cdot TN - FP \cdot FN}{\sqrt{(TP+FP) \cdot (TP+FN) \cdot (TN+FP) \cdot (TN+FN)}}$ |

Looking carefully at the table, it is important to focus on the meaning of the individual measures with reference to melanoma detection. The True Positive rate, also known as Sensitivity, concerns the portion of positives melanoma images that are correctly identified. This provide important information because highlights the skill to identify images containing skin lesions and contributes to increase the degree of robustness of result. The same concept is true for the True Negative rate, also known as Specificity, which instead measures the portion of negatives, not containing skin lesions, that have been correctly identified. The Positive and Negative Predictive values, also known as Precision and Recall, respectively, are probabilistic measures that indicate whether an image with a positive or negative melanoma test may or may not have a skin lesion. In essence, Recall expresses the ability to find all relevant instances in the dataset, Precision expresses the proportion of instances that the framework claims to be relevant were actually relevant. Accuracy, a well-known performance measure, is the proportion of true results among the total number of cases examined. In our case, it provides an overall analysis, certainly a rough measurement compared to the previous ones, about the skill of a classifier to distinguish a skin lesion from an image without lesions. $F_1-$Score measure combines the Precision and Recall of the model, as the harmonic mean, in order to find an optimal blend. The choice of the harmonic mean instead of a simple mean concerns the possibility of eliminating extreme values. Finally, Matthew's correlation coefficient is another overall well-known quality measure. It takes into account True/False Positives/Negatives values and is generally regarded as a balanced measure which can be adopted even if the classes are of very different sizes.

The Tables 4 and 5 describe the comparison with existing skin cancer classification methods (we referred with the results which appear in the corresponding papers). The provided performance can be considered satisfactory compared to competitors. In terms of accuracy, although it provides a rough measurement, we have provided the best result for MED-NODE and the second for Skin-lesion (only surpassed by BIBS). Differently, predictive positive value and negative positive value give good indications on the classification ability. True positive rate, a measure that provides greater confidence about addressed problem, is very high for both datasets. Otherwise, true negative rate, which also provides a high degree of sensitivity related to the absence of tumors within the image, is the best value for both datasets. Regarding the remaining measures, $F_1^p$, $F_1^n$ and Matthew's Correlation Coefficient, considerable values were obtained but, unfortunately, not available for all competitors. We can certainly attribute the satisfactory performance to two main aspects. First, the deep learning features, which even if abstract, are able to best represent the images. Furthermore, the framework provides multiple representation models that certainly constitute a different starting point than a standard approach, in which a single representation is provided. This aspect is relevant for improving performance. A non negligible issue, the normalization of the image size, with respect to the request of the first layer of the neural network, before the features extraction phase, does not produce a performance degradation. In other cases, normalization causes loss of quality of the image content and a consequent degradation of details. Otherwise, the weak point is the computational load even if pretrained networks include layers with already tuned weights. Surely, the time required for training is long but less than a network created from scratch. Second, the classification scheme, which provides multiple choices in decision making. In fact, at each iteration, the framework chooses which classifier is suitable for recognizing melanoma in the images included in the proposed set. Certainly, this approach is more computationally expensive but produces better results than a single classifier.

**Table 4.** Experimental results on the MED-NODE dataset.

| Method | TPR | TNR | PPV | NPV | ACC | $F_1^p$ | $F_1^n$ | MCC |
|---|---|---|---|---|---|---|---|---|
| MED-NODE annoted [18] | 0.78 | 0.59 | 0.56 | 0.80 | 0.66 | 0.65 | 0.68 | 0.36 |
| Spotmole [39] | 0.82 | 0.57 | 0.56 | 0.83 | 0.67 | 0.67 | 0.68 | 0.39 |
| Barhoumi and Zagrouba [40] | 0.46 | 0.87 | 0.70 | 0.71 | 0.70 | 0.56 | 0.78 | 0.37 |
| MED-NODE color [18] | 0.74 | 0.72 | 0.64 | 0.81 | 0.73 | 0.69 | 0.76 | 0.45 |
| MED-NODE texture [18] | 0.62 | 0.85 | 0.74 | 0.77 | 0.76 | 0.67 | 0.81 | 0.49 |
| Jafari et al. [24] | 0.90 | 0.72 | 0.70 | 0.91 | 0.79 | 0.79 | 0.80 | 0.61 |
| MED-NODE combined [18] | 0.80 | 0.81 | 0.74 | 0.86 | 0.81 | 0.77 | 0.83 | 0.61 |
| Nasr Esfahani et al. [19] | 0.81 | 0.80 | 0.75 | 0.86 | 0.81 | 0.78 | 0.83 | 0.61 |
| Benjamin Albert [20] | 0.89 | 0.93 | 0.92 | 0.93 | 0.91 | 0.89 | 0.92 | 0.83 |
| Pereira et [21] ght/svm-smo/f23-32 | 0.45 | 0.92 | - | - | 0.73 | - | - | - |
| Pereira et [21] ght/svm-smo/f1-32 | 0.56 | 0.86 | - | - | 0.74 | - | - | - |
| Pereira et al. [21] lbpc/svm-smo/f23-32 | 0.49 | 0.93 | - | - | 0.75 | - | - | - |
| Pereira et al. [21] lbpc/svm-smo/f1-32 | 0.58 | 0.91 | - | - | 0.78 | - | - | - |
| Pereira et al. [21] ght/svm-sda/f23-32 | 0.66 | 0.83 | - | - | 0.76 | - | - | - |
| Pereira et al. [21] ght/svm-sda/f1-32 | 0.66 | 0.86 | - | - | 0.78 | - | - | - |
| Pereira et al. [21] lbpc/svm-isda/f23-32 | 0.69 | 0.83 | - | - | 0.77 | - | - | - |
| Pereira et al. [21] lbpc/svm-isda/f1-32 | 0.65 | 0.88 | - | - | 0.79 | - | - | - |
| Pereira et al. [21] ght/ffn/f23-32 | 0.63 | 0.84 | - | - | 0.76 | - | - | - |
| Pereira et al. [21] ght/ffn/f1-32 | 0.63 | 0.84 | - | - | 0.76 | - | - | - |
| Pereira et al. [21] lbpc/ffn/f23-32 | 0.64 | 0.83 | - | - | 0.75 | - | - | - |
| Pereira et al. [21] lbpc/ffn/f1-32 | 0.66 | 0.86 | - | - | 0.77 | - | - | - |
| Sultana et al. [22] | 0.73 | 0.86 | 0.77 | 0.83 | 0.81 | - | - | - |
| Ge, Yunhao and Liet al. [23] | 0.94 | 0.93 | - | - | 0.92 | - | - | - |
| Mandal et al.[41] Case 1 | 0.61 | 0.65 | 0.74 | 0.87 | 0.65 | - | - | - |
| Mandal et al.[41] Case 2 | 0.80 | 0.73 | 0.74 | 0.87 | 0.71 | - | - | - |
| Mandal et al.[41] Case 3 | 0.84 | 0.66 | 0.68 | 0.86 | 0.71 | - | - | - |
| Jafari et al. [42] | 0.82 | 0.71 | 0.67 | 0.85 | 0.76 | - | - | - |
| T. Do et al. [25] Color | 0.81 | 0.73 | 0.66 | 0.85 | 0.75 | - | - | - |
| T. Do et al. [25] Texture | 0.66 | 0.85 | 0.75 | 0.79 | 0.78 | - | - | - |
| T. Do et al. [25] Color and Texture | 0.84 | 0.72 | 0.70 | 0.87 | 0.77 | - | - | - |
| E. Nasr-Esfahani et al. [19] | 0.81 | 0.80 | 0.75 | 0.86 | 0.81 | - | - | - |
| Resnet50+Resnet18 | 0.80 | 1.00 | 1.00 | 0.83 | 0.90 | 0.88 | 0.90 | 0.81 |
| **Resnet50+Googlenet+Alexnet** | 0.90 | 0.97 | 0.97 | 0.90 | 0.93 | 0.93 | 0.94 | 0.87 |

**Table 5.** Experimental results on the Skin-lesion dataset.

| Method | TPR | TNR | PPV | NPV | ACC | $F_l^p$ | $F_l^n$ | MCC |
|---|---|---|---|---|---|---|---|---|
| Texture analysis [17] | 0.87 | 0.71 | 0.76 | - | 0.75 | - | - | - |
| HLIFs [16] | 0.96 | 0.73 | - | - | 0.83 | - | - | - |
| BIBS [15] | 0.92 | 0.88 | 0.91 | - | 0.90 | - | - | - |
| Decision Support [14] | 0.84 | 0.79 | - | - | 0.81 | - | - | - |
| Color pigment boundary [13] | 0.95 | 0.88 | 0.92 | - | 0.82 | - | - | - |
| R. Amelard et al. [43] Asymmetry $F_C$ | 0.73 | 0.64 | - | - | 0.69 | - | - | - |
| R. Amelard et al. [43] Proposed HLIFs | 0.79 | 0.68 | - | - | 0.75 | - | - | - |
| R. Amelard et al. [43] Cavalcanti feature set | 0.84 | 0.78 | - | - | 0.82 | - | - | - |
| R. Amelard et al. [43] Modified $F_C$ | 0.86 | 0.75 | - | - | 0.72 | - | - | - |
| R. Amelard et al. [43] Combined $F_{MC}$ $F_A^{HLIFS}$ | 0.91 | 0.80 | - | - | 0.86 | - | - | - |
| **Resnet50+Resnet18** | 0.84 | 0.92 | 0.91 | 0.85 | 0.88 | 0.87 | 0.88 | 0.76 |
| Resnet50+Googlenet+Alexnet | 0.87 | 0.65 | 0.71 | 0.84 | 0.76 | 0.78 | 0.73 | 0.54 |

## 5. Conclusions and Future Works

The challenge in the discrimination of melanoma and nevi has resulted to be very interesting in recent years. The complexity of the task is linked to different factors such as the large amount of types of melanomas or the difficulties for digital phase acquisition (noise, lighting, angle, distance and much more). Machine learning classifiers suffer greatly these factors and inevitably reflect on the quality of the results. In support, the convolutional neural networks give a big hand for both classification and features extraction phases. In this context, we have proposed a framework that combines standard classifiers and features extracted with convolutional neural networks using a transfer learning approach. The results produced certainly support the theoretical thesis. A multiple representation of the image compared to a single one is a high discrimination factor even if the features adopted are completely abstract. The extensive experimental phase has shown how the proposed approach is competitive, and in some cases surpassing, with respect to state-of-the-art methods. Certainly, the main weak point concerns the computational complexity relating to features extraction phase, as it is known, takes a long time especially when the data to be processed grows. Future work will certainly concern the study and analysis of convolutional neural networks still unexplored for this type of problem, the application of the proposed framework to additional datasets (such as $PH^2$ [44]) and alternative tasks from the melanoma detection. Finally, also interesting are different dataset balancing approaches, such as proposed in [45] where all the melanom images are duplicated by including zero-mean Gaussian noise with variance equal to 0.0001.

**Author Contributions:** Both authors conceived the study and contributed to the writing of the manuscript and approved the final version. All authors have read and agreed to the published version of the manuscript.

**Funding:** This research received no funding.

**Acknowledgments:** Our thanking is for Alfredo Petrosino. He followed us during the first steps towards the Computer Science, through a whirlwind of goals, ideas and, especially, love and passion for the work. We will be forever grateful great master.

**Conflicts of Interest:** The authors declare no conflict of interest.

## References

1. Codella, N.; Cai, J.; Abedini, M.; Garnavi, R.; Halpern, A.; Smith, J.R. Deep learning, sparse coding, and SVM for melanoma recognition in dermoscopy images. In *International Workshop on Machine Learning in Medical Imaging*; Springer: Berlin/Heidelberg, Germany, 2015; pp. 118–126.
2. Mishra, N.K.; Celebi, M.E. An overview of melanoma detection in dermoscopy images using image processing and machine learning. *arXiv* **2016**, arXiv:1601.07843.

3. Binder, M.; Schwarz, M.; Winkler, A.; Steiner, A.; Kaider, A.; Wolff, K.; Pehamberger, H. Epiluminescence microscopy: A useful tool for the diagnosis of pigmented skin lesions for formally trained dermatologists. *Arch. Dermatol.* **1995**, *131*, 286–291. [CrossRef] [PubMed]
4. Barata, C.; Celebi, M.E.; Marques, J.S. A survey of feature extraction in dermoscopy image analysis of skin cancer. *IEEE J. Biomed. Health Inform.* **2018**, *23*, 1096–1109. [CrossRef] [PubMed]
5. Celebi, M.E.; Kingravi, H.A.; Uddin, B.; Iyatomi, H.; Aslandogan, Y.A.; Stoecker, W.V.; Moss, R.H A methodological approach to the classification of dermoscopy images. *Comput. Med. Imaging Graph.* **2007**, *31*, 362–373. [CrossRef]
6. Tommasi, T.; La Torre, E.; Caputo, B. Melanoma recognition using representative and discriminative kernel classifiers. In *International Workshop on Computer Vision Approaches to Medical Image Analysis*; Springer: Berlin/Heidelberg, Germany, 2006; pp. 1–12.
7. Pathan, S.; Prabhu, K.G.; Siddalingaswamy, P. A methodological approach to classify typical and atypical pigment network patterns for melanoma diagnosis. *Biomed. Signal Process. Control.* **2018**, *44*, 25–37. [CrossRef]
8. Krizhevsky, A.; Sutskever, I.; Hinton, G.E. Imagenet classification with deep convolutional neural networks. *Advances in Neural Information Processing Systems*; ACM Digital Library: New York, NY, USA, 2012 , pp. 1097–1105.
9. Ronneberger, O.; Fischer, P.; Brox, T. U-net: Convolutional networks for biomedical image segmentation. In *International Conference on Medical Image Computing and Computer-Assisted Intervention*; Springer: Berlin/Heidelberg, Germany, 2015, pp. 234–241.
10. Yu, L.; Chen, H.; Dou, Q.; Qin, J.; Heng, P.A. Automated melanoma recognition in dermoscopy images via very deep residual networks. *IEEE Trans. Med Imaging* **2016**, *36*, 994–1004. [CrossRef]
11. Shie, C.K.; Chuang, C.H.; Chou, C.N.; Wu, M.H.; Chang, E.Y. Transfer representation learning for medical image analysis. In Proceedings of the 2015 37th annual international conference of the IEEE Engineering in Medicine and Biology Society (EMBC), Milan, Italy, 25–29 August 2015; pp. 711–714.
12. Shin, H.C.; Roth, H.R.; Gao, M.; Lu, L.; Xu, Z.; Nogues, I.; Yao, J.; Mollura, D.; Summers, R.M. Deep convolutional neural networks for computer-aided detection: CNN architectures, dataset characteristics and transfer learning. *IEEE Trans. Med. Imaging* **2016**, *35*, 1285–1298. [CrossRef]
13. Jayant Sachdev, Shashank Shekhar, D.S.I. Skin Lesion Images Classification Using New Color Pigmented Boundary Descriptors. In Proceedings of the 3rd International Conference on Pattern Recognition and Image Analysis (IPRIA 2017), Shahrekord, Iran, 19–20 April 2017.
14. Amelard, R.; Glaister, J.; Wong, A.; Clausi, D.A. Melanoma Decision Support Using Lighting-Corrected Intuitive Feature Models. In *Computer Vision Techniques for the Diagnosis of Skin Cancer*; Springer: Berlin/Heidelberg, Germany, 2013, pp. 193–219.
15. Mahdiraji, S.A.; Baleghi, Y.; Sakhaei, S.M. BIBS, a New Descriptor for Melanoma/Non-Melanoma Discrimination. In Proceedings of the Iranian Conference on Electrical Engineering (ICEE), Mashhad, Iran, 8–10 May 2018, pp. 1397–1402.
16. Amelard, R.; Glaister, J.; Wong, A.; Clausi, D.A. High-Level Intuitive Features (HLIFs) for Intuitive Skin Lesion Description. *IEEE Trans. Biomed. Eng.* **2015**, *62*, 820–831. [CrossRef]
17. Karabulut, E.; Ibrikci, T. Texture analysis of melanoma images for computer-aided diagnosis. In Proceedings of the International Conference on Intelligent Computing, Computer Science & Information Systems (ICCSIS 16), Pattaya, Thailand, 28–29 April 2016; Volume 2, pp. 26–29.
18. Giotis, I.; Molders, N.; Land, S.; Biehl, M.; Jonkman, M.; Petkov, N.MED-NODE: A Computer-Assisted Melanoma Diagnosis System using Non-Dermoscopic Images. *Expert Syst. Appl.* **2015**, *42*, [CrossRef]
19. Nasr-Esfahani, E.; Samavi, S.; Karimi, N.; Soroushmehr, S.M.R.; Jafari, M.H.; Ward, K.; Najarian, K. Melanoma detection by analysis of clinical images using convolutional neural network. In Proceedings of the 2016 38th Annual International Conference of the IEEE Engineering in Medicine and Biology Society (EMBC), Orlando, FL, USA, 16–20 August 2016; pp. 1373–1376.
20. Albert, B.A. Deep Learning From Limited Training Data: Novel Segmentation and Ensemble Algorithms Applied to Automatic Melanoma Diagnosis. *IEEE Access* **2020**, *8*, 31254–31269. [CrossRef]

21. Pereira, P.M.; Fonseca-Pinto, R.; Paiva, R.P.; Assuncao, P.A.; Tavora, L.M.; Thomaz, L.A.; Faria, S.M. Skin lesion classification enhancement using border-line features—The melanoma vs. nevus problem. *Biomed. Signal Process. Control* **2020**, *57*, 101765, [CrossRef]
22. Sultana, N.N.; Puhan, N.B.; Mandal, B. DeepPCA Based Objective Function for Melanoma Detection. In Proceedings of the 2018 International Conference on Information Technology (ICIT), Bhubaneswar, India, 19–21 December 2018; pp. 68–72.
23. Ge, Y.; Li, B.; Zhao, Y.; Guan, E.; Yan, W. Melanoma segmentation and classification in clinical images using deep learning. In Proceedings of the 2018 10th International Conference on Machine Learning and Computing, Macau, China, 26–28 February 2018; pp. 252–256.
24. Jafari, M.H.; Samavi, S.; Karimi, N.; Soroushmehr, S.M.R.; Ward, K.; Najarian, K. Automatic detection of melanoma using broad extraction of features from digital images. In Proceedings of the 2016 38th Annual International Conference of the IEEE Engineering in Medicine and Biology Society (EMBC), Orlando, FL, USA, 16–20 August 2016; pp. 1357–1360.
25. Do, T.; Hoang, T.; Pomponiu, V.; Zhou, Y.; Chen, Z.; Cheung, N.; Koh, D.; Tan, A.; Tan, S. Accessible Melanoma Detection Using Smartphones and Mobile Image Analysis. *IEEE Trans. Multimed.* **2018**, *20*, 2849–2864. [CrossRef]
26. Astorino, A.; Fuduli, A.; Veltri, P.; Vocaturo, E. Melanoma detection by means of Multiple Instance Learning. *Interdiscip. Sci. Comput. Life Sci.* **2020**, *12*, 24–31. [CrossRef] [PubMed]
27. Vocaturo, E.; Zumpano, E.; Giallombardo, G.; Miglionico, G. DC-SMIL: A multiple instance learning solution via spherical separation for automated detection of displastyc nevi. In Proceedings of the 24th Symposium on International Database Engineering & Applications, Incheon (Seoul), South Korea, August 12–18, 2020; pp. 1–9.
28. Fuduli, A.; Veltri, P.; Vocaturo, E.; Zumpano, E. Melanoma detection using color and texture features in computer vision systems. *Adv. Sci. Technol. Eng. Syst. J.* **2019**, *4*, 16–22. [CrossRef]
29. Corinna Cortes, V.V. Support-vector networks. *Mach. Learn.* **1995**, *20*, 273–297. [CrossRef]
30. Kobayashi, T.; Watanabe, K.; Otsu, N. Logistic label propagation. *Pattern Recognit. Lett.* **2012**, *33*, 580–588. [CrossRef]
31. Dasarathy, B.V. *Nearest Neighbor (NN) Norms: NN Pattern Classification Techniques*; IEEE Computer Society Press: Los Alamitos, CA, USA 1991; ISBN 978-0-8186-8930-7
32. Likas, A.; Vlassis, N.; Verbeek, J.J. The global k-means clustering algorithm. *Pattern Recognit.* **2003**, *36*, 451–461. [CrossRef]
33. Shorten, C.; Khoshgoftaar, T.M. A survey on image data augmentation for deep learning. *J. Big Data* **2019**, *6*, 60. [CrossRef]
34. Szegedy, C.; Liu, W.; Jia, Y.; Sermanet, P.; Reed, S.; Anguelov, D.; Erhan, D.; Vanhoucke, V.; Rabinovich, A. Going deeper with convolutions. In Proceedings of the IEEE Conference on Computer Vision and Pattern Recognition, Boston, MA, USA, 7–12 June 2015; pp. 1–9.
35. LeCun, Y.; Boser, B.; Denker, J.S.; Henderson, D.; Howard, R.E.; Hubbard, W.; Jackel, L.D. Backpropagation applied to handwritten zip code recognition. *Neural Comput.* **1989**, *1*, 541–551. [CrossRef]
36. He.; Kaiming.; Zhang, X.; Ren, S.; Sun, J. Deep residual learning for image recognition. In Proceedings of the IEEE Conference on Computer Vision and Pattern Recognition, Las Vegas, NV, USA , 27–30 June 2016; pp. 770–778.
37. Deng, J.; Dong, W.; Socher, R.; Li, L.J.; Li, K.; Fei-Fei, L. Imagenet: A large-scale hierarchical image database. In Proceedings of the 2009 IEEE Conference on Computer Vision and Pattern Recognition, Miami, FL, USA, 20–25 June 2009; pp. 248–255.
38. Russakovsky, O.; Deng, J.; Su, H.; Krause, J.; Satheesh, S.; Ma, S.; Huang, Z.; Karpathy, A.; Khosla, A.; Bernstein, M.; et al. Imagenet large scale visual recognition challenge. *Int. J. Comput. Vis.* **2015**, *115*, 211–252. [CrossRef]
39. Munteanu, C.; Cooclea, S. Spotmole Melanoma Control System. 2009. https://play.google.com/store/apps/details?id=com.spotmole&hl=en=AU (accessed on 18 November 2020).
40. Zagrouba, E.; Barhoumi, W. A prelimary approach for the automated recognition of malignant melanoma. *Image Anal. Stereol.* **2004**, *23*, 121–135, [CrossRef]

41. Mandal, B.; Sultana, N.; Puhan, N. Deep Residual Network with Regularized Fisher Framework for Detection of Melanoma. *IET Comput. Vis.* **2018**, *12*, [CrossRef]
42. Jafari, M.H.; Samavi, S.; Soroushmehr, S.M.R.; Mohaghegh, H.; Karimi, N.; Najarian, K. Set of descriptors for skin cancer diagnosis using non-dermoscopic color images. In Proceedings of the 2016 IEEE International Conference on Image Processing (ICIP), Phoenix, AZ, USA, 25–28 September 2016; pp. 2638–2642.
43. Amelard, R.; Wong, A.; Clausi, D.A. Extracting high-level intuitive features (HLIF) for classifying skin lesions using standard camera images. In Proceedings of the 2012 Ninth Conference on Computer and Robot Vision, Toronto, ON, Canada, 28–30 May 2012; pp. 396–403.
44. Mendonca, T.; Celebi, M.; Mendonca, T.; Marques, J. Ph2: A public database for the analysis of dermoscopic images. *Dermoscopy Image Anal.* **2015**, 419–439, doi:10.1201/b19107-14.
45. Barata, C.; Ruela, M.; Francisco, M.; Mendonça, T.; Marques, J.S. Two systems for the detection of melanomas in dermoscopy images using texture and color features. *IEEE Syst. J.* **2013**, *8*, 965–979. [CrossRef]

MDPI
St. Alban-Anlage 66
4052 Basel
Switzerland
Tel. +41 61 683 77 34
Fax +41 61 302 89 18
www.mdpi.com

*Journal of Imaging* Editorial Office
E-mail: jimaging@mdpi.com
www.mdpi.com/journal/jimaging

www.ingramcontent.com/pod-product-compliance
Lightning Source LLC
LaVergne TN
LVHW070133170726
843515LV00007B/1784